Vigilance and Performance in Automatized Systems/
Vigilance et Performance de l'Homme dans les Systèmes Automatisés

NATO ASI Series

Advanced Science Institutes Series

A Series presenting the results of activities sponsored by the NATO Science Committee, which aims at the dissemination of advanced scientific and technological knowledge, with a view to strengthening links between scientific communities.

The Series is published by an international board of publishers in conjunction with the NATO Scientific Affairs Division

A Life Sciences **B Physics**	Plenum Publishing Corporation London and New York
C Mathematical and Physical Sciences **D Behavioural and Social Sciences** **E Applied Sciences**	Kluwer Academic Publishers Dordrecht, Boston and London
F Computer and Systems Sciences **G Ecological Sciences** **H Cell Biology**	Springer-Verlag Berlin, Heidelberg, New York, London, Paris and Tokyo

Series D: Behavioural and Social Sciences - Vol. 49

Vigilance and Performance in Automatized Systems/ Vigilance et Performance de l'Homme dans les Systèmes Automatisés

edited by

A. Coblentz
Université René Descartes,
Laboratoire d'Anthropologie et d'Ecologie Humaine,
Paris, France

Kluwer Academic Publishers

Dordrecht / Boston / London

Published in cooperation with NATO Scientific Affairs Division

Proceedings of the NATO Advanced Research Workshop on
Vigilance and Performance in Automatized Systems
Paris, France
19–23 September 1988

Library of Congress Cataloging in Publication Data

Vigilance and performance in automatized systems : proceedings / NATO advanced research workshop, Paris, 1988 ; edited by A. Coblentz = Vigilance et performance de l'homme dans les systèmes automatisés : comptes rendus / journées d'études prospectives OTAN, Paris, 19-23 septembre 1988 ; edité par A. Coblentz.

p. cm. -- (NATO ASI series. Series D, Behavioural and social sciences ; vol. 49)

English and French.

At head of title: NATO International Scientific Exchange Programme.

"C'est avec le soutien de la Division des affaires scientifiques de l'Organisation du Traité de l'Atlantique Nord que ces journées d'études ont été organisées".

Includes index.

ISBN-13: 978-94-010-6930-4

1. Man-machine systems--Congresses. 2. Fatigue--Congresses. I. Coblentz, A. M. (Alex M.), 1930- . II. NATO International Scientific Exchange Programme. III. North Atlantic Treaty Organization. Scientific Affairs Division. IV. Title: Vigilance et performance de l'homme dans les systèmes automatisés. V. Series: NATO ASI series. Series D, Behavioural and social sciences ; no. 49.

T59.72.V54 1989
620.8'2--dc20 89-11061

ISBN-13: 978-94-010-6930-4 e-ISBN-13: 978-94-009-0981-6

DOI: 10.1007/978-94-009-0981-6

Published by Kluwer Academic Publishers,
P.O. Box 17, 3300 AA Dordrecht, The Netherlands.

Kluwer Academic Publishers incorporates the publishing programmes of
D. Reidel, Martinus Nijhoff, Dr W. Junk and MTP Press.

Sold and distributed in the U.S.A. and Canada
by Kluwer Academic Publishers,
101 Philip Drive, Norwell, MA 02061, U.S.A.

In all other countries, sold and distributed
by Kluwer Academic Publishers Group,
P.O. Box 322, 3300 AH Dordrecht, The Netherlands.

Printed on acid free paper

C'est avec le soutien de la Division des Affaires Scientifiques de l'Organisation du Traité de l'Atlantique Nord que ces Journées d'Etudes ont été organisées.

Le Docteur Craig SINCLAIR, Directeur des Programmes des Journées d'Etudes Prospectives de l'OTAN, a constamment témoigné un vif intérêt pour notre projet ; qu'il soit remercié pour son aide efficace sans laquelle notre réunion n'aurait pu se tenir.

Le Professeur Louis AUQUIER, Président de l'Université René Descartes accepta, avec la plus grande bienveillance, d'accueillir le colloque dans les locaux de notre Université.

Nous tenons à lui exprimer ici notre reconnaissance.

SOMMAIRE

CONTENTS

SOMMAIRE
CONTENTS

PREFACE

PREFACE

Il y a une ou deux décennies les ingénieurs ont cru ou voulu faire croire que l'automatisation des systèmes de production et de contrôle permettrait de s'affranchir entièrement de ce maillon faillible que constitue l'Homme.On a même vu une réunion scientifique se réunir sur le thème "L'homme élément limitatif de l'utilisation des systèmes d'armes". Une telle attitude montre avant tout que l'on n'a pas su intégrer l'utilisateur dans la conception du système considéré. Elle consacre la faillite d'un concept.

On est revenu à une notion plus réaliste de la situation. En effet, même dans les systèmes dits "entièrement automatisés" on trouve toujours un ou plusieurs hommes, indispensables pour surveiller le déroulement du processus, intervenir lorsque se produit une défaillance, voire reprendre en main l'ensemble des opérations lorsque survient un incident.

Le développement des automatismes modifie évidemment la tâche de l'opérateur. L'homme contribue moins directement à la production elle-même, en ce sens qu'on n'exige plus de lui un effort physique. On lui confie, dans des conditions de travail encore inhabituelles et auxquelles il n'est pas spécialement préparé, des systèmes dont les coûts sont très élevés et qui risquent d'être détériorés s'il commet la moindre faute. Sans compter les risques qu'il est alors susceptible de faire courir à ses concitoyens.

Le bon sens reprenant le dessus, on s'intéresse maintenant à la place de l'Homme confronté à ces situations qu'il rencontre de plus en plus fréquemment dans l'industrie.

La monotonie de bon nombre de postes de travail modernes conduit à se poser bien des questions sur la sélection des personnels, leur formation initiale et en cours de carrière, la simulation des tâches, celle des pannes et incidents, etc. En outre maints postes de travail bousculent les rythmes biologiques sans que les conséquences en soient véritablement identifiées et analysées.

Il faut reconnaître que les scientifiques ne savent pas encore très bien résoudre toutes ces questions, relativement nouvelles, mais qui se posent avec beaucoup d'acuité du fait de leur universalité.

En première approche on peut dire qu'il existe deux familles de situations dégradées :

- celles pour lesquelles l'homme doit intervenir immédiatement, de façon réflexe, pour rétablir la situation,
- celles pour lesquelles au contraire l'intervention de l'opérateur peut être différée plus ou moins longtemps, lui donnant la possibilité de réfléchir avant d'agir. Il faut attirer l'attention sur le fait que ce n'est pas dans ce genre de situations que les erreurs sont les moins fréquentes.

Face à ces nouvelles contraintes, que l'on ne sait pas toujours affronter de façon constructive, il est apparu nécessaire de confronter les travaux des meilleures équipes travaillant sur ces sujets difficiles.

L'objectif était double :

- en premier lieu faire le point des acquis sans attendre que de nouveaux résultats soient publiés. Actuellement, les publications rapportent des résultats obtenus deux ou trois ans avant leur parution dans la littérature ;

- examiner les protocoles proposés par les différentes équipes de manière à les harmoniser si possible et, en outre, aplanir un certain nombre de difficultés sémantiques : les mêmes mots ne possédant pas forcément la même signification quand ils sont utilisés par des équipes différentes.

On peut, je crois, admettre que ces deux objectifs ont été atteints. Certes, bien des questions sont restées sans réponse car la situation se révèle très complexe, mais nous avons constaté que les sujets d'étude développés par les différentes équipes représentées :
- étaient abordés avec des procédures faisant appel simultanément à la recherche de base et à l'expérimentation sur le terrain,
- utilisaient des techniques identiques ou au moins comparables, tant pour l'acquisition que pour l'exploitation des données,
- couvraient bon nombre de situations rencontrées depuis peu dans l'industrie et les transports.

Dans ces conditions, on peut espérer des progrès rapides pour les différentes équipes, malgré la diversité des problèmes affrontés.

L'ensemble des communications présentées à ces journées de Septembre 1988 se trouvent réunies dans ce volume. Elles ont été classées, indépendamment de la spécificité des équipes, physiologie, neurophysiologie, psychologie et pharmacologie, en fonction des trois thèmes suivants :
- méthodes d'évaluation de la vigilance et de la performance,
- évaluation de la charge de travail dans les systèmes totalement ou partiellement automatisés,
- incidence des substances médicamenteuses sur la vigilance et la performance.

Il me paraît impossible de terminer sans dire combien notre équipe a été honorée de recevoir les scientifiques qui ont accepté notre invitation. Pendant ces quelques jours une ambiance chaleureuse s'est établie et développée, n'excluant pas des discussions parfois âpres, mais attestant de la volonté de chacun de faire progresser les connaissances sur des sujets difficiles qui constituent une épreuve nouvelle pour l'homme et qu'il faudra résoudre pour qu'il puisse bénéficier des acquis du progrès technique.

Un nouvel humanisme se développe, peut-on imaginer qu'il se mette en place au détriment de ceux qui le créent et qui contribuent à son épanouissement ?

Professeur A. COBLENTZ

VIGILANCE AND PERFORMANCE IN AUTOMATIZED SYSTEMS

LISTE DES PARTICIPANTS

LIST OF PARTICIPANTS

BEATTY J.
Department of Psychology
and Brain Research Institute
University of California
Los Angeles, California 90024
(ETATS-UNIS)

BERTIN M.
Electricité de France (E.D.F.)
Comité de Radioprotection
3, rue de Messine
75384 PARIS Cedex 08
(FRANCE)

BLOMBERG R.D.
Dunlap and Associates, Inc.
17 Washington Street
Norwalk, Connecticut 06854
(ETATS-UNIS)

BORREDON J.C.
Direction des Recherches,
Etudes et Techniques (D.R.E.T.)
Biologie et Sciences humaines
26, Boulevard Victor
75015 PARIS
(FRANCE)

BROUGHTON R.J.
Division of Neurology
Ottawa General Hospital and
University of Ottawa
501 Smyth Rd
Ottawa, K1H 8L6
(CANADA)

CABON Ph.
Laboratoire d'Anthropologie Appliquée
45, rue des Saints-Pères
75270 PARIS Cedex 06
(FRANCE)

CARMIGNIANI V.
Direction Générale de l'Aviation Civile
Service de la Formation Aéronautique et du
Contrôle Technique
246, rue Lecourbe
75732 PARIS Cedex 15
(FRANCE)

COBLENTZ A.
Université René Descartes
Laboratoire d'Anthropologie et d'Ecologie Humaine
45, rue des Saints-Pères
75270 PARIS Cedex 06
(FRANCE)

EYSENCK M.W.
Department of Psychology
Royal Holloway and Bedford New College
University of London
Egham Hill, Egham
Surrey, TW 20 OEX
(GRANDE BRETAGNE)

FOLKARD S.
MRC/ESRC Social and Applied Psychology Unit
University of Sheffield
Sheffield S10 2TN
(GRANDE BRETAGNE)

FOUILLOT J.P.
Laboratoire de Physiologie
Faculté de Médecine Cochin
24, rue du Faubourg Saint-Jacques
75014 PARIS
(FRANCE)

FULLER R.G.C.
Department of Psychology
Trinity College
25 - 28 Westland Row
Dublin 2
(IRLANDE)

GAILLARD A.W.K.
TNO Institute for perception
Tilburg University
P.O. Box 23
3769ZG Soesterberg
(PAYS BAS)

GRAEBER R.C.
Aerospace Human Factors Research Division
NASA - Ames Research Center
Moffett Field, California 94035
(ETATS-UNIS)

GUILLEMINAULT C.
Stanford University
Sleep Disorders Center
701 Welch Rd, suite 2226
Palo Alto, Ca 94304
(ETATS-UNIS)

HOCKEY G.R.J.
MRC/ESRC Social and Applied Psychology Unit
University of Sheffield
Sheffield S10 2TN
(GRANDE BRETAGNE)

IGNAZI G.
Centre National de la Recherche Scientifique
(C.N.R.S.)
Laboratoire d'Anthropologie Appliquée
45, rue des Saints-Pères
75270 PARIS Cedex 06
(FRANCE)

INOMATA O.
Railway Technical Research Institute
2.8.38 Ikari-cho, Kokubunji-shi
Tokyo 185
(JAPON)

JOUFFRAY L.
Centre d'Etudes du Bouchet
Le Bouchet
B.P. 13
91710 Vert-Le-Petit
(FRANCE)

JOURET R.
Société Nationale des Chemins de Fer Français
(S.N.C.F.)
Direction du Personnel
Place Louis Armand
75012 PARIS
(FRANCE)

KOREISHI K.
T.R.D.I. Japan Defense Agency
2.2.1. Nakameguro, Meguro Ku
Tokyo-To 153
(JAPON)

LAVIE P.
Sleep Disorders Center,
Faculty of Medicine
Guthwirth Bldg, Technion City
Haifa 32000
(ISRAEL)

LEONARD J.P.
Boeringer Ingelheim KG
Labor für Experimentelle
Pharmakopsychologie und EEG
Abt Medizin/HPZ
6507 Ingelheim
(R.F.A.)

MOLLARD R.
Centre National de la Recherche Scientifique
(C.N.R.S.)
Laboratoire d'Anthropologie Appliquée
45, rue des Saints-Pères
75270 PARIS Cedex 06
(FRANCE)

MUZET A.
Laboratoire de Physiologie
et de Psychologie Environnementales
21, rue Becquerel
67087 STRASBOURG Cedex
(FRANCE)

O'DONNELL R.D.
NTI, Incorporated
4130 Linden Avenue
Suite 235
Dayton, Ohio 45432
(ETATS-UNIS)

PASCOE P.A.
Royal Air Force Institute of Aviation Medicine
Farnborough
Hampshire GU14 6SZ
(GRANDE BRETAGNE)

PRONOST A.
Direction Générale de l'Aviation Civile
Service de la Formation Aéronautique et du
Contrôle Technique
246, rue Lecourbe
75732 PARIS Cedex 15
(FRANCE)

ROSCOE A.H.
Britannia Airways
Luton
Bedforshire LU2 9ND
(GRANDE BRETAGNE)

Mrs Nadine SABLOWSKI-SARTER
183, West Pacemont Avenue
Colombus, Ohio 43202
(ETATS-UNIS)

SHINGLEDECKER C.A.
NTI, Incorporated
4130 Linden Avenue
Suite 235
Dayton, Ohio 45432
(ETATS-UNIS)

SINCLAIR C.
Division des Affaires Scientifiques
Organisation du Traité de l'Atlantique Nord
(OTAN)
1110 Bruxelles
(BELGIQUE)

SPENCER M.B.
Royal Air Force Institute of Aviation Medicine
Farnborough
Hampshire GU14 6SZ
(GRANDE BRETAGNE)

SPEYER J.J
Flight Division
Airbus Industrie
1, Rond-Point Maurice Bellonte
31707 BLAGNAC
(FRANCE)

STAMPI C.
Human Neurosciences Research Unit
University of Ottawa
451 Smyth Rd
Ottawa, K1H 8M5
(CANADA)

TATTERSALL A.J.
MRC/ESRC Social and Applied Psychology Unit
University of Sheffield
Sheffield S10 2TN
(GRANDE BRETAGNE)

UGAJIN H.
Railway Technical Research Institute
2.8.38 Ikari-cho, Kokubunji-shi
Tokyo 185
(JAPON)

YAMASAKI K.
T.R.D.I. Japan Defense Agency
2.2.1. Nakameguro, Meguro Ku
Tokyo-To 153
(JAPON)

VIGILANCE AND PERFORMANCE IN AUTOMATIZED SYSTEMS

LIST OF LECTURERS

BEATTY J.
Department of Psychology
and Brain Research Institute
University of California
Los Angeles, California 90024
(ETATS-UNIS)

BLOMBERG R.D.
Dunlap and Associates, Inc.
17 Washington Street
Norwalk, Connecticut 06854
(ETATS-UNIS)

BROUGHTON R.J.
Division of Neurology
Ottawa General Hospital and
University of Ottawa
501 Smyth Rd
Ottawa, K1H 8L6
(CANADA)

CABON Ph.
Laboratoire d'Anthropologie Appliquée
45, rue des Saints-Pères
75270 PARIS Cedex 06
(FRANCE)

EYSENCK M.W.
Department of Psychology
Royal Holloway and Bedford New College
University of London
Egham Hill, Egham
Surrey, TW 20 OEX
(GRANDE BRETAGNE)

FOLKARD S.
MRC/ESRC Social and Applied Psychology Unit
University of Sheffield
Sheffield S10 2TN
(GRANDE BRETAGNE)

FOUILLOT J.P.
Laboratoire de Physiologie
Faculté de Médecine Cochin
24, rue du Faubourg Saint-Jacques
75014 PARIS
(FRANCE)

FULLER R.G.C.
Department of Psychology
Trinity Collège
25-28 Westland Row
Dublin 2
(IRLANDE)

GAILLARD A.W.K.
TNO Institute for Perception
Tilburg University
P.O. Box 23
3769ZG Soesterberg
(PAYS BAS)

GRAEBER R.C.
Aerospace Human Factors Research Division
NASA -Ames Research Center
Moffett Field, California 94035
(ETATS-UNIS)

GUILLEMINAULT C.
Stanford University
Sleep Disorders Center
701 Welch Rd, suite 2226
Palo Alto, Ca 94304
(ETATS-UNIS)

HOCKEY G.R.J.
MRC/ESRC Social and Applied Psychology Unit
University of Sheffield
Sheffield S10 2TN
(GRANDE BRETAGNE)

INOMATA O.
Railway Technical Research Institute
2.8.38 Ikari-cho, Kokubunji-shi
Tokyo 185
(JAPON)

KOREISHI K.
T.R.D.I. Japan Defense Agency
2.2.1. Nakameguro, Meguro Ku
Tokyo-To 153
(JAPON)

LAVIE P.
Sleep Disorders Center,
Faculty of Medicine
Guthwirth Bldg, Technion City
Haifa 32000
(ISRAEL)

LEONARD J.P.
Boeringer Ingelheim KG
Labor für Experimentelle
Pharmakopsychologie und EEG
Abt Medizin/HPZ
6507 Ingelheim
(R.F.A.)

MUZET
Laboratoire de Physiologie
et de Psychologie Environnementales
21, rue Becquerel
67087 STRASBOURG Cedex
(FRANCE)

O'DONNELL R.D.
NTI, Incorporated
4130 Linden Avenue
Suite 235
Dayton, Ohio 45432
(ETATS-UNIS)

PASCOE P.A.
Royal Air Force Institute of Aviation Medicine
Farnborough
Hampshire GU14 6SZ
(GRANDE BRETAGNE)

ROSCOE A.H.
Britannia Airways
Luton
Bedforshire LU2 9ND
(GRANDE BRETAGNE)

Mrs Nadine SABLOWSKI-SARTER
183, West Pacemont Avenue
Colombus, Ohio 43202
(ETATS-UNIS)

SHINGLEDECKER C.A.
NTI, Incorporated
4130 Linden Avenue
Suite 235
Dayton, Ohio 45432
(ETATS-UNIS)

SPENCER M.B.
Royal Air Force Institute of Aviation Medicine
Farnborough
Hampshire GU14 6SZ
(GRANDE BRETAGNE)

SPEYER J.J.
Flight Division
Airbus Industrie
1, Rond-Point Maurice Bellonte
31707 BLAGNAC
(FRANCE)

STAMPI C.
Human Neurosciences Research Unit
University of Ottawa
451 Smyth Rd
Ottawa, K1H 8M5
(CANADA)

UGAJIN H.
Railway Technical Research Institute
2.8.38 Ikari-cho, Kokubunji-shi
Tokyo 185
(JAPON)

YAMASAKI K.
T.R.D.I. Japan Defense Agency
2.2.1. Nakameguro, Meguro Ku
Tokyo-To 153
(JAPON)

I

METHODES D'EVALUATION DE LA VIGILANCE ET DE LA PERFORMANCE

METHODS OF VIGILANCE AND HUMAN PERFORMANCE EVALUATION

NEUROPHYSIOLOGY OF SUSTAINED ATTENTION

JACKSON BEATTY
Department of Psychology
and
Brain Research Institute
University of California, Los Angeles
Los Angeles, California 90024
USA

ABSTRACT. Sustained attention requires the elaborated processing of sensory information by an alert and aroused nervous system. The spectrum of the electroencephalogram provides useful information concerning the state of CNS arousal as related to vigilance performance. Similarly, task-evoked pupillary responses reflect the intensity of cognitive processing, which declines during prolonged sensory monitoring. Such effects are not confined to simple sensory tasks, but are also observed in monotonous tests of complex reasoning.

1. Sustained Attention

"There is no such thing as voluntary attention sustained for more than a few seconds at a time," wrote William James in his classic textbook, *The Principles of Psychology* (James, 1890). "What is called sustained voluntary attention is a repetition of successive efforts which bring back the topic to the mind [p. 421]."

As usual, James's insightful analysis has stood up well to the test of time. He recognized clearly the deleterious effects of monotony on attentive behavior: "No one can possibly attend continuously to an object that does not change [p. 421]." And he emphasized ever so clearly and elegantly the importance of mental alertness if attention is to be successfully maintained: "The *conditio sine qua non* of sustained attention to a given topic of thought is that we should roll it over and over incessantly and consider different aspects and relations of it in turn....And now we can see why it is that what is called sustained attention is the easier, the richer in acquisitions and the fresher and more original the mind. In such minds, subjects bud and sprout and grow. At every moment, they are pleased by a new consequence and rivet the attention afresh [p. 423]."

Today—nearly a century later—we continue to hold James's view of the psychology of sustained attention. Monotony brings the demise of vigilance. Mental activity holds

A. Coblentz (ed.), Vigilance and Performance in Automatized Systems/ Vigilance et Performance de l'Homme dans les Systèmes Automatisés, 3–12.

the key to successful performance in monotonous environments. Alertness provides the essential substrate for maintaining attention to signals that rarely, if ever, change.

Such issues become increasingly important with the emergence of large automated man/machine systems, in which the human is often assigned only monitoring functions while the machine performs more interesting tasks. Examples of such automated systems may be found in many aspects of contemporary life, including radar and sonar operation, the piloting of large aircraft, industrial inspection, and medical patient monitoring.

The Jamesian approach to understanding sustained attention has had implications for neuroscience as well. By emphasizing the relations between attention and alertness, the contribution to the study of vigilance that could be made by neurophysiology is suggested. Judiciously employed, physiological indicators can provide relevant converging information to aid in the specification of unobservable psychological processes (Posner *et al.*, 1987). The study of sustained attention is a particularly appropriate topic of a joint behavioral and biological analysis, as physiological indicators of both mental alertness and the intensity of cognitive processing are available (Beatty, 1986a).

2. EEG Spectra Reflect CNS Alertness

The electroencephalogram [EEG] recorded from the human scalp is the most useful physiological indicator of central nervous system activation and alertness (Lindsley, 1987). Long regarded as the definitive CNS indicator of the sleep-waking cycle, the EEG has also proven capable of measuring more subtle shifts of alertness that naturally occur during waking.

Belyavin and Wright (1987), for example, recently have examined the changes in EEG spectra that occur during the course of a pair of digit and letter discrimination tasks that were performed over a 15-hour period. EEG recorded between P3-O1 was spectrally analysed over several minute epochs. Comparisons were then made between averaged spectra with very high error rates (75 percent or greater) and all other epochs. A canonical variate analysis of these spectra suggested that periods of low vigilance were marked by an increase in delta (1-3 Hz) and theta (4-7 Hz) activity that is coupled with a decrease in beta (14-21 Hz) activity. Further analyses suggested that it was the increase in theta and the decrease in beta activity that contributed most substantially to Belyavin and Wright's results.

In a similar task, Winsum, Sergeant, and Geuze (1984) examined the effect of presentation rate upon both performance and the suppression of alpha frequency activity induced by task-relevant stimuli. They found that both the extent and duration of alpha suppression increase at slow presentation rates. These changes in alpha suppression were associated with a decline in accuracy and an increase in reaction time. Their data may be interpreted as indicative of decreased general alertness during the slow event rate task.

Such findings are similar to those previously reported in a series of experiments from my own laboratories, although our analyses were somewhat less sophisticated. For example, Beatty and Kornfeld (1977) measured EEG spectra as seven individuals performed a continuous tracking task over a 75-minute period. In that task, errors systematically increased as a function of time in the task. In five of these individuals, there were significant positive correlations between performance decrement and increases of occipital theta activity. There was also an accompanying increase in power in the high alpha frequency range.

In a related experiment, O'Hanlon and Beatty (1977) examined the relation between EEG spectra and sustained attentive performance in twenty college students performing a simulated sea-surveillance radar monitoring task. Using a CRT resembling a standard radar Planned Position Indicator, subjects were required to monitor the display for occasional targets displayed against a background of visual noise. On the first day subjects were trained to perform the monitoring task. On a subsequent testing day, each subject engaged in a two-hour sustained vigilance test in which five targets were presented in each of the four thirty-minute quarters of the task. This sustained monitoring period was preceded and followed by two brief (five-minute) alerted tests in which eight targets were presented. One channel of EEG (P3-O1) was recorded during the sustained attention task.

Under these conditions, there were significant correlations between EEG spectra and task performance reflecting a decline in alertness. Theta and alpha power both increased as performance declined, whereas the power of higher frequency beta activity decreased. Such findings are congruent with those reported by Belyavin, Kornfeld, and others (see O'Hanlon and Beatty, 1977).

The relation between EEG and alertness has also been demonstrated outside the physiological laboratory. Recently, Torsvall and Ålerstedt (1987) measured continuous EEG (O2-P4) spectra in eleven train drivers during a four-and-one-half hour trip through southern Sweden, once in daylight and another time at night. Subjective sleepiness ratings were obtained for each of the seven segments of the journey.

Torsvall and Ålerstedt provided clear evidence that reported sleepiness, the occurance of slow eye movements, and both alpha and theta power increase across segments of the journey at night, but not in the daytime. Lapses of attention, as indicated by missing a signal, were preceded by increased low frequency EEG activity. Significant and large intra-individual correlations between rated sleepiness and EEG delta, alpha, and theta activity were also present.

Thus, there is considerable evidence that EEG spectral patterns may reflect CNS alertness under conditions requiring sustained attention. Under such conditions, increases in EEG delta and theta activity are related to impaired performance, as are decreases in beta frequency activity. For the alpha frequency band, results are less consistent.

3. Task-Evoked Pupillary Responses Reflect Processing Intensity

A second approach to studying the neurophysiology of vigilance examines the responses of the peripheral nervous system to task-relevant stimuli. This line of research exploits the fact that stimulus-evoked changes in autonomic activity reflect the extent of cognitive processing evoked by a stimulus. The most widely studied autonomic indicator of cognitive processing is the task-evoked pupillary response (Beatty, 1982; 1986b), although similar results have been obtained using cardiovascular, skin conductance, and spinal reflex indicators.

Task-evoked pupillary responses [TEPRs] are pupillary dilations elicited by a sensory stimulus or motor response. In this respect, the TEPR is the pupillary equivalent of the EEG's event-related potential. Produced jointly by transient sympathetic activation and parasympathetic inhibition, these small dilations begin within .1-.2 sec following the onset of a meaningful stimulus and are maintained during a brief period of cognitive processing, returning to baseline as processing is complete. Like voluntary attention, TEPRs are characteristically brief.

Perhaps most important is that the amplitude of this phasic pupillary response reflects the degree of processing load imposed by the cognitive task. For example, in a digit span task in which subjects hear and then repeat a string of between 3 and 7 digits, pupillary diameter increases as each successive digit is stored in memory and decreases as each digit is subsequently spoken, returning to baseline at the completion of the response. Peak pupillary diameter is a monotonic function of the number of digits momentarily stored in memory (Kahneman and Beatty, 1966). Similar results have been reported for a wide variety of cognitive tasks, leading to the general conclusion that the task-evoked pupillary response provides a physiological index of the extent of mental processing evoked by meaningful sensory stimuli (Beatty, 1986b).

4. Task-Evoked Pupillary Responses in Vigilance

If the TEPR reflects the extent of task-related stimulus processing and—as James asserted—the essence of sustained attention is active stimulus processing, then task-evoked pupillary dilations could provide a physiological means of tracking vigilance and its decline in monitoring performance.

Beatty (1982b) measured both the task-evoked pupillary response and monitoring performance in a sustained auditory detection task. Eleven university students listened to a series of 1 kHz tone bursts presented at 3.2 sec intervals for a period of 48 minutes. Their task was to indicate the occurrence of infrequent target tones, which were slightly less intense than the standards.

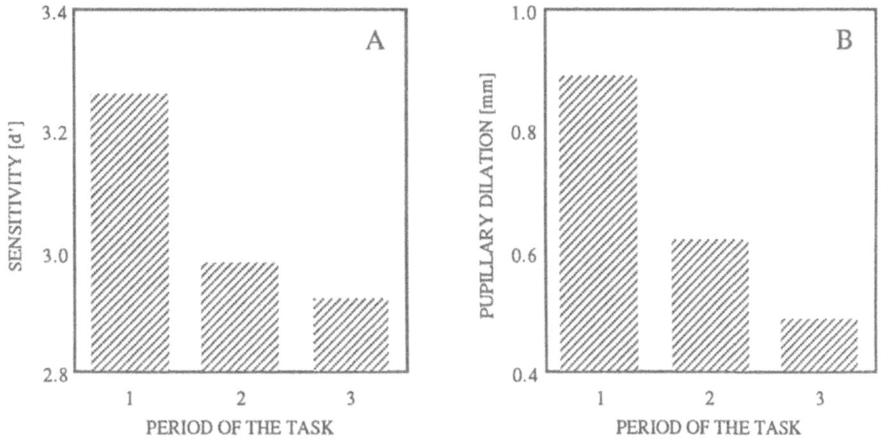

Figure 1. Performance and pupillary data averaged across eleven subjects as a function of period in an auditory vigilance task. A] Sensitivity, as measured by the signal detection statistic d', declines from the first through the third period of the task. B] Similarly, there is a decline in the amplitude of the task-evoked pupillary response. Tonic pupillary diameter remains constant over the course of the task. (After Beatty, 1982b.)

Under these conditions, detection performance declined as a function of time in the task. This vigilance decrement resulted from both a conservative shift of decision criteria and—more importantly—from an absolute decline in listener sensitivity, as determined by signal detection analysis. A loss of sensitivity is the classic definition of a vigilance decrement produced by a decline of sustained attention.

The vigilance decrement was reflected pupillometrically by a corresponding reduction in the amplitude of the phasic task-evoked pupillary. TEPR amplitude decreased more than 40 percent from the first to the last third of the task. In the interpretation of the TEPR first suggested by Kahneman, Beatty, and Pollack (1967), the extent of processing of the task-relevant auditory stimuli had undergone a substantial decline during the course of the monitoring task. The pupillary data suggest quite clearly that the vigilance decrement results from the failure to sustain sufficient cognitive processing of each and every monotonous stimulus as the duration of the watch unfolds. It is task-related cognitive processing, as reflected in the TEPR, that declines and with it brings the demise of vigilance.

5. Monitoring in Complex Systems: Examples from Medicine

Simple monitoring tasks, such as that employed by Beatty (1982b), have formed the basis of most contemporary work on vigilance. Such studies are derived from experimental work that began after World War II and that was motivated by a need to understand the increasing failure of radar operators to detect enemy aircraft as time in the watch increased.

Norman Mackworth (1948) first employed a laboratory analog of the radar monitoring task to quantitatively analyse the breakdown of vigilance in sustained visual search. The visual search task—with infrequently presented signals against a background of noise—had a number of distinct experimental advantages: it was easy to implement, easy to analyse, and perhaps most important, it had practical relevance. Radar was the key to air defense in the developing postwar world and operator error was a major factor in determining system reliability.

But much has changed in the four decades since Mackworth's first publication. Automated systems have become much more complex and are controlled to large extent by computer systems that would have been unimaginable only a decade ago. Have these cured the problem of the vigilance decrement? Unfortunately not, for sustaining vigilance is a *human*, not a *machine*, problem.

Difficulty in sustaining attention for prolonged periods, particularly in the face of sleep loss, continues to present real problems. This is particularly apparent in the high technology world of modern medicine, where extremely sophisticated systems of medical monitoring and treatment are increasingly common. For example, consider the highly trained and highly selected anesthesiologists, who operate intricate, largely automated, medical monitoring systems for prolonged periods during surgery. Here, the anesthesiologist is responsible not only for attending to all vital signs, but also for serving as the primary physician involved in medicating the patient. Often, a number of drugs and their interactions are involved. Here the vigilance tasks range from simple monitoring to complex reasoning.

Beatty, Ahern, and Katz (1977) examined the effects of decreased alertness in anesthesiology residents using three different tests of cognitive performance. Six anesthesiology residents were tested on two separate mornings, one following a full night's sleep and the other after working through most of the night.

The simplest monitoring task was a version of Neisser's letter-search test (Neisser, 1963), in which each anesthesiologist scanned columns of letters searching for and marking selected target letters. One, two, or four targets were employed on different trials. In this test, which constitutes the least complicated type of monitoring, search time is a linear function of the number of targets held in memory.

The second task simulated surgical monitoring of six vital signs—heart rate, blood pressure, tidal volume, arterial p[O_2], and arterial p[CO_2]—for fifty minutes using a computer driven display. Occasionally the display would be blanked and the subject would be asked if a particular vital sign was within normal bounds. This procedure

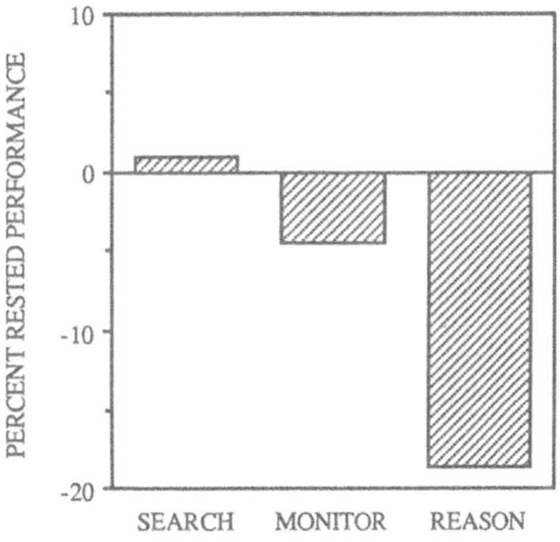

Figure 2. Performance change in 6 surgical anesthesiologists as a function of one night's sleep loss. Neisser's visual letter search task showed no performance decrement. Similarly, little effect of sleep loss was seen in a simulated surgical monitoring task. However, significant and profound difficulties appeared in a test of higher cognitive function, Baddeley's grammatical reasoning task.

was modelled upon the type of patient monitoring actually employed in surgery at the UCLA medical center.

The third test—first suggested by Baddeley (1968)—measured abstract reasoning. Here, the anesthesiologists were required to judge the truth or falsity of simple sentences with exemplars, of the form "A follows B: AB." Individual sentences could be active, passive, or negated. Although seemingly simple, the lack of content of these sentences, coupled with their highly schematized representation of information, provides a serious challenge to anyone's reasoning ability.

One night of sleep loss had no effect on the speed or accuracy of a low-level memory search task and had only moderate and inconsistent effects on a multichannel visual monitoring task. But the decline of alertness had significant and profound effects on Baddeley's mental reasoning task. Sleeploss prevented the anesthesiologists from sustaining a mental life that would permit concepts to "bud and sprout and grow"; thus, the failure of complex reasoning. In complex cognition, the vigilance decrement finally and decisively appeared.

Such findings are not without practical consequence. Today, for example, a major lawsuit is being pressed in New York State concerning the effects of sleep loss in the case of a patient who died as the consequence of an unexpected drug interaction not detected by a sleep deprived resident. This case provides a clear indication of the importance of vigilance on effective abstract reasoning in real life medical monitoring.

6. Conclusion

Sustained attention requires the engagement of the cognitive system at a level commensurate with the demands of the task. When the task is intrinsically interesting and motivating, sustaining attention is never a problem. But when the task is not compelling, when it is prolonged excessively, or when the nervous system is not alert, sustaining attention becomes difficult.

Physiological indicators provide an important approach to the experimental study of sustained attention, in that they permit assessment of both the general level of central nervous system activation and the specific cognitive activity that is evoked by task-relevant stimuli. The judicious use of these techniques may make significant contributions to understanding vigilance in the context of large automated systems, by providing information concerning internal—and otherwise unobservable—states and processes.

7. References

Baddeley, A.D. A three-minute reasoning test based on grammatical transformation. *Psychonomic Science*, 1968, **10**, 341-342.

Beatty, J. Computation, control, and energetics: A biological perspective. In G.R.J. Hockey, A.W.K. Gaillard, and M.G.H. Coles (Eds.), *Energetics and Human Information Processing*. Dordrect, Netherlands: Martinus Nijhoff, 1986a, 43-52.

Beatty, J. The pupillary system. In M.G.H. Coles, S.M. Porges, and E. Donchin (Eds.), *Psychophysiology: Systems, Processes, and Applications*. New York: Guilford Press, 1986b, 43-50.

Beatty, J. Task-evoked pupillary responses, processing load, and the structure of processing resources. *Psychological Bulletin*, 1982a, **91** (2), 276-292.

Beatty, J. Phasic not tonic pupillary responses vary with auditory vigilance performance. *Psychophysiology*, 1982b, **19**, 167-172.

Beatty, J., Ahern, S.K., and Katz, R. Sleep deprivation and the vigilance of anesthesiologists during simulated surgery. In R. Mackie (Ed.), *Vigilance: Theory, Operational Performance, and Psychophysiological Correlates*. New York: Plenum Press, 1977, 189-202.

Belyavin, A., and Wright, N.A. Changes in electrical activity of the brain with vigilance. *Electroencephalography and Clinical Neurophysiology*, 1987, **66**, 137-144.

James, W. *The Principles of Psychology, Vol. I*. New York: Henry Holt and Company, 1890.

Kahneman, D., and Beatty, J. Pupil diameter and load on memory. *Science*, 1966, **154**, 1583-1585.

Kahneman, D., Beatty, J., and Pollack, I. Perceptual deficit during a mental task. *Science*, 1967, **157**, 218-219.

Kornfeld, C.M., and Beatty, J. EEG spectra during a long-term compensatory tracking task. *Bulletin of the Psychonomic Society*, 1977, **10** (1), 46-48.

Lindsley, D.B. Activation, arousal, alertness, and attention. In G. Adelman (Ed.), *Encyclopedia of Neuroscience, Vol. I*. Boston: Birkhäuser, 1987, 3-6.

Mackworth, N.H. The breakdown of vigilance during prolonged visual search. *Quarterly Journal of Experimental Psychology*, 1948, **1**, 6-21.

Neisser, U., Novick, R., and Lazar, R. Searching for ten targets simultaneously. *Perceptual and Motor Skills*, 1963, **17**, 955-961.

O'Hanlon, J.F., and Beatty, J. Concurrence of electroencephalographic and performance changes during a simulated radar watch and some implications for the arousal theory of vigilance. In R. Mackie (Ed.), *Vigilance: Theory, Operational Performance, and Psychophysiological Correlates*. New York: Plenum Press, 1977, 189-202.

Posner, M.I., Inhoff, A.W., Friedrich, F.J., and Cohen, A. Isolating attentional systems: A cognitive-anatomical analysis. *Psychobiology*, 1987, **15** (2), 107-121.

Torsvall, L., and Åkerstedt, T. Sleepiness on the job: Continuously measured EEG changes in train drivers. *Electroencephalography and Clinical Neurophysiology*, 1987, **66**, 502-511.

Winsum, W.V., Sergeant, J., and Geuze, R. The functional significance of event-related desynchronization of alpha rhythm in attentional and activating tasks. *Electroencephalography and Clinical Neurophysiology*, 1984, **58**, 519-524.

Acknowledgment

Research supported in part by grant MH 37430-07 from the National Institute of Mental Health.

THE MAINTENANCE OF VIGILANCE DURING AUTOMATED MONITORING

G. Robert J. Hockey and Andrew J. Tattersall
MRC/ESRC Social and Applied Psychology Unit
University of Sheffield
S10 2TN
England

ABSTRACT. This paper is concerned with the maintenance of vigilance in modern, complex human-machine systems. It questions the applicability of models of the human operator derived from tradional studies of watchkeeping, since these assume a ´passive´ attentional state. It argues instead for a model in which the operator actively regulates his cognitive state to adapt to changes in prevailing task and internal demands, through the adoption of strategies which permit a trade-off between energetical costs and performance benefits.

1. INTRODUCTION

1.1. Conceptions of Vigilance

The term "vigilance" has traditionally been defined as a state of heightened awareness or alertness. French use of the term has emphasised the broader concept of a state of general arousal and enhanced mental activity, in line with Head´s (1926) definition: "a state of high-grade efficiency of the nervous system". In contrast, since Mackworth´s (1950) work on radar detection after the second World War, British and American psychology has focussed on the narrower perceptual readiness aspects of this state, in relation to the detection of small, infrequent changes in displays. This distinction takes on a renewed importance when we consider the problems of sustained attention in automated human-machine systems. We would argue that the problems facing the operator cannot usefully be thought of in terms of monitoring for well-defined environmental changes. Instead, the vigilance required of him or her is of the French kind: He or she must be generally alert, and be actively involved in searching, problem-solving, predicting and planning.

1.2. Active and Passive Vigilance

The essence of vigilance is sustained alertness: But this may involve a number of components. We distinguish in this paper two general

A. Coblentz (ed.), Vigilance and Performance in Automatized Systems/ Vigilance et Performance de l'Homme dans les Systèmes Automatisés, 13–22.

models of sustained alertness: passive and active. In the classical watchkeeping task the operator is required to sustain a state which allows him to detect occasional visual targets - a passive receiver/detector mode. In semi-automated human-machine systems, such as air traffic control or chemical production, however, sustaining attention is emphatically an active process. In addition to monitoring for faults, the operator needs to be constantly updating his information about system changes, to be planning contingency actions - depending on the outcome of current operational steps, and to be considering the implications of current changes for future required states (Sheridan, 1981).

Even the more clearly perceptual aspects of this task are different from the classical (passive) model of the radar operator. In such complex systems monitoring involves active search for configurations of symptoms of plant state (Rasmussen, 1986), rather than waiting for discrete isolated events. It may, of course, be argued that the passive model is inappropriate even for classical vigilance tasks, though this only becomes apparent when the operator's job is made more effortful by demands on working memory or by rapid decision-making (Davies & Parasuraman, 1982). As we shall see below, the classical watchkeeping situation does not usually provide either a sufficient flow of information or the necessary feedback from task actions to permit an active mode to be consistently successful.

Maintaining vigilance in automated jobs thus involves more than a state of perceptual readiness. It demands maintenance of alertness, execution of the appropriate level of control, and regulation of effort, both across different component activities and over the duration of the task (Sheridan, 1981). Where a breakdown in performance occurs the problem is not only the classical one of momentary lapses of detection efficiency (whether sensory or decision-based). Disruption may also arise through ineffective use of strategies for information search, inadequate planning, inability to manage high levels of workload, or a failure to deploy resources efficiently over the complete period of work.

2. STRATEGIES OF CONTROL

2.1. Controllability in Human-Machine Systems

What are the implications of the deployment of active control strategies for effectiveness and health? Increased task involvement (operator-generated activity) has long been considered an important strategy for maintaining effectiveness in monotonous tasks (eg, McBain, 1970). In analogues of complex human-machine systems, a number of studies have shown a clear advantage for active (manual, or 'in-the-loop') control over passive control (monitoring or supervising the performance of the automatic controller), in detecting changes in system functioning (see Sheridan, 1987). The superiority of active participation in these studies may not, however, be universal. Detection may be both faster and more reliable in the automatic monitoring mode when either high-order transfer functions are involved or when there are several different sources of information to be monitored.

A central issue here is that of controllability - the potential within the human-machine system for the exercising of discretion in the planning and execution of work. Current models of occupational stress (eg, Karasek, 1979; Frankenhaeuser, 1986) emphasise the fundamental role of controllability in buffering the individual against the effects of heavy workload. In the present context, we would predict that a system which permitted flexibility in the operator's handling of task demands would result in greater overall system performance than one which did not. This means not only a higher level of output or behavioural effectiveness, but effectiveness achieved at acceptable costs to the operator, in terms of strain, discomfort and physiological disturbance. Note that the use of an active control strategy when it is required by the task structure would not necessarily be desirable, or even effective. In a controllable work environment, the operator is able to regulate his own actions, including electing to operate passively for a period.

Below, we outline a model of state regulation which emphasises the costs and benefits of different strategies for coping with disruptions in cognitive balance. Before doing so, however, it is necessary to consider the nature of control in such tasks more directly.

2.2. Modes of Action

2.2.1. Open- and closed-loop control. Complex sequences of actions may be controlled on a number of different levels, and in several modes of feedback use. Rasmussen (1986), for example, distinguishes three kinds of action of process operators in relation to task events: the 'skill' level, the 'rule' level, and the 'knowledge' level, corresponding to increasing involvement of active problem-solving. In terms of use of feedback, the primary distinction is between open-loop and 'closed-loop' modes of control.

Closed-loop control operates on the basis of feedback from the mismatch between required and achieved performance parameters. Actions are designed to achieve particular sub-goals of the task (identifying an annunciator signal; executing the procedure to restore hydraulic system pressure). This kind of control is typical of less experienced operators, or those experiencing unfamiliar tasks (Umbers, 1979), though it is also thought to be more appropriate for dealing with 'difficult' problems (when unexpected or rare faults are diagnosed, or during high workload periods; eg, Bainbridge, 1974). In conditions such as these closed-loop control is, in fact, often preferred, even by experienced process operators (Umbers, 1979). It allows them to take one step at a time, and attend fully to the feedback to each action.

Open-loop strategies, by contrast, enable the operator to control sequences of actions by reference to a plan of the broad pattern of required actions over a longer time period (carrying out the filtration stage of a batch of dyestuffs; maintaining the appropriate system state for take-off). This kind of control typically involves a form of ´feedforward´ through prediction of future task states. It can only be acquired through experience with a particular system, and requires the development of extensive operational knowledge. The advantage to operators is the possibility of more efficient management of the unfolding workload of the job, through planning, scheduling and temporal grouping of activities.

2.2.2. Choice of control mode. In general terms, open-loop control may thus be seen as more skilled and more efficient. It would be a mistake, however, to assume that such control is necessary or even desirable in all cases. As several theorists have pointed out (Broadbent, 1977; Rasmussen, 1986; Reason, 1984), control may be exercised on different levels, depending on the task context, the operator´s state, and so on. For the same sequence of actions by an operator, the use of an open-loop/ feedforward programme (ie, one involving prediction and planning) will normally make greater overall demands on working memory and executive functions than a succession of closed-loop acts, since it entails (a) the maintenance of a task model or plan in the foreground of memory, and (b) the monitoring of mismatches between the model´s predictions and environmental events, to permit ´fine-tuning´ of the plan (Summers, 1981). By contrast, the demands of closed-loop operations, although more obviously deliberate, are short-lived and localised, though more frequent control actions are needed.

Studies of process control (eg, Umbers, 1979) have shown that while open-loop control is a general characteristic of skilled work, operators are likely to ´fall back´ on closed loop strategies when unfamiliar fault states are detected, or when several concurrent activities were being carried out. In general, as we shall argue, continued activity, at any level, which makes excessive demands on executive processes is likely to result in cognitive fatigue and strain. An assertion put forward here is that operators have a need to exercise control over their mode of control! One way they can do this is by trading off their deployment of slowly-changing general decision-making strategies (open-loop) for those of fast turn-over, specific control operations (closed-loop), whenever one becomes too difficult or inappropriate. We now consider how this regulation of cognitive state might be achieved, and what the energetical costs to the operator are.

3. A MODEL OF COGNITIVE STATE REGULATION

3.1. Elements of state regulation

In cybernetic models of action the maintenance of organised, goal-oriented behaviour is treated as the problem of maintaining a state of stable equilibrium. The system regulates its own behaviour by the use of negative feedback from observed discrepancies between desired and achieved outputs. In the context of sustained work and stress the general state control model has been developed in a number of previous papers (eg, Hockey & Hamilton, 1983; Hockey, 1986), so that we need outline only the central elements here.

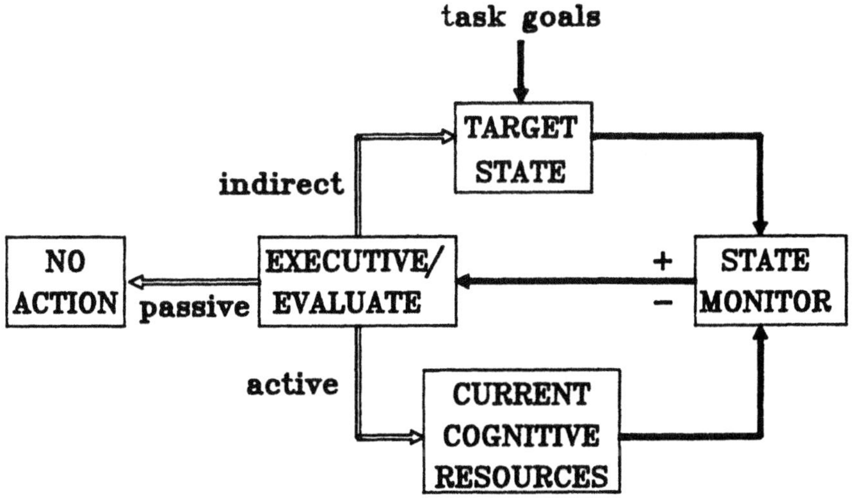

Figure 1. Strategies of state regulation for maintaining alertness.

The main components of state regulation are a state monitor and executive process: see Figure 1. For any well-practiced task a target state is set up in memory to serve as a reference pattern and guide action over the duration of the task. This includes specifications of task goals, the functional levels of relevant information processing resources, priming of component skills, setting of error tolerance limits and decision biases, and so on. The state monitor checks for significant discrepancies between target state and current cognitive activity, resulting from either environmental demands or internal changes. The output is fed to an executive/ evaluation system, which regulates the response to task demands. A monitoring facility of this kind is assumed to be available for both open- and closed-loop control activity, though higher levels of the goal hierarchy will be involved in the case of the former.

3.2. State Management Strategies

Maintenance of the appropriate processing state over prolonged periods of work requires adaptive regulation on the part of the operator. Several different regulatory strategies are illustrated in Figure 1. In general, these are independent of whether the operator is employing open- or closed-loop control, though, as we note below, there are likely to be boundary conditions for such independence. The most direct adaptation involves recruitment of additional processing resources; ie, changing the supply of resources until it matches the required level. This active mode retains task goals, but is effortful and attracts costs, both psychological and physiological (Hockey, 1986). When an active strategy is required by the job structure, or the operator persists with it in the face of heavy workload, costs may become excessive, and the subjective state of effortful control change to one of strain.

The trade-off between performance and costs is central to the analysis of performance in sub-optimal states. For example, in considering the effects of environmental stressors on task performance, high levels of activation may be interpreted as the result rather than the cause of successful maintenance of task orientation (Hockey & Hamilton, 1983; Hockey, 1986). Equally, the observed correlation between performance level and the maintenance of activation over a prolonged monitoring session (eg, O'Hanlon & Beatty, 1976) is interpreted in this model as the result of compensatory effort on the part of the operator, rather a passive consequence of task or environmental factors or of intrinsic arousal modulations.

The second kind of strategy, here labelled indirect, achieves equilibrium by modification of the target state. This may take the form of a reduction in the required speed of search, for example, or a reduced commitment to planning ahead. Task goals are modified downwards to match the supply of resources. The pattern of trade-off here is biased towards reducing costs, with a consequent reduction in performance. This strategy may also be thought of as adaptive, since it provides a way of reducing costs when they become too high, while resolving the problem of mismatch. The third strategy indicated in Figure 1 is essentially passive. In this case the discrepancy is tolerated (or ignored, or not noticed) rather than being resolved. The passive strategy attracts the lowest costs, of course, since no corrective action is required, though performance is likely to suffer from the failure to make use of feedback. This is more likely to occur in the case of open-loop control, since feedback is less direct, and infrequent. Such a strategy may be also be accompanied by negative subjective reactions such as boredom and detatchment. Why, then, should this strategy ever be adopted? One answer is that it may be an inevitable consequence of the design of the human-task system.

3.3. Regulatory Strategies in Sustained Operations

Active state regulation relies on the state monitor to provide a steady flow of information about the efficacy of current actions. This can only occur when there is regular feedback from the environment, in the form of normal task events, or faults and targets. Without dynamic human-task interaction there is no clear input into the evaluation/ executive system. Active control, in such a case, has to be achieved by a top-down strategy, relying extensively on feedforward skill. Maintaining such a state over prolonged periods in the absence of feedback is highly effortful, since executive processes have to be kept in constant use to guide action. We would therefore anticipate that a more manageable passive strategy would eventually replace active control in this kind of situation.

In the case of traditional vigilance tasks there may well be a difference in the nature of control between high and low event rate situations. Vigilance tasks of the ´pure´ variety, involving very low event/signal rates, are known to suffer from shifts of decision criterion, rather than a loss of perceptual discriminability (Davies & Parasuraman, 1982). In terms of state control this represents an adaptive change of target state parameters towards values which may be more ´natural´ for the individual, and more easily maintained in the absence of suitable task information. High event rate tasks, on the other hand, permit regular feedback to occur (albeit of a limited kind), so that active control may be attempted. The limitation on performance in high event rate tasks is typically a loss of discriminability (Davies & Parasuraman, 1982). In terms of state regulation, this is consistent with expectations from the continuous need for paced decision-making, giving rise to brief periods of cognitive fatigue (Broadbent, 1977). The observations of similar brief disruptions or "lapses" in sleep deprivation (Williams, Lubin & Goodnow, 1959), and prolonged work in continuous serial responding (Broadbent, 1963) are associated with tasks having this kind of temporal structure.

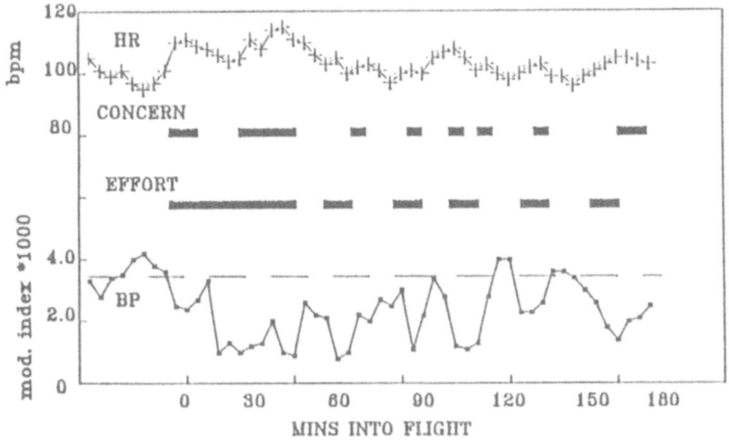

Figure 2. Variations in BP band energy over a 3-hour simulated flight.

Only in tasks which encourage continued high involvement are active strategies likely to be readily maintained, through the use of effortful executive activities. Figure 2 illustrates the pattern of effort seen in such tasks, using preliminary data from a study of workload management in trainee air engineers. It shows energy in the blood pressure (BP) or mid-frequency band (around 0.07 - 0.14 Hz) of the cardiac interval signal (Mulder, 1980), over successive 4-min periods of a 3-hour simulated flight for one trainee. Reduced power in BP band (below the baseline level shown by the dotted line) is indictative of suppression of vagal control, responsible for short-term regulation of arterial pressure. Figure 2 also shows periods of high workload, (derived from analysis of a video film of the session) in the form of either effort-demands (concentration and time pressure) or concern/ anxiety, and imposed 'critical events' of the training session. In addition to the generally lower level of BP band energy during the 'flight', suppression is seen to be associated with periods of high effort, whether or not coincident with critical events. Heart rate, which is also shown, tends to correlate more with concern (e.g. take off, landing and alarms).

Despite the preliminary nature of these data, the potential value of spectral analyses for measuring effort investment in moment-to-moment control of processing state is considerable. We are currently developing methods for studying the relationship between such short-term regulatory activity and costs associated with the longer-term maintenance of task orientation over prolonged work. This involves, for example, increased tonic levels of catecholamines (O'Hanlon & Beatty, 1976), as well as possible chronically low levels of vagal control. These kind of issues become increasingly important as automation makes monitoring work more cognitively demanding on the one hand, and less controllable on the other.

4. MAINTAINING VIGILANCE IN AUTOMATED SYSTEMS

4.1. Problems of Automation

In jobs with extremely high levels of automation, where the human has effectively been 'designed out' of the control process, the operator is required to 'supervise' or monitor the performance of the control computer, which is largely responsible for controlling the system state. This inevitably means a severe reduction in controllability. In supervisory systems the operator/computer dialogue is machine-centred, leaving the human little or no opportunity to initiate actions or make use of personal cognitive management skills (Sheridan, 1981). The resulting detatchment from the task prevents the operator from maintaining an effective mental model of the plant or process (Rasmussen, 1986; Wickens, 1984), and forces him to adopt a passive strategy.

As Wickens and others have argued, this 'out-of-the-loop' role may seriously impair the operator's capability of diagnosing and dealing with unexpected faults, since he has no easy way of keeping track of changes in the system when it is behaving as predicted. Such problems are evident in arguments about the 'sterile cockpit' (Wiener, 1985), where the reduced opportunities for direct control of flying are implicated in a number of accidents.

4.2. Possibilities of Automation

The kinds of difficulty normally associated with automatic control systems are by no means universal, and some degree of automation seems to be highly desirable in many applications. Where they do occur, such difficulties appear to arise from fundamental misconceptions about the nature of the human operator. First, there is a failure to recognise that the automatic system should be used to support human decision-making wherever possible, rather than replace it. Decision support systems can be provided which the operator may consult if he wishes, or which provide independent decisions (eg, Sheridan, 1987). Second, discussions of the allocation of function (eg, Chapanis, 1970) have tended to take the view that humans should be given those jobs which computers cannot do well. As Bainbridge (1987) has pointed out, this means that the operator is not only prevented from taking an active role in the control process, but is expected to be able to switch from being a passive monitor into an active problem-solving mode whenever a problem occurs which the automatic controller cannot handle.

Instead, we would propose that the design of automated systems be based on the principle of controllability. Wherever a human is needed in the system (and this is true for all practical purposes), he or she must be provided with the possibility of regulating the level of control. This means being able to adopt whatever level of interaction with the system is necessary, in order to manage their cognitive and energetical resources effectively. A system which prevents active involvement will lead to boredom and dissatisfaction, as well as a loss of monitoring efficiency over time. On the other hand, one which demands a high level of active processing (high workload tasks involving multiple displays and complex control dynamics, or where all commands must be initiated by the operator) is likely to cause a breakdown in vigilance through fatigue and strain. Naturally, some 'hard-wired' division of function may often be necessary, but allowing operators controllability wherever necessary will not only make their job more satisfying, but allow them to operate more safely and effectively.

5. REFERENCES

Bainbridge, L. (1974). 'Problems in the assessment of mental load'. Le Travail Humain, **37**, 279-304.

Bainbridge, L. (1987). 'Ironies of automation'. In J. Rasmussen, K. Duncan & J. Leplat (Eds.), New Technology and Human Error. Chichester: Wiley.

Broadbent, D. E. (1963). 'Differences and interactions between stresses'. Quarterly Journal of Experimental Psychology, **15**, 205-211.

Broadbent, D. E. (1977). 'Levels, hierarchies and the locus of control'. Quarterly Journal of Experimental Psychology, **29**, 181-201.

Chapanis, A. (1970). 'Human factors in system engineering'. In K.B. DeGreene (Ed.), Systems Psychology. New York: McGraw Hill.

Davies, D. R. & Parasuraman, R. (1982). The Psychology of Vigilance. London: Academic Press.

Frankenhaeuser, M. (1986). ´A psychobiological framework for research on human stress and coping´. In M. H. Appley & R. Trumbull (Eds.), Dynamics of Stress. New York: Plenum.
Head, H. (1926). Aphasia and Kindred Disorders of Speech. New York: Macmillan.
Hockey, G. R. J. (1986). ´A state control theory of adaptation and individual differences in state management´. In G. R. J. Hockey, A. W. K. Gaillard & M. G. H. Coles (Eds.), Energetics and Human Information Processing. Dordrecht: Martinus Nijhoff.
Hockey, G. R. J. & Hamilton, P. (1983). ´The cognitive patterning of stress states´. In G. R. J. Hockey (Ed.), Stress and Fatigue in Human Performance. Chichester: Wiley.
Karasek, R. A. (1979). ´Job demands, job decision latitude, and mental strain: Implications for job redesign´. Administrative Science Quarterly, **24**, 285-308.
McBain, W. N. (1970). ´Arousal, monotony and accidents in line driving´. Journal of Applied Psychology, **54**, 509-519.
Mackworth, N. H. (1950). ´Researches on the measurement of human performance´. Medical Research Council Special Report, **No. 268.** London: HMSO.
Mulder, G. (1980). The Heart of Mental Effort. Groningen, The Netherlands: Groningen University Press.
O´Hanlon, J. F. & Beatty, J. (1976). ´Catecholamine correlates of radar monitoring performance´. Biological Psychology, **4**, 293-304.
Reason, J. T. (1984). ´Lapses of attention in everyday life´. In R. Parasuraman & D. R. Davies (Eds.), Varieties of Attention. New York: Academic Press.
Summers, J. J. (1981). ´Motor programs´. In D. H. Holding (Ed.), Human Skills. Chichester: Wiley.
Rasmussen, J. (1986). Information Processing and Human-Machine Interaction. New York: North-Holland.
Sheridan, T. B. (1981). ´Understanding human error and aiding human diagnostic behavior in nuclear power plants´. In J. Rasmussen & W. B. Rouse (Eds.), Human Detection and Diagnosis of System Failures. New York: Plenum.
Sheridan, T. B. (1987). ´Supervisory control´. In G. Salvendy (Ed.), Handbook of Human Factors. Chichester: Wiley.
Umbers, I. G. (1979). ´Models of the process operator´. International Journal of Man-Machine Studies, **11**, 263-284
Wickens, C. D. (1984). Engineering Psychology and Human Performance. Columbus, Ohio: Merrill.
Wiener, E. L. (1985). ´Beyond the sterile cockpit´. Human Factors, **27**, 75-90.
Williams, H. L. Lubin, A. & Goodnow, J. J. (1959). ´Impaired performance with acute sleep loss´. Psychological Monographs, **73**, No. 14, Whole No. 484.

THE MULTIPLE SLEEP LATENCY TEST: A MEASURE OF DAYTIME SLEEPINESS

Christian Guilleminault
Stanford University Sleep Disorders Center
701 Welch Road, Suite 2226
Palo Alto, CA 94304, U.S.A.

ABSTRACT

Methods such as subjective rating, performance tests, electronic pupillogram, or cerebral evoked potentials had shown some limitations for the assessment of sleepiness. Thus, there was the need to develop a measure of physiological sleep tendency that avoids the need for 24 hour continuous electrophysiological recordings. The Multiple Sleep Latency Test (MSLT) was designed with this goal in mind. It consists of five scheduled naps during which the EEG, EMG, and ECG are recorded. Subjects are given instructions that encourage falling asleep. The mean latency between lights-out time and sleep onset is calculated for the overall test. This value gives an indication of the sleepiness of the subject : lower scores indicate greater sleepiness and higher scores the contrary. The MSLT has proved to be a useful tool in determining excessive sleepiness associated with syndromes such as narcolepsy. It was also used to study the effect of aging, sleep loss, time of day, and hypnotic drugs on sleep tendancy. Study of sleepiness in a low stimulus environment should also be an important issue because of its potential impact on industry.

A. Coblentz (ed.), Vigilance and Performance in Automatized Systems/ Vigilance et Performance de l'Homme dans les Systèmes Automatisés, 23–30.

The daily progression of sleep and wakefulness is a continuous feature of human behavior. Little attention, however, has been paid to the alertness/sleepiness continuum, whose range is wide. Its dimensions cannot be appropriately measured by such criteria as performance, attention, or fatigue testing. Sleepiness and alertness have been seen as opposites, but there is still no clear evidence, as emphasized by Dement and Carskadon (1), that they are two reciprocal mechanisms. An accurate, objective measurement of sleepiness would therefore be very desirable. In a 1961 review article (2), Kamiya described four ways to assess drowsiness/sleepiness: (a) measures of reduced activity, (b) attenuation of 'best effort' on tests of performance, (c) reports of sleepiness on subjective rating scales, and (d) readiness to fall asleep. We tried to measure sleepiness quantitatively, using the Stanford Sleepiness Scale: a seven-point scale giving information on the subjective state of sleepiness (3). It became clear, however, that there were discrepancies between subjective ratings and the behavior that we observed in normal subjects and patients presenting disorders of alertness such as narcolepsy or obstructive sleep apnea. A different approach was needed.

It was possible, using the tools of chemical neurophysiology, to determine sleepiness by measuring the variations of pupil diameter or by studying the late components of cerebral evoked potentials. Yoss et al. (4) described the electronic pupillogram (EPG) as a method to measure decreased levels of alertness. The technique is based on the association of peripheral autonomic manifestations with states of arousal/excitation as well as with sleep. The pupil, an index of autonomic activity, constricts during sleep (5,6). A normal, alert individual sitting quietly in total darkness can maintain a stable pupil diameter that is usually well above 7 mm for at least 10 minutes, without subjective difficulty or pupillary oscillation. In a sleepy subject, however, pupillary diameter is unstable when adapting to the dark. There are problems and limitations with this technique: a subject may experience lid drooping and closure, and investigators may have difficulty in interpreting the data if conditions are not optimal or standardized (6).

The late components of auditory evoked potentials have been explored in drowsy subjects. The changes reported for drowsiness have consisted mainly of reduced amplitudes of these late components (N_1-P_1). These findings have been confirmed for all of the main sensory modalities, including visual and somatosensory (7,8). A marked increase in amplitude of the late component P_2 and especially N_2 during well-established stages 2, 3, and 4 NREM sleep has been attributed to an "enhanced cortical responsiveness" (9) at sleep onset and during sleep.

Despite these available approaches, there was still the need to develop a measure of physiological sleep tendency in the absence of alerting factors. In the late 1970s, a method was designed that eliminated the need for 24-hour continuous monitoring with electrophysiological techniques (10). Called the Multiple Sleep Latency Test, it consists of five (sometimes six) scheduled naps, usually at 10:00, 12:00, 14:00, 16:00, and 18:00 hours, during which a subject is polygraphically monitored in a comfortable, soundproof, dark room. Because the test is currently given under various clinical conditions, a committee of sleep researchers has recently published recommendations for its proper administration (11,12). Sleep onset is defined according to the international criteria of Rechtschaffen and Kales (13), and the latency between lights-out time and sleep onset is calculated for each nap.

Electro-encephalographic electrodes placed on the subject's scalp, using the C_3/A_2-C_4/A_1 and O_2-O_1 electrode placements from the 10-20 international EEG electrode placement system. Electro-oculogram and chin electromyogram are monitored simultaneously. Usually one ECG derivation (V_2) is also monitored. The recording is made on a polygraph, or, as has recently been done, with the 9-channel Oxford Medilog^{TM} recorder which monitors the data on tape.

The polygraphic recording is performed as indicated in the international manual of Rechtschaffen and Kales (13). Paper speed is usually 10 mm/second in North America, with 30-second epochs; and 15 mm/second in Europe, with 20-second epochs. Epoch length difference has been a problem and the 30-second length is now favored (13). Sleep onset is defined as the first 30-second epoch in which sleep occupies more than 50 percent of the epoch. This definition requires careful scoring of sleep records and may include any non-rapid eye movement (NREM) or rapid eye movement (REM) sleep stages or state. If no sleep occurs during any one of the 20-minute monitoring periods, that particular nap is discontinued. If the subject falls asleep during the test, the test is continued for 15 minutes after the first epoch of sleep. At the end of each test, the subject leaves the monitoring room but is kept awake until the following scheduled nap. Efforts are made to facilitate sleep during the test. If the patient smokes, smoking should be avoided for 30 minutes before each test. Vigorous activity by the subject should be suspended 15 minutes before each test, and the subject should be encouraged to go to the bathroom during these last 15 minutes before testing (12). The MSLT gives the latency of each nap (i.e., the time from lights-out until the onset of sleep, defined as the first epoch scored as sleep) and the mean sleep latency for the overall test. It also indicates whether REM sleep is present during the first 15 minutes after sleep onset (an abnormal feature, since REM usually occurs about 90 minutes after sleep onset) in any of the naps. The MSLT was initially designed to remove as many of the alerting factors as possible so that the underlying sleep tendency could be measured. It is based on the assumption that an alert person will not fall asleep, while a drowsy person will fall asleep quickly. This test eliminates many of the factors that could mask, at least temporarily, the physiological sleep tendency: light, noise, activity level, motivation, anxiety, excitement, attention, and room temperature.

Five minutes before the actual start of the test, the subject, wearing street clothes, removes shoes, loosens constrictive clothing, and lies down on the comfortable bed where testing is done. During these 5 minutes the subject is given standard instructions to encourage falling asleep ('close your eyes, try to fall asleep,' etc.).

The underlying assumption of the MSLT is that lower scores indicate greater sleepiness and vice-versa. The MSLT has already been used in many studies, all of which have demonstrated that a mean daily score of less than 5 minutes indicates a pathological level of daytime sleepiness (15-22). This level is associated with impaired performance in patients and sleep-deprived normal controls. It is linked to reports of falling asleep while driving, while on the job, while eating, etc. These impairments usually cause the subject concern over the ability to perform adequately in daily life. Adult normal control volunteers usually score in the range of 10 to 20. Scores between 5 and 8 have been associated with impaired performance, but subjects presenting these scores have not necessarily recognized the impairment and may have denied excessive sleepiness. They may have reported 'feelings of fatigue' or noted that in situations experienced as boring and quiet, such as going to church or to a concert, watching a non-exciting program on television, or doing monotonous work, they may have fallen asleep, particularly after lunch or in the late afternoon, especially if seated in a comfortable armchair. It is interesting to note that although the subject denies subjective sleepiness, the environment (for example, the subject's spouse) may consider the subject's behavior to be 'sleepy'. The MSLT therefore contradicts the subjective ratings but confirms the decreased ability to maintain alertness.

Scores between 8 and 10 have been considered a grey zone. There is not sufficient currently available data to allow clear evaluation of the performance impairments linked to these scores.

The MSLT has been used in several types of investigation. It has been used as a clinical tool to recognize syndromes in need of medical treatment. With this in mind, it has been recommended that the MSLT be used for such purposes as obtaining sleep/wake diaries kept for 1 to 2 weeks preceding the MSLT, to give an idea of the subject's routine (shift-work, mealtimes, way of life, etc.); performing one night of polysomnography in accordance with the subject's habitual schedule; and carefully reviewing, and eventually withdraw subjects from, drug intake (particularly stimulent medication).

The MSLT has proved to be of great help in determining excessive sleepiness associated with syndromes such as narcolepsy, obstructive sleep apnea, CNS hypersomnia, and periodic leg movements during sleep, that impair daytime alertness. The test has also been used with other goals in mind: Carskadon has performed many studies on normal individuals, ranging from pre-pubertal children to elderly subjects, whose day/night cycles were manipulated in various ways (14-19, 23,24). The first finding was that, even though nocturnal sleep time remained virtually unchanged at slightly over 9 hours per night, there was a striking shift toward daytime sleepiness that accompanied puberty. At no time in life will anyone be more alert than an 8- to 11-year old pre-pubertal child, yet the mid-pubescent adolescents whom we followed longitudinally over 5 to 6 years presented bouts of daytime somnolence that were also clearly marked in college students (23,24). An important change in subject alertness thus occurs during the second decade. When adult groups were assessed for further changes later in life, the MSLT data of Carskadon were very informative. An example was the evaluation, in similar experimental settings under standardized conditions, of three groups of normal adults: a 'young' adult group (17 to 25 years old); a 'middle-aged' group (30 to 48 years old); and an 'elderly' group (62 to 77 years old) (19). The available nocturnal time was 8 hours, with bedtime held constant at midnight for the young and middle-aged groups; and 8 1/2 hours for the elderly group with bedtime at 23:30 hours. Data indicated that there was a moderate decrease in total nocturnal sleep time noted in the elderly group (438 ± 7 and 441 ± 12 minutes versus 416 ± 12 minutes). But it was found that MSLT scores were different for each group, with a progressive decline in daytime alertness with age. This decline was significant when young and elderly subjects were compared, but the trend was also clear in the middle-aged group. Moreover, one-third of the young adults, two-thirds of the middle-aged adults, and nine-tenths of the elderly subjects were scored on at least one test with a sleep latency below 5 minutes. The MSLT has also allowed evaluation of the circadian variation of sleep tendency in diverse age groups. To study sleep tendency as measured by the MSLT, Richardson et al. (25) controlled the dominant influence of time-of-day and inter-subject variation by correcting the individual MSLT values, using a linear model to account for the contributions of these factors. In the case of young and elderly subjects, a very clear biphasic pattern of sleep tendency, indicated by MSLT results, occurred on a normal schedule. This biphasic pattern of MSLT scores did not change significantly with aging. Mid-day (and expected early morning) minima were associated with morning and late afternoon-early evening maxima. The pattern was consistent across subjects and across days; it persisted despite limited sleep loss or sleep extension (25). The late afternoon and early evening decline in sleep tendency without intervening sleep period was considered as evidence that sleepiness was not a simple function of time awake. The mid-day dip, which has been noted even in the crews of jetliners crossing multiple time-zones, is not related to food intake. The MSLT thus appears to have provided information on spontaneous individual sleep tendency, quite apart from other known measures of behavioral functioning. This may be an important issue, as the tendency to fall asleep in a low-stimulus environment has not been thoroughly explored, despite its potential impact on industry. Undoubtedly, the MSLT should be only part of a test battery, but it has proven to be a sensitive tool in studying the following-day effects of drugs, particularly hypnotic drugs with various half-lives, such as the two benzodiazepines triazolam and flurozepam (26,27); the effects of partial or total sleep loss; the effect of aging; and the effect of pathology. The MSLT should be given in association with the classic performance tests, where it may be able to identify subjects with greater or smaller biphasic sleepiness responses or spontaneous daytime scores. The MSLT should probably be associated with 24-hour monitorings to evaluate the effect of daytime challenges on vulnerability to daytime sleepiness. Many techniques have been proposed to evaluate the combined and opposite effects of daytime sleepiness and daytime challenges. One possibility is the determination of micro-sleeps, using EEG criteria and continuous EEG monitoring on ambulatory recorders such as the Medilog 9000. Santamaria and Chiappa recently published a monograph on the EEG of drowsiness (28,29). The most interesting part of their experimental work was the emphasis placed on the importance and value of eye-movement recordings and the changes in ocular patterns helpful in the electrophysiological recognition of drowsiness. Their article also gave a detailed analysis of patterns associated with drowsiness. The application of these findings to the field, using micro-electrodes on the eyelids and carrying out MSLT investigations, may give us new tools to explore the human vulnerability to sleepiness and also, possibly, early warning of drowsiness in subjects whose professional activities place them in 'at risk' situations.

References

1. Dement WC, Carskadon MA. 'Current perspectives on daytime sleepiness:the issues'. Sleep 1982;5:S56-S66.
2. Kamiya J. 'Behavioral, subjective and physiological aspects of drowsiness and sleep'. In: Fiske DW, Maddi ST, eds: Functions of varied experience. Homewood, Illinois: Dorsey Press, 1961, pp. 145-174.
3. Hoddes E, Zarcone V, Smythe H, Phillips R, Dement WC. 'Quantification of sleepiness: a new approach'. Psychophysiology 1973;10:431-436.
4. Yoss RE, Mayer NJ, Ogle KN. 'The pupillogram and narcolepsy'. Neurology 1969;19:921-928.
5. Berlucci G, Moruzzi G, Salvi G, Stratta P. 'Pupil behavior and ocular movements during synchronized and desynchronized sleep'. Arch Ital Biol 1964;102:230.

6. Schmidt HS, Fortin LD. 'Electronic pupillography in disorders of arousal'. In: Guilleminault C, ed. Sleep and waking disorders: indications and techniques. Menlo Park, California: Addison-Wesley, 1982, pp. 127-141.
7. Broughton R. 'Performance and evoked potential measures of various states of daytime sleepiness'. Sleep 1985;5:S135-S146.
8. Pressman MR, Spielman AJ, Pollack CP, Weitzman ED. 'Long latency auditory evoked responses during sleep deprivation and in narcolepsy'. Sleep 1985;5:S147-S156.
9. Fruhstorfer H, Bergstrom RM. 'Human vigilance and auditory evoked responses'. Electroencephal Clin Neurophysiol 1969;27:346-355.
10. Richardson GS, Carskadon MA, Flagg W, van den Hoed J, Dement WC, Mitler MM. 'Excessive daytime sleepiness in man: multiple sleep latency measurements in narcoleptic and control subjects'. Electroencephal Clin Neurophysiol 1978;45:621-627.
11. Carskadon MA, Dement WC. 'The Multiple Sleep Latency Test: what does it measure?' Sleep 1982;5:S67-S72.
12. ASDC Task Force on Daytime Sleepiness. 'Guidelines for the Multiple Sleep Latency Test (MSLT): a standard measure of sleepiness'. Sleep 1986;9:519-524.
13. Rechtschaffen A, Kales A, eds. A manual of standardized terminology, techniques and scoring systems for sleep stages of human subjects. Brain Information Service/Brain Research Institute. University of California, Los Angeles, 1968.
14. Carskadon MA, Harvey K, Dement WC. 'Acute restriction of nocturnal sleep in children'. Percept Mot Skills 1981;53:103-112.
15. Carskadon MA, Dement WC. 'Nocturnal determinants of daytime sleepiness'. Sleep 1982 5;S73-S81.
16. Carskadon MA, Dement WC. 'Sleep loss in elderly volunteers'. Sleep 1985;8:207-221.
17. Carskadon MA, Dement WC. 'Sleepiness in the normal adolescent'. In: Guilleminault C, ed. Sleep and its disorders in children. New York: Raven Press, 1987, pp. 53-66.
18. Carskadon MA, Brown ED, Dement WC. 'Sleep fragmentation in the elderly: Relationship to daytime sleep tendency'. Neurobiol Aging 1982;3:321-327.
19. Carskadon MA, Dement WC. 'Daytime sleepiness: Quantification of a behavioral state'. Behav Rev 1987;11:217.
20. Stepanski E, Lamphere J, Badia P, Zorick F, Roth T. 'Sleep Fragmentation and daytime sleepiness'. Sleep 1984;7:18-26.
21. Seidel WF, Dement WC. 'Chronic insomnia and daytime sleepiness'. Sleep Res 1982;11:117.
22. Roehrs TA, Tietz EI, Zorick FJ, Roth T. 'Daytime sleepiness and antihistamines'. Sleep 1984;7:137-141.
23. Carskadon MA, Harvey K, Duke P, Anders TF, Litt IF, Dement WC. 'Pubertal changes in daytime sleepiness'. Sleep 1980;2:453-460.
24. Carskadon MA, Keenan S, Dement WC. 'Nighttime sleep and daytime sleep tendency in preadolescents'. In: Guilleminault C, ed. Sleep and its disorders in children. New York: Raven Press, 1987, pp. 43-52.
25. Richardson GS, Carskadon MA, Orav EJ, Dement WC. 'Circadian variation of sleep tendency in elderly and young adult subjects'.

Sleep 1982;5:S82-S94.
26. Seidel WF, Dement WC. 'Sleepiness in insomnia: Evaluation and treatment'. Sleep 1982;5:S182-S190.
27. Dement WC, Richardson GS, Prinz P, Carskadon M, Kripke D, Czeisler C. 'Changes of sleep and wakefulness with age'. In: Finch C, Schneider EL, eds. Handbook of the biology of aging, 2nd edition. New York: Van Nostrand Reinhold, 1985, pp. 692-717.
28. Santamaria J, Chiappa KH. 'The EEG of drowsiness in normal adults'. J Clin Neurophysiol 1987;4:327-382.
29. Santamaria J, Chiappa KH. The EEG of drowsiness. New York: Demos Publications, 1987, 202 pp.

INDIVIDUAL DIFFERENCES IN VIGILANCE PERFORMANCE

Michael W. Eysenck
Department of Psychology
Royal Holloway and Bedford New College
University of London
Egham Hill, Egham
Surrey, TW20 0EX
U. K.

ABSTRACT. For many practical purposes it is of major importance to consider individual differences in vigilance performance. Much of the relevant research has investigated individual differences in intelligence and in temperament or personality. Of the various dimensions which have received systematic attention, the most consistent effects have been obtained with the personality dimension of introversion-extraversion. Introverts typically outperform extraverts in vigilance performance; they also show less of a vigilance decrement over time. The differences between introverts and extraverts have conventionally been ascribed to the allegedly higher level of arousal in introverts than in extraverts. Limitations in this view are discussed, and an alternative view is presented. According to this view, extraverts are more affected than introverts by immediate environmental stimuli and by their current internal state. In contrast, introverts are more inclined than extraverts to formulate plans and goals, and to make use of compensatory, effortful activities to ensure that their performance goals are attained.

1. Vigilance Tasks

Over the years there have been literally hundreds of laboratory studies of vigilance in which a very wide range of tasks has been used. An obvious problem is that performance on these tasks typically correlates poorly from one task to another, indicating that so-called vigilance tasks differ substantially in the demands which they impose on the subject. The pattern of correlations indicates the necessity for a taxonomy of vigilance tasks, but there is as yet little sign of any consensus in this direction.

Of course, there are various clear-cut differences among vigilance tasks in terms of the situation which is presented to the subject. These differences occur in the modality of stimulus presentation, the predictability of the timing of signal stimuli, the discriminability of the signal-present and the signal-absent environments, and the rate of signal presentation. Correlations between vigilance tasks which are similar to each other in these characteristics are generally much

A. Coblentz (ed.), Vigilance and Performance in Automatized Systems/ Vigilance et Performance de l'Homme dans les Systèmes Automatisés, 31–40.

higher than they are between vigilance tasks which differ in terms of these parameters. This indicates that all of these factors do influence the psychological processes involved.

Some theorists (e.g., Davies & Parasuraman, 1982) have argued that there is an important distinction between speed and closure tasks. Speed tasks can be defined as those in which subjects must detect either an increment or a decrement in the intensity (or duration) of the signal stimulus, whereas closure tasks require the detection of signal stimuli within larger stimulus events. Some of the research on individual differences in vigilance performance has successfully made use of this particular distinction, as we will see later.

An alternative approach is to consider the different ways in which the vigilance decrement can occur. According to signal-detection theory, this can be due either to decreased perceptual sensitivity or to an increase in the criterion level for responding. Which of these factors is of greater importance appears to depend on the specific task demands of the vigilance task. Tasks in which there is a high rate of signal stimuli (approximately 30 - 60 per min) or in which the current stimulus needs to be compared with stimulus representations stored in memory typically lead over time to a loss of sensitivity, whereas most other vigilance tasks lead to increased caution of responding. This increased caution of responding presumably reflects a strategic change of some kind.

Do the tasks which have traditionally been used in studies of vigilance or sustained attention have any real relevance for practical situations? Of course, laboratory vigilance tasks were originally introduced in order to simulate the practical task of maintaining radar watch for submarines, and there are undoubtedly various other practical situations (e.g., detecting faulty goods on a conveyor belt) which resemble in some important way standard laboratory tasks. However, there are important ways in which these tasks appear to differ from many real-life vigilance situations in the work environment. At the risk of over-simplification, the participant in a traditional vigilance task is typically placed in a rather passive and unchanging role, and there are probably far fewer workers than there were in years gone by who work under such conditions. Technological innovation and changing attitudes to work have led to an increased emphasis on permitting the worker to be more interactively involved with his or her working environment, and to structure that environment in ways which increase efficiency and job satisfaction.

2. Theories of Vigilance

There is an important distinction between those theories of vigilance which emphasize the role of decreasing arousal and/or motivation in producing performance decrements on vigilance tasks and those theories which focus on information-processing constructs (e.g., expectancy matching or probability matching). Since individual differences in vigilance have often been interpreted within an arousal framework, it is worth considering some of the limitations of that approach at this point. One of the central (and often unrecognized) difficulties with the arousal perspective concerns the interpretation of what is typically correlational evidence: good performance tends to occur at the same time as relatively high levels of arousal, whereas poor performance occurs concurrently with low levels of arousal. Traditionally, it has been assumed that the level of arousal causally affects the level of performance. However, as Hockey (1988) has argued persuasively, it is probably more accurate to argue that the level of arousal reflects the extent to which the individual has endeavoured to maintain vigilance performance through effort expenditure.

Another crucial weakness of the arousal approach is that it can do no more than account for general patterns of performance (e.g., impaired performance over time as the arousal level decreases). It has essentially nothing to say about the specific details of performance, such as the finding that the vigilance decrement sometimes stems from reduced perceptual sensitivity and sometimes from increased cautiousness of responding.

A final major weakness of arousal theory concerns the amorphous construct of 'arousal' itself. For example, it has been claimed that a variety of manipulations (e.g., white noise; amphetamine; incentives) all lead to increased arousal. However, detailed examination of the behavioural effects associated with these and other manipulations indicates that they all differ in some respects (Eysenck, 1984; Hockey & Hamilton, 1983). The implication is that there are several qualitatively different activation states rather than a single arousal state (Hockey & Hamilton, 1983).

3. Individual Differences in Vigilance

3.1. INTELLIGENCE

Most attempts to investigate individual differences in vigilance performance have considered either intelligence or personality/temperament. The typical finding so far as intelligence is concerned is that intelligence-test performance correlates either modestly or not at all with various aspects of vigilance-task performance. The same relative failure to predict vigilance performance has been found when the predictor variable consists of a reasoning task, memory span, or a specific aptitude test. Thus, neither intelligence per se nor more specific components of intelligence appear to be associated with vigilance performance. However, this statement may need to be qualified somewhat. There are cases of impaired intellectual functioning (e.g., Alzheimer's disease) in which there are marked reductions in sustained attention (Parasuraman & Nestor, 1986). It is doubtful, however, whether a direct causal relationship is involved in this case.

3.2. TRAIT ANXIETY

There are grounds for supposing that individual differences in trait anxiety or the related personality dimension of neuroticism should be predictive of vigilance performance. Eysenck (1982, 1988) has discussed evidence indicating that there are several differences in attentional functioning between those high and low in trait anxiety. More specifically, individuals high in trait anxiety or neuroticism exhibit high susceptibility to distraction, generally poor concentration, and a high level of attentional selectivity. In addition, they show reduced available working memory capacity (especially of the central executive), but this appears to occur primarily under moderately stressful conditions.

There are only a few studies which have been concerned with the relationship between the personality dimension of trait anxiety or neuroticism and vigilance performance. The typical finding in these studies is that there is either a weak or no correlation between these two variables. One possible explanation for these non-significant effects is that groups of high and low trait anxious subjects do not differ in state anxiety under the rather monotonous and non-evaluative conditions prevailing in most vigilance studies. Since the level of state anxiety was not assessed in these studies, it is not possible to be confident that there were no group differences in state anxiety. However, it has often been found that state anxiety is interactively determined by trait anxiety and by the amount of stress in the situation, with state anxiety differences being more pronounced under condtions of moderate than of low stress.

Even if high and low trait anxious groups differ little in state anxiety in most vigilance-task situations, it is still not clear that this would account for the absence of any relationship between trait anxiety and vigilance performance. The reason is that those high in trait anxiety appear to manifest greater distractibility, poorer concentration, and increased attentional selectivity even in situations involving minimal stress or threat. In other words, individuals high and low in trait anxiety differ in various aspects of attentional functioning regardless of the level of situational stress. The implication therefore seems to be that those attentional processes which are important in the performance of numerous cognitive tasks are of little of no relevance in the performance of most vigilance tasks.

3.3 INTROVERSION-EXTRAVERSION

Of the various dimensions of individual differences which have been related to vigilance performance, there is clearer evidence of the predictive power of introversion-extraversion than of any other dimension. The research literature has been discussed by Davies and Parasuraman (1982) and by Eysenck (1982), and so only a summary of the findings will be presented here. There are 12 experimental studies in which introverts and extraverts were compared on overall vigilance-task performance. In five of those studies, it was found that introverts performed the vigilance task significantly better than extraverts. In all of the remaining studies, there was a non-significant difference in performance between the introverts and the extraverts. These findings are not dramatic, but it is of interest that all of the significant differences were in the same direction. Thus, introverts generally outperform extraverts across a variety of vigilance tasks, although it is probable that the magnitude of the performance difference between groups depends upon the precise requirements of the task in question.

In some of the experiments there was a more fine-grained analysis of performance differences as a function of introversion-extraversion. In seven cases, the number of false alarms was considered. Extraverts made significantly more false alarms than introverts in five of those studies, and the two personality groups did not differ in the remaining two studies. There were three studies in which perceptual sensitivity was indexed by the d' measure of signal-detection theory. Introverts had significantly greater sensitivity than extraverts in two of those studies, and there was no difference in the remaining one. Thus, it appears that introverts differ from extraverts in having both a higher level of sensitivity by means of the administration of intense white noise. It has also been found that the addition of a secondary task to a primary auditory vigilance task reduced somewhat the performance of introverts but actually improved that of extraverts (Bakan, 1959). In one of the most interesting studies, Keister and McLaughlin (1972) reported that caffeine administration did not affect the vigilance performance of introverts, but improved that of extraverts substantially so that they performed at the same level as introverts. This pattern of results led Keister and McLaughlin (1972) to the following conclusion: "The effect of caffeine was to increase the cortical arousal of the extraverts and make them equal to the introverts who characteristically function at a high arousal level" (p. 10).

4. Theories of Introversion-Extraversion

4.1. AROUSAL THEORY

Historically, differences in performance between introverts and extraverts have often been accounted for by means of arousal theory (e. g., H. J. Eysenck, 1967). According to this theoretical position, introverts are chronically more cortically aroused than extraverts, and they are also more arousable. These differences allegedly occur because of individual differences in the level of activity in the ascending reticular activating system. If, in addition, it is assumed that there is an inverted curvilinear relationship between arousal and performance, it then becomes possible to make some predictions about the levels of performance expected from introverts and from extraverts. More specifically, the relatively monotonous and unarousing conditions characteristic of most vigilance situations might mean that extraverts are further below the optimal level of arousal than introverts. This line of argument could be extended to account for the greater vigilance decrement typically shown by extraverts than by introverts. However, in the latter case a number of additional (and possibly implausible) assumptions would need to be made.

The findings which apparently provide the strongest evidence for arousal theory are the interaction effects discussed earlier. If one assumes that extraverts perform poorly on vigilance tasks because their level of arousal is below the optimal level, then it follows that manipulations designed to"increase arousal level should benefit their performance. Various factors (i.e., white noise; caffeine; a secondary task) which might be regarded as producing increases in arousal level all had the predicted effect of either reducing or eliminating the inferiority of extraverts to introverts in vigilance performance.

While there may be a grain of truth in arousal theory, it is undoubtedly a substantial over-simplification of a complex reality. It is important to distinguish between arousal level and arousability, with the former reflecting chronic arousal and the latter reflecting the response to stimulation. Psychophysiological differences between introverts and extraverts have recently been reviewed by Stelmack and Geen (in press). They concluded as follows: "Notwithstanding the difficulty in determining a truly resting, stimulation free condition, the exception being a sleep state, there is little evidence to indicate that introverts and extraverts differ in tonic or basal levels of physiological activity. There has been considerable success in demonstrating the enhanced response to stimulation of introverts with various modes of stimulation and methods of response recording." In other words, introverts and extraverts differ in terms of arousability but not in terms of arousal level, which is only partially in line with the predictions of arousal theory.

The notion that there is a curvilinear relationship between arousal and performance is supported by much less evidence that is often assumed. Some of the limitations in the evidence were pointed out by Eysenck (1982). He noted that there are six possible orderings of three conditions varying in arousal level so far as performance is concerned, and only two of those orderings are inconsistent with the predictions of arousal theory. Therefore, two-thirds of experiments including three levels of arousal would apparently support arousal theory by chance alone.

Close examination of the evidence leads to the identification of additional problems for arousal theory. Extraverts, who are allegedly lower in arousal level than introverts, are typically less cautious in their responding in vigilance situations. In apparent conflict with that finding, arousal generally falls during the course of a vigilance session, but the cautiousness of responding tends to increase. In other words, low arousal is sometimes associated with incautious responding and sometimes with cautious responding, which suggests that the arousal construct is inadequate to account for the data.

As predicted by arousal theory, white noise, caffeine, and the introduction of a secondary task all improve the performance of extraverts. However, what is more problematic from the perspective of arousal theory is that the performance of introverts is relatively unaffected by all of these manipulations. This can only be explained by arousal theory in an entirely post hoc fashion.

Finally, if differences between introverts and extraverts in vigilance performance occur because of the unarousing conditions of most vigilance tasks, it follows that performance differences as a function of introversion-extraversion should tend to be greatest when the task is least stimulating. In fact, the opposite seems to be the case, since it is tasks with a high event frequency which produce the clearest differences between introverts and extraverts.

4.2. CONTROL THEORY

Broadbent (1971), Eysenck (1982), and Hockey (1986) have all argued that the effects of arousing and/or stressing factors on performance can be understood adequately only when the existence of a high-level control system is postulated. Within such a conceptualization, it is assumed that the effects of arousal on performance are typically indirect, being influenced by the extent to which the control system instigates effortful activities to compensate for the sub-optimal internal state produced by the arousing or stressful conditions. Hockey (1986) has presented the most developed version of such a theory, and his views will therefore be used as the basis for understanding the effects of introversion-extraversion on vigilance performance.

According to Hockey (1986), there is a comparison mechanism or state monitor which detects mismatches between the current cognitive state and the required or target state. Mismatches may occur for many different reasons, including the introduction of arousing and/or stressful manipulations. The executive system can respond to mismatch detection in various different ways: the target state can be altered (e.g., by reducing performance goals); the mismatch can be ignored; or effort can be applied in the attempt to make the current cognitive state correspond more closely to the target state. Effort is involved, "whenever an attempt to resolve mismatch of target and current state takes the form of active manipulation of cognitive resources resources" (Hockey, 1986, p. 291).

Of greatest relevance here is the effortful response of the executive control system to the detection of mismatches. As a first approximation, it will be assumed that introverts are often more likely than extraverts to make use of effort in order to reduce the discrepancy between the current cognitive state and the target state. Effort and motivation are clearly of relevance to vigilance performance. Nachreiner (1977) pointed out that vigilance decrements are much more commonly observed under laboratory than under naturalistic conditions, and argued that this difference occurred because motivation is much more likely to diminish over time in the laboratory. Under laboratory conditions, Nachreiner (1977) and Lucaccini, Freedy, and Lyman (1968) discovered that motivating instructions produced significant improvements in vigilance performance. O'Hanlon (1965) discovered on a very long visual vigilance task that the concentration of adrenalin remained constant among those subjects who did not show a vigilance decrement, but decreased among those whose performance deteriorated over time. Since effort expenditure is typically associated with the production of adrenalin, it may be that those subjects whose performance remained constant over time continued to expend a relatively steady level of effort throughout the task.

The control theory account of introversion-extraversion provides a potential explanation for the finding that the vigilance performance of introverts is affected much less than that of extraverts by manipulations such as white noise, caffeine, the introduction of a secondary task, or the passage of time on a vigilance task. All of these manipulations affect the current cognitive state and the discrepancy between that state and the target state, but they do not affect the performance of introverts because of their active control system. In contrast, extraverts make less use of the executive control system, and so manipulations affecting the current cognitive state have fairly direct effects on performance. In other words, arousal theory claims that introverts are always more aroused than extraverts, i.e., there is a direct effect of personality on arousal level. In contrast, control theory argues that introverts are more aroused than extraverts only under certain conditions, i.e., there is an indirect effect of personality on arousal which depends on the extent to which the current and target states differ from each other.

The control theory can also help to account for the fact that introverts and extraverts differ in arousability rather than in their chronic levels of arousal. If many of the observed differences in arousal level between introverts and extraverts reflect differential use of the executive control system, then there is no particular reason why the two groups should differ in arousal level under resting conditions. It is in the response to challenging conditions that the groups should differ in arousal level, and that is in general terms what has been found.

There are other soures of evidence which are consistent with control theory. The manipulation which most clearly leads to increased use of the executive control system and effort expenditure is incentive. According to the theory, introverts typically make more extensive use of the executive control system even in the total or relative absence of external incentives. It thus seems reasonable to predict that the performance of extraverts should be affected more than that of introverts by the introduction of incentives. Corcoran (1962) discovered with a letter-cancellation task that incentive had no effect on the performance of introverted subjects, but improved that of extraverted subjects by approximately 80 %. On the five-choice serial reaction task under sleep-deprived conditions, Corcoran (1962) found that there was a correlation of +.59 between extraversion and the beneficial effect of incentive on performance.

One of the interesting characteristics of the control theory of introversion-extraversion is that it remains true to the basic meanings of 'extraversion' (turning outwards) and 'introversion' (turning inwards). That is to say, external stimuli and events (e.g., white noise) have a relatively greater impact on the performance of extraverts than of introverts. In contrast, internally generated plans and goals tend to have more influence on introverts than on extraverts. A major prediction of control theory is that the performance of introverts should be more stable than that of extraverts under changing circumstances and also over time. Recent research by Hepburn and Eysenck (submitted) has kndicated that differences in stability over time also apply to mood states. They investigated mood (positive affect; negative affect; and overall mood based on an amalgam of positive and negative affect) three times a day on 15 days over a three-week period. Extraversion correlated .50 with variability of positive affect, .60 with variability of negative affect, and .65 with overall mood variability. Thus, extraverts are considerably more volatile in mood than introverts.

Despite the potential value of control theory, it is still clearly at an early stage of development, and requires detailed testing and extensions. One obvious deficiency of control theory is that it does not account for the characteristically greater cautiousness of responding of introverts than of extraverts. This difference between introverts and extraverts can be considered within the theoretical framework proposed by Gray (1973). According to him, introverts are more susceptible to punishment than extraverts, whereas extraverts are more susceptible to reward than introverts. It may be that the perceived costs or negative consequences of making false alarms are higher for introverts than for extraverts, whereas the perceived gains or rewards associated with making correct responses are higher for extraverts than for introverts. These tendencies would lead introverts to be more cautious about responding than extraverts.

An alternative theoretical position has been adopted by Brebner and Cooper (1978). They claimed that introverts are "geared to inspect," whereas extraverts are "geared to respond." Their strongest evidence in favour of this viewpoint was obtained in an experiment in which subjects had the choice of either inspecting a visual stimulus or of responding in an attempt to change it. As predicted, introverts spent much more time inspecting the stimuli, whereas extraverts responded more than twice as often as introverts. The notion that extraverts are "geared to respond" would lead directly to low cautiousness of responding, but the problem with the entire theoretical approach of Brebner and Cooper is that it is descriptive rather than explanatory.

References

Bakan, P. (1959). 'Extraversion-introversion and improvement in an auditory vigilance task.' **British Journal of Psychology, 50,** 325-332.

Brebner, J., & Cooper, C. (1978). 'Stimulus- or response-produced excitation: A comparison of the behaviour of introverts and extraverts.' **Journal of Research in Personality, 12,** 306-311.

Broadbent, D. E. (1971). **Decision and Stress.** London: Academic Press.

Corcoran, D. W. J. (1962). 'Individual differences in performance after loss of sleep.' Unpublished Ph.D. thesis, University of Cambridge.

Davies, D. R., & Hockey, G. R. J. (1966). 'The effects of noise and doubling the signal frequency on individual differences in visual vigilance performance.' **British Journal of Psychology, 57,** 381-389.

Davies, D. R., & Parasuraman, R. (1982). **The Psychology of Vigilance.** London: Academic Press.

Eysenck, H. J. (1967). **The Biological Basis of Personality.** Springfield: Thomas.

Eysenck, M. W. (1982). **Attention and Arousal: Cognition and Performance.** Berlin: Springer.

Eysenck, M. W. (1984). **A Handbook of Cognitive Psychology.** London: Erlbaum.

Eysenck, M. W. (1988). 'Individual differences, arousal, and monotonous work.' In J. P. Leonard (Ed.), **Vigilance: Methods, Models, and Regulation.** Frankfurt: Lang.

Hepburn, L., & Eysenck, M. W. (Submitted). 'Personality, average mood, and mood variability.'

Hockey, G. R. J. (1986). 'A state control theory of adaptation and individual differences in stress management.' In G. R. J. Hockey, A. W. K. Gaillard, and M. G. H. Coles (Eds.), **Energetics and Human Information Processing.** Dordrecht: Martinus Nijhoff.

Hockey, G. R. J., & Hamilton, P. (1983). 'The cognitive patterning of stress states. ' In G. R. J. Hockey (Ed.), **Stress and Fatigue in Human Performance.** Chichester: Wiley.

Hockey, G. R. J. (1988). 'The maintenance of vigilance: A state control analysis.' In J. P. Leonard (Ed.), **Vigilance: Methods, Models, and Regulation.** Frankfurt: Lang.

Keister, M. E., & McLaughlin, R. J. (1972). 'Vigilance performance related to extraversion-introversion and caffeine.' **Journal of Experimental Research in Personality, 6,** 5-11.

Lucaccini, L. F., Freedy, A., & Lyman, J. (1968). 'Motivational factors in vigilance: The effects of instructions on performance in a complex vigilance task.' **Perceptual and Motor Skills, 26,** 783-786.

Nachreiner, F. (1977). 'Experiments on the validity of vigilance experiments.' In R. R. Mackie (Ed.), **Vigilance: Theory, Operational Performance, and Physiological Correlates.** London: Plenum.

O'Hanlon, J. (1965). 'Adrenaline and noradrenaline: Relation to performance in a visual vigilance task.' **Science, 150,** 507-509.

Parasuraman, R., & Nestor, P. (1986). 'Energetics of attention and Alzheimer's disease.' In G. R. J. Hockey, A. W. K. Gaillard, and M. G. H. Coles (Eds.), **Energetics and Human Information Processing.** Dordrecht: Martinus Nijhoff.

Stelmack, R. M., & Geen, R. G. (In press) 'The psychophysiology of extraversion.' In A. Gale and M. W. Eysenck (Eds.), **Handbook of Individual Differences: Biological Perspectives.** Chichester: Wiley.

EFFETS DE L'ALTERNANCE D'UNE TACHE MONOTONE ET DE TACHES COMPLEXES SUR LA VIGILANCE ET LA PERFORMANCE : ETUDE EXPLORATOIRE

A. COBLENTZ , P. CABON , R. MOLLARD
Laboratoire d'Anthropologie et Ecologie Humaine
Université R. Descartes-Paris V
45, rue des Saints- Pères - 75270 Paris Cedex 06

ABSTRACT. Vigilance researches will find certainly more and more applications in the area of industry and of transport operations because of the increased automatisation of work. This evolution often creates monotonous situations that can complicate the task of the operator especially when he has to take a decision or achieve a manual action after a long and monotonous period of monitoring (*e.g.* in transport operation). Thus, the aim of this research is to study the effects on vigilance of an alternation of a monotonous task (Vigilance task : VIG) with complex tasks (sustained Attention Task : TA). The method and preliminary results are presented. Continuous recording of EEG and of performance parameters show, that when TA are present, there is a resulting decrease in performance for subjects who present a marked ultradian rhythm of performance ; no significant effect were observed for the subject who presented no variation of performance. Furthermore, subjects with important variations of performance show, at the same time, decreases of power in alpha band which coincide with the presentations of TA. Both theoretical and practical implications are discussed.

1. Introduction

L'application des recherches sur la vigilance et la performance de l'opérateur humain devrait connaître un essor important en raison du niveau élevé d'automatisation vers lequel évoluent un grand nombre d'activités professionnelles. Cependant, l'application de ces recherches requièrt une meilleure connaissance des facteurs susceptibles de perturber la vigilance dans les systèmes automatisés. Mais au delà de cette connaissance approfondie, il est nécessaire que soit abandonnée l'idée, qualifiée de mythe par Wiener (1987), que "l'automatisation des fonctions élimine ou même diminue le rôle de la vigilance humaine".

Les premières recherches menées sur ce thème ont débuté au cours de la dernière guerre mondiale. Ces études avaient à l'époque pour but principal d'améliorer l'efficience des opérateurs radar. C'est Macworth (1948) qui fut le premier à définir une situation de laboratoire (test de l'horloge) permettant de mettre en évidence une diminution progressive de la vigilance des sujets, en se fondant sur les temps de réponse et le pourcentage de détection. La situation et la méthode sont devenues classiques : une tâche de vigilance est définie par la faible probabilité d'apparition de signaux critiques (signaux à détecter), une durée élevée et par la monotonie qu'elle contribue à créer. Bien que depuis le test de l'horloge la plupart des tâches aient été fondées sur le même principe, l'interprétation des résultats a connu un certain nombre d'évolutions. Ainsi, la baisse de performance a été selon les auteurs expliquée par une augmentation de la sévérité du critère de détection du signal (Broadbendt et coll., 1963), ou par une simple absence de motivation (Smith et coll., 1969).

A. Coblentz (ed.), Vigilance and Performance in Automatized Systems/ Vigilance et Performance de l'Homme dans les Systèmes Automatisés, 41–54.

Parallèlement, de nombreuses théories se sont développées, influencées par les données soit de la neurophysiologie : théorie de l'inhibition (Macworth, 1950) et théorie de l'activation (Macworth., 1970), soit de la psychologie expérimentale : théorie des filtres (Broadbendt, 1958, 1971), théorie de l'expectative (Baker, 1959), théorie de la détection du signal (Broadbendt, 1963).

La plupart des recherches ont été entreprises dans le but d'étudier les facteurs susceptibles d'influencer la vigilance. Parmi les facteurs qui améliorent la vigilance on peut citer : la connaissance des résultats (Lucaccini et coll., 1968), l'augmentation du nombre de signaux critiques (Macworth, 1968), et l'interruption de la tâche par de courtes pauses (Macworth, 1970). Les facteurs qui au contraire perturbent la vigilance sont : le bruit (Wilkinson, 1963), l'augmentation du nombre de signaux non critiques (Jerison et coll., 1964), la privation de sommeil, et l'alcool à forte dose (Wilkinson et coll., 1969).

Les tâches de vigilance en laboratoire évoluent actuellement vers des tâches plus complexes n'impliquant plus seulement les capacités sensorielles des sujets comme la plupart des tâches inspirées du modèle de Mackworth. Dans les situations expérimentales classiques, un seul type de signal critique est en général proposé au sujet qui doit fournir une réponse unique à ces signaux. Depuis quelques années, les situations mettent en jeu plusieurs signaux dont la combinaison constitue le signal ou les signaux critiques auxquels le sujet doit apporter plusieurs réponses différentes (Craig, 1984).

Cette évolution apporte en partie une réponse au nombre important de critiques formulées à l'encontre des situations classiques depuis plusieurs années. Il leur a été reproché de n'être, dans le meilleur des cas qu'une approximation de situations réelles et donc, de ne pas prendre en compte un certain nombre de variables susceptibles de modifier les résultats comme la motivation, le réalisme de la tâche et la faible intensité des signaux à détecter. Cette critique a reçu des confirmations à la suite d'observations effectuées sur le terrain (Smith et coll., 1969). Ces observations montrent que les manifestations du déclin de la vigilance en situation réelle sont moins nombreuses qu'on le pensait et que la fonction de décrement telle que l'a décrite Macworth ne serait qu'une réalité de laboratoire. La transposition des résultats de ces recherches aux situations réelles a donc été remise en cause (Nachreiner, 1977). Cependant, il n'est pas certain que les tâches de vigilance futures reflèteront plus fidèlement les situations réelles en raison de l'importance des problèmes théoriques restant à résoudre pour les tâches les plus simples.

Les modifications progressives des tâches de vigilance qui ont été évoquées, s'expliquent surtout par l'évolution technique dans le sens d'une automatisation croissante du travail. Parallèlement à la réduction de la pénibilité physique, les tâches pour lesquelles les capacités requises se sont déplacées du domaine purement sensoriel au domaine cognitif ont pris une ampleur considérable.

Cette prise en charge par la machine d'un nombre important d'activités entraînera des effets négatifs ou positifs, selon la nature de la tâche, la qualification et des caractéristiques propres à l'opérateur, notamment des facteurs de personnalité. Dans certains cas, cette automatisation va augmenter la difficulté des opérateurs à maintenir une vigilance stable au cours du temps. En effet, les systèmes automatisés, si fiables soient-ils, peuvent s'avérer inadaptés lors de situations extrêmes. On observe, par exemple, une augmentation de la part de responsabilité revenant à l'homme dans les accidents aériens, alors que le nombre total d'accidents a connu une diminution (Mollard et coll.,1986).

La prise de conscience progressive du rôle des facteurs humains dans le fonctionnement et la sécurité des systèmes automatisés (Coblentz, 1986), a contribué à renforcer l'application des recherches sur la vigilance, soit en étudiant directement la situation sur le terrain (Mackie et coll., 1977; Lille et coll., 1982; Akerstedt et coll., 1985), soit en utilisant des simulations (O'Hanlon et coll., 1977). Ce dernier type de recherche a permis notamment d'étudier des systèmes de sécurité comme ceux utilisés pour la conduite des trains. Coblentz et coll. (1985) ont montré, en plaçant des conducteurs dans une situation simplifiée de conduite monotone, une relation entre des diminutions de vigilance attestées par l'électro-encéphalogramme (EEG), des actions biomécaniques altérées sur la commande de sécurité (pédale) et une augmentation des temps de réponse et des omissions pour la détection des signaux. Peter (1983) a montré sur un système de sécurité similaire (SIFA) que les diminutions de vigilance observées sur l'EEG n'empêchaient pas la manipulation correcte du système. Dans des conditions réelles, une telle situation peut conduire à des accidents parfois très graves.

Ces baisses de vigilance enregistrées dans des situations de laboratoire ont été confirmées sur le terrain par divers auteurs : pour le transport ferroviaire, les travaux de Torsvall et coll. (1985) ont mis en évidence l'apparition de somnolences et leurs incidences sur la conduite ; dans le domaine aéronautique, Nicholson et coll.(1984, 1985), Wegmann et coll. (1986) ont étudié l'influence de certains facteurs perturbant la vigilance des pilotes : décalages horaires, cycles repos-activités irréguliers...

Si l'automatisation contribue, comme nous l'avons vu, à accroître la difficulté de certaines tâches, il est important de connaître les mécanismes en jeu. L'hypothèse de ce travail est que l'automatisation d'une partie des activités, par la monotonie qu'elle occasionne, diminue la préparation de l'opérateur dans les décisions et les actions qu'il doit déclencher au cours de tâches non automatisées. La situation choisie est donc une situation semi-automatique comme le sont la plupart des activités réelles (dans les transports, par exemple) où l'homme continue à jouer un rôle essentiel.

L'objectif de cette recherche réside dans l'étude des effets sur la vigilance de l'alternance d'une tâche simple et monotone, caractéristique d'une surveillance de dispositifs automatisés, et de tâches complexes représentant un mode manuel ou un fonctionnement dégradé.

Il est possible que ces effets soient différents selon les sujets. Coblentz et coll. (1985) ont reporté plusieurs types de profils électro-encéphalographiques et de performance chez des sujets réalisant des tests de performance : certains sujets présentent des variations ultradiennes très faibles de la performance au cours du temps, d'autres présentent au contraire une grande amplitude dans la variation de ces rythmes. Ces derniers semblent les moins adaptés à ce type d'activité. L'hypothèse de cette étude est que pour des sujets dits stables, l'introduction au cours d'une tâche de vigilance monotone de tâches nécessitant une attention et une activité plus soutenue aura peu d'effets sur leurs performances. Au contraire, des sujets dont la vigilance et la performance présentent une rythmicité marquée verront leur performance se dégrader en raison d'une préparation insuffisante et d'une somnolence due à la monotonie.

Le protocole élaboré permet de faire alterner ces deux types de tâches qui ne diffèrent que dans le rythme d'apparition des signaux critiques et dans la réponse du sujet. La phase monotone correspond à un mode automatique, les phases d'attention soutenue représentent un mode manuel ou dégradé. Il est à noter que ces dernières ne constituent pas une tâche secondaire mais font partie intégrante de l'activité.

Nous décrivons ci-après la méthode expérimentale mise en oeuvre ainsi que les premiers résultats obtenus sur un petit groupe de sujets. Les acquis de cette phase exploratoire sont destinés à être utilisés pour la définition d'un protocole de recherche sur les possibilités de réactivation des opérateurs par un changement de contenu de la tâche.

2. Méthode

Le sujet est assis dans une cabine d'expérimentation face à un écran sur lequel lui est présentée la tâche de surveillance gérée par un micro-ordinateur situé à l'extérieur de la cabine. Il dispose d'une poignée de commande munie d'un bouton poussoir.

2.1 Description de la tâche

Nous présentons au sujet un cadre dans lequel se déplace aléatoirement un curseur. Aux extrémités droite et gauche de ce cadre, se trouvent deux bornes. Le signal critique à détecter est le passage du curseur devant l'une de ces deux bornes.

Cette tâche est composée de deux niveaux :

-un niveau très monotone (tâche de vigilance) : le curseur se déplace très lentement, les signaux à détecter sont rares (trois pour une période de 10 minutes) et l'action motrice du sujet demeure faible : le sujet doit appuyer sur un bouton lorsqu'un signal critique apparaît ,

-un niveau où l'attention du sujet est sollicitée d'une manière plus importante (tâche d'attention soutenue) : la vitesse de déplacement du curseur et la fréquence des signaux critiques augmente (50 par 10 minutes), l'action motrice devient plus importante puisque le sujet doit ramener le curseur jusque dans la zone centrale à l'aide de la poignée de commande.

Dans ces deux tâches, il n'y a pas, au sens strict de *signaux non critiques*, mais des *périodes non critiques* durant lesquelles le sujet ne doit pas répondre. Ces périodes constituent le temps qui sépare deux signaux critiques.

2.2. Description du protocole (figure n°1)

Deux conditions sont comparées :

2.2.1. *Condition expérimentale.* Le niveau dit "Tâche d'Attention soutenue" (TA) est présenté au sujet par périodes de 10 minutes :

-au début, afin d'éviter un éventuel effet d'apprentissage, TAE, puis pour évaluer la performance de base du sujet : TA REF,

-à la suite d'une longue période de vigilance : TA1, TA2, TA3, TA 4,

-à la fin de l'expérimentation TA FIN, afin d'évaluer l'effet des TA successives sur la vigilance du sujet.

L'épreuve de vigilance VIG est présentée :

-pendant 45 minutes au début (VIG 1),

-pendant respectivement 5, 10 et 20 minutes (V5, V10,V20) entre les tâches d'attention soutenue,

-pendant 30 minutes après les tâches d'attention soutenue.

2.2.2. *Condition contrôle.* TA est présentée :

-au début (TA REF),

-à la fin (TA FIN).

La tâche de vigilance est présentée pendant 2h45 sans interruption.

Chaque sujet passe les deux conditions et constitue ainsi son propre contrôle. Au total, l'expérimentation dure 3h30 dans les deux conditions.

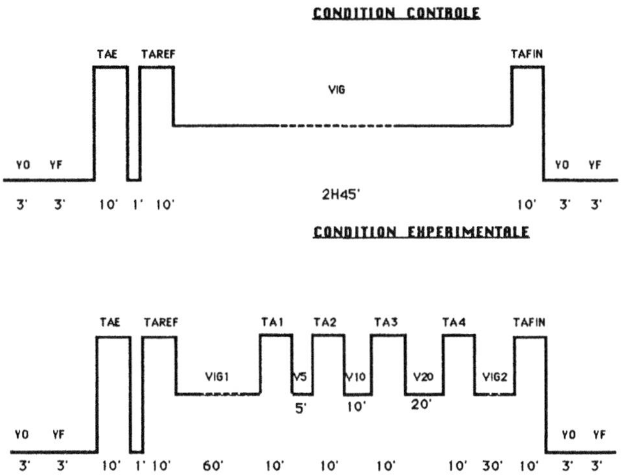

Figure n°1. *Protocole*

2.3. Paramètres recueillis

2.3. 1. *Paramètres de performance.* Les deux paramètres pris en compte sont :
- -les temps de réponse,
- -les omissions.

2.3.2. *Paramètres physiologiques.* L'électro-encéphalogramme (EEG) est enregistré en continu par un système de cartographie cérébrale (C3 ALVAR). L'enregistrement est effectué grâce à 16 électrodes placées sur le cuir chevelu du sujet rendant possible une analyse par zones anatomiques. Le montage utilisé est celui de la norme 10/20 internationale.

L'électro-oculogramme (EOG) est restitué en continu sur le polygraphe. La fréquence des mouvements oculaires est considérée comme un bon indice des niveaux de vigilance.

2.4. *Echantillon.* Il est composé de dix sujets agés de 21 à 30 ans. Chaque sujet a réalisé l'expérimentation dans les deux conditions : contrôle et expérimentale

3. Premiers résultats

Les résultats portant sur quatre sujets ont été analysés, et on retiendra à titre d'exemple les résultats caractéristiques d'un sujet stable (AB), et d'un sujet fluctuant (PL).

3.1. Performances

3.1.1. *Tâche de vigilance.* Si l'on compare pour chaque sujet les résultats dans les deux conditions, on note une assez grande reproductibilité : le sujet qui présente des performances stables (sujet AB) dans une condition, conserve un profil similaire dans l'autre condition à une semaine d'intervalle.

Sur les quatre sujets, trois présentent des variations importantes de leurs performances dans les deux conditions. Chez ces sujets, les baisses de performance sont bien réparties au cours de l'expérimentation et se produisent aussi bien au début qu'à la fin. Le quatrième sujet (AB) ne présente pas de baisses de performance dans la condition contrôle et très peu dans la condition expérimentale.

Si l'on compare l'évolution des temps de réponse des sujets AB et PL (figure n°2) on observe :

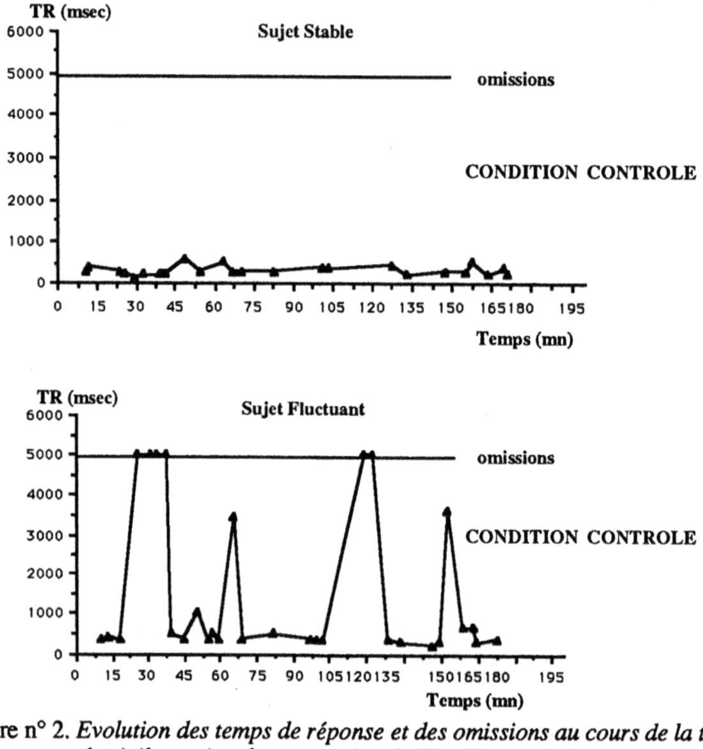

Figure n° 2. *Evolution des temps de réponse et des omissions au cours de la tâche de vigilance (mode automatique).(TR : Temps de Réponse).*

- *condition contrôle* :

-AB : des temps de réponse très stables et une absence d'omission,

-PL : un profil très fluctuant, des temps de réponse élevés et des omissions apparaissant très tôt au cours de l'expérimentation ; ce sujet présente un décrément de performance dans la première demi-heure,

- *condition expérimentale :*

-AB: à l'exception de deux temps de réponse élevés au début de l'expérimentation et une omission à la fin, les performances de ce sujet restent stables,

-PL : les omissions et les temps de réponse élevés semblent plus nombreux à la fin de l'expérimentation (à la suite des TA) qu'ils ne l'étaient dans la condition contrôle.

3.1.2. *Tâches d'Attention soutenue : comparaison de TAREF et TAFIN dans les deux conditions.* Les temps de réponse ont été séparés en deux classes : temps de réponse inférieurs et supérieurs à 500 msec. Les histogrammes représentés sur la figure n°3 correspondent au nombre de temps de réponse dans chaque classe pour le sujet stable et fluctuant, dans les deux conditions. Ce type de traitement des résultats permet une meilleure analyse des profils individuels que la comparaison de temps de réponse moyens. Un Khi ^{2} a été calculé sur ces deux classes.

-Sujet AB :

Dans la condition contrôle, les temps de réponse ont diminué dans TA FIN, mais le degré de significativité est faible (.05<p<.1). Les différences ne sont pas significatives dans la condition expérimentale.

-Sujet PL :

La performance s'est améliorée (.001<p<.01) entre TA REF et TA FIN dans la condition contrôle. Dans la condition expérimentale, les temps de réponse ont, au contraire, significativement augmenté (.02<p<.05).

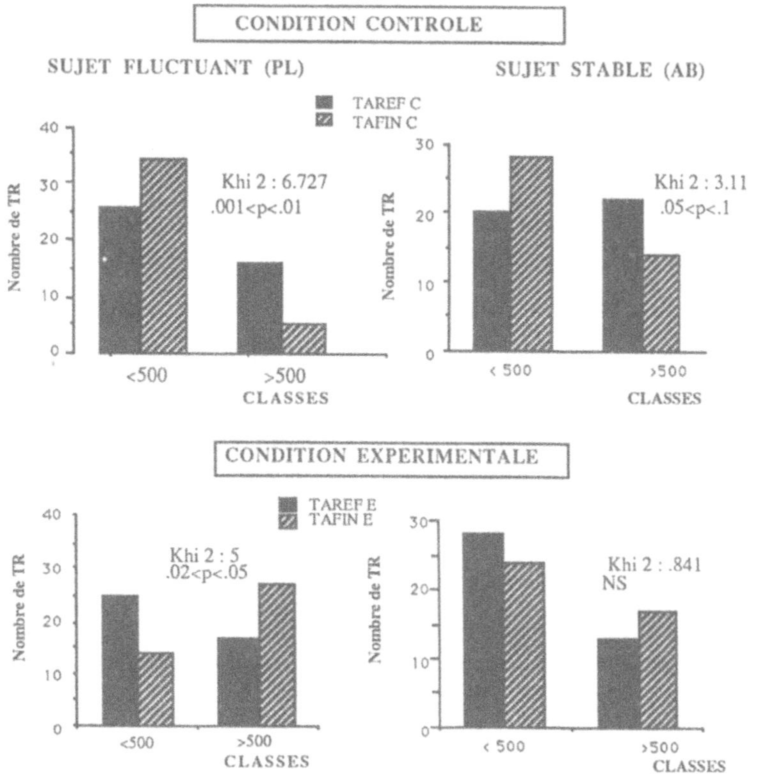

Figure n° 3. *Comparaison des nombres de temps de réponse par classes pour la Tâche d'Attention Soutenue de début (TA REF) et de fin (TA FIN) pour le sujet stable et le sujet fluctuant. En haut : condition contrôle ; en bas : condition expérimentale. (TR : Temps de réponse; NS : Non Significatif).*

Dans la condition contrôle, on observe donc une amélioration pour tous les sujets. Par contre, les interruptions de monotonie, dans la condition expérimentale, provoquent une dégradation significative des performances uniquement pour les sujets fluctuants.

3.2. Paramètres physiologiques

3.2.1. *EEG.* Les résultats des cinq zones (occipitale, frontale, pariétale et les deux régions temporales) différant assez peu, seuls les résultats de la zone occipitale ont été retenus pour une première analyse (figure n°4).

Les variations inter et intra-individuelles sont du même type que pour les paramètres de performance. On observe :

-des variations individuelles : trois sujets présentent d'importantes fluctuations de leurs rythmes EEG, et un sujet présente une relative stabilité EEG tout au long de l'expérimentation,

-que ces profils individuels sont comparables dans les deux conditions pour chaque sujet.

L'analyse en pourcentage des rythmes alpha et thêta fait apparaître pour les sujets fluctuants une rythmicité ultradienne, très marquée pour le rythme alpha.

Chez le sujet stable (AB), les rythmes alpha et thêta ne présentent pas cette rythmicité ultradienne. La puissance alpha est, en pourcentage, toujours supérieure à celle du thêta, la différence étant de 10 à 20%.

3.2.2. *EOG.* L'analyse des mouvements oculaires a été réalisée de façon manuelle. On constate une diminution des mouvements oculaires coïncidant avec des augmentations de la bande alpha ou thêta. Par ailleurs, cette analyse a permis de montrer qu'à plusieurs occasions les sujets avaient les yeux ouverts parallèlement à des augmentations de la bande alpha, alors que celui-ci apparaît le plus souvent à la fermeture des yeux.

3.3. Relations entre les paramètres de performance et les données physiologiques

3.3.1 *Performances et EEG.* Sur le plan des variations inter et intra-individuelles, les données convergent assez bien : le sujet qui présente de nombreuses diminutions des performances est celui qui a les plus grandes fluctuations EEG avec notamment une rythmicité ultradienne des rythmes alpha. Chez le sujet fluctuant, il faut noter que dans la condition expérimentale, l'apparition des TA provoque une diminution très nette du rythme alpha (cf. figure n°4). La fin des TA entraîne une augmentation de l'alpha. Le sujet stable dont les baisses de performances sont peu nombreuses présente, au contraire, de très faibles variations EEG.

Pour les sujets fluctuants, les indices EEG varient d'une manière assez différente : dans les TA, les augmentations des temps de réponse sont surtout accompagnées d'une diminution de la bande alpha ; dans la tâche de vigilance, les variations sont plus complexes puisque les augmentations des temps de réponse coïncident tantôt avec des augmentations, tantôt avec des diminutions du rythme alpha.

Si l'on compare l'activité alpha et thêta dans les TA de début et de fin dans les deux conditions, on note :

-pour le sujet fluctuant, des diminutions importantes dans TA FIN, plus marquées dans la condition expérimentale (figure n° 5),

-pour le sujet stable, de faibles différences et ceci, dans les deux conditions (figure n° 5).

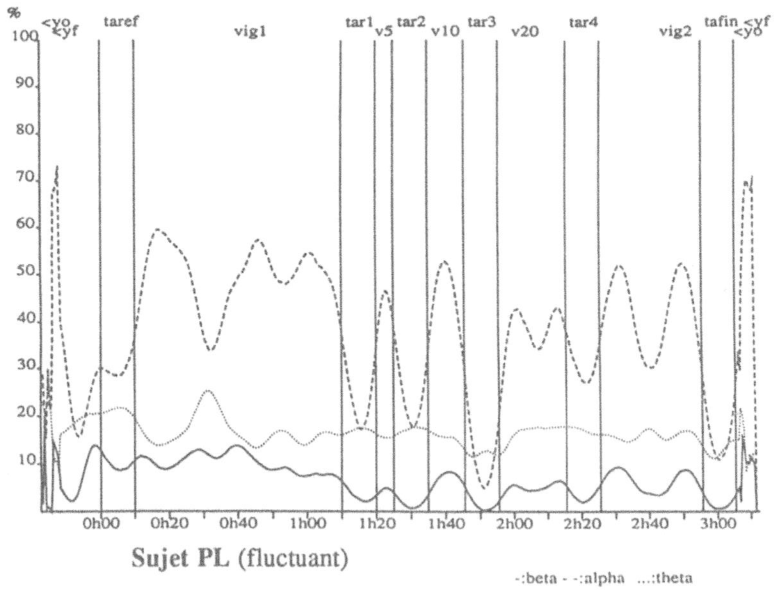

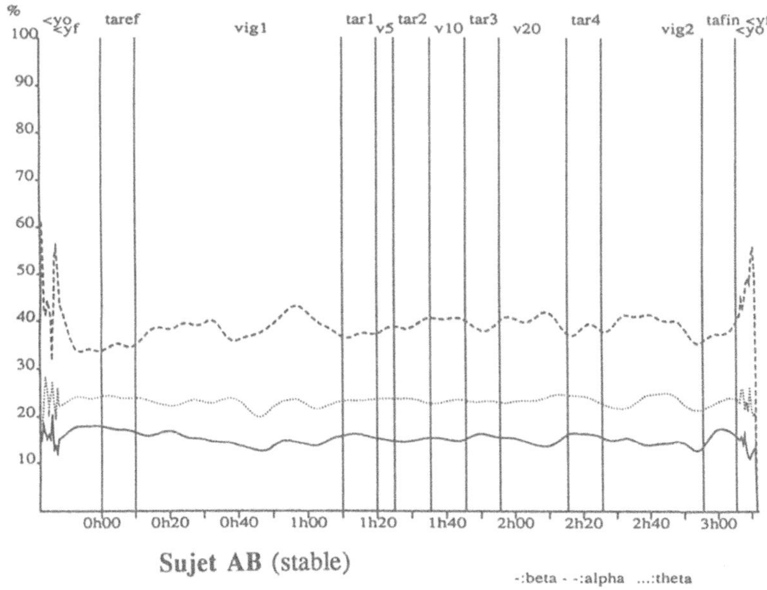

Figure n°4. *Pourcentage des fréquences EEG alpha, thêta et bêta au cours de la tâche de vigilance monotone (mode automatique) et des Tâches d'Attention soutenue successives (mode manuel).*

SUJET FLUCTUANT (PL) SUJET STABLE (AB)

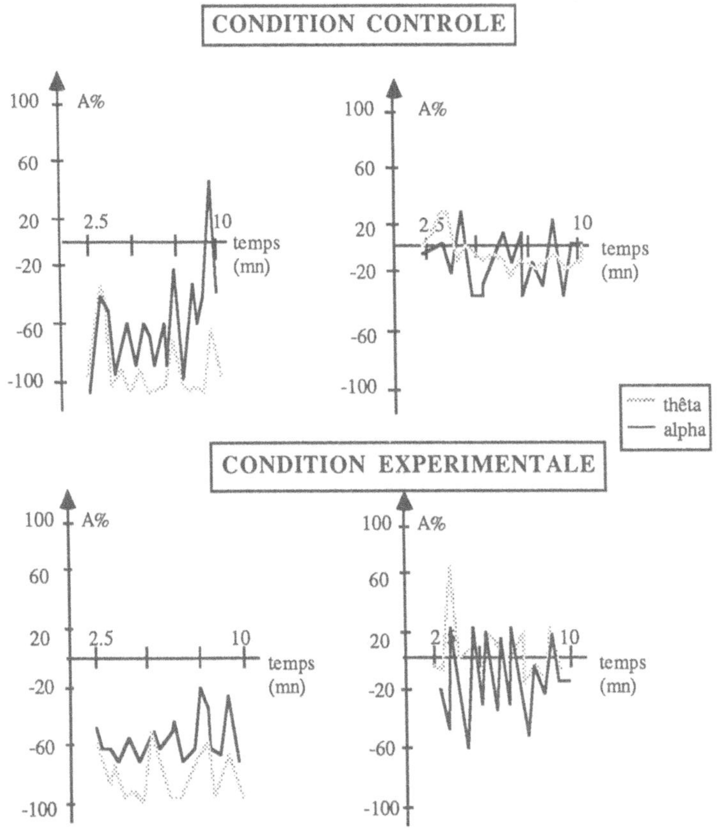

Figure n°5. *Rapports de spectres calculés sur 30 secondes des bandes alpha et thêta entre TA REF et TA FIN dans la condition contrôle et la condition expérimentale pour le sujet stable et le sujet fluctuant. (A% : pourcentage d'augmentation ou de diminution des puissances alpha et thêta).*

Ces résultats corroborent assez bien les données concernant la performance : les différences de performance sont faibles chez le sujet stable, importantes chez le sujet fluctuant. Par ailleurs, les sujets fluctuants sont plus sensibles aux changements de tâches : du point de vue EEG, leur vigilance augmente lors de la présentation des TA, avec notamment une diminution de l'alpha et du thêta.

3.3.2. *EOG et performances.* L'analyse conventionnelle de l'EOG montre que les périodes pour lesquelles les sujets présentent des baisses de performances importantes sont les périodes où le nombre de clignements tend à diminuer.

4. Discussion

Malgré la faiblesse évidente de l'échantillon, la convergence des paramètres de performance et des mesures physiologiques permet de formuler les remarques suivantes :

-les sujets présentent des profils physiologiques et de performances reproductibles d'un jour à l'autre, ce qui confirme des résultats obtenus antérieurement (Coblentz et coll.,1985),

-les performances s'améliorent dans la condition contrôle entre le début et la fin, pour tous les sujets, et se dégradent dans la condition expérimentale pour les sujets fluctuants,

-enfin, les interruptions de monotonie entraînent chez les sujets fluctuants une diminution de la puissance alpha (réaction d'arrêt), ce qui n'a pas été observé pour le sujet stable.

Les fluctuations des rythmes EEG et notamment des fréquences alpha et thêta se sont révélées être des indices assez fiables, car ils ont permis de distinguer les sujets dont les niveaux d'activation fluctuent au cours du temps de sujets dont l'activation reste stable. Cette typologie se conserve pour les performances, ce qui permet de penser que l'EEG peut être considéré comme un prédicteur de la performance globale. Cependant, ce pouvoir prédicteur s'atténue lorsque l'on tente de corréler un temps de réponse à un signal, avec la variation EEG concomittante. O'Hanlon et coll. (1977) ont montré que l'apparition du rythme alpha pouvait être interprétée à la fois comme un indicateur d'un haut ou d'un bas niveau de vigilance. La disparition de ce rythme a été également mise en relation avec une activité motrice augmentée, une réaction à un stimulus nouveau (Mulholland et coll., 1962), ou avec l'exploration d'un champ visuel non uniforme (Lehtonen et coll., 1972). Ces interprétations s'appliquent assez bien à nos résultats puisque nous avons observé pour les sujets fluctuants un blocage de l'alpha au moment de la présentation des TA qui sont des tâches pour lesquelles les activités motrices et visuelles requises s'intensifient. Dans le contexte de notre expérimentation, ce blocage de l'alpha peut être considéré comme une réaction d'arrêt, témoin d'une augmentation de la vigilance.

Le résultat essentiel concerne l'effet différentiel des deux conditions. Alors que dans la condition contrôle les quatre sujets améliorent leurs performances dans TA FIN, l'introduction de TA (condition expérimentale) provoque une diminution des performances chez les sujets fluctuants et n'a que peu ou pas d'effet sur les temps de réponse du sujet non fluctuant. Dans la condition contrôle, il y aurait donc eu apprentissage malgré l'entraînement donné aux sujets.

Ceci constitue un résultat surprenant, dans la mesure où un apprentissage est le plus souvent établi lorsque la tâche est répétée ce qui n'est justement pas le cas dans la condition contrôle. Si l'on admet qu'il y a eu apprentissage dans la condition contrôle, les interruptions de monotonie semblent avoir empêché cet apprentissage dans la condition expérimentale puisque la performance a diminué dans cette condition.

Il est pour l'instant difficile de proposer une interprétation pour ce résultat car il ne s'agit ici que d'une phase exploratoire et, avant tout d'un travail de méthode. Des expérimentations supplémentaires s'avèrent nécessaires pour confirmer ces premières tendances et pour proposer une interprétation.

Si ces premiers résultats se vérifiaient, leurs applications seraient à la fois théoriques et pratiques. Théoriques, car ils montreraient que des interruptions de monotonie par des tâches plus complexes engendrent des effets différents selon le comportement des sujets. Les applications pratiques de ce type d'étude concernent en premier lieu la sélection et la formation des personnels puisque nous avons observé un effet différentiel des conditions sur des sujets présentant des profils EEG et de performances différents. Si ce résultat était reproduit, il signifierait que certains sujets, en raison d'une vigilance fluctuante, sont peu adaptés à des situations dans lesquelles des tâches monotones alternent avec des tâches complexes. Ces situations devenant de plus en plus fréquentes, il sera essentiel de posséder des critères permettant une sélection appropriée des personnels. La deuxième application concerne les possibilités d'aménagement des tâches requérant une performance stable au cours de longues durées par des moyens artificiels. Des travaux anciens (Baker, 1960) et plus récents (Maningat et coll., 1986) permettent de penser que des solutions de ce type sont envisageables.

RESUME. L'objectif de cette recherche est d'étudier l'effet sur la vigilance de l'alternance d'une tâche simple et monotone (tâche de vigilance : VIG) et de tâches complexes représentant un mode manuel (Tâche d'Attention : TA). La méthode expérimentale et les premiers résultats portant sur un petit groupe de sujets sont présentés. L'enregistrement en continu de l'EEG et de paramètres de performance montrent que chez des sujets "fluctuants" caractérisés par une une rythmicité ultradienne importante des performances, la présence des TA dégrade les performances ; par contre, aucun effet significatif n'a été mis en évidence pour des sujets "stables". Les implication théoriques et pratiques de ces résultats sont discutés.

Références

Akerstedt, T., Torsvall, L., Gillberg, M., 1985, 'Sleepiness : continuous EEG and EOG measures in the field and laboratory', in the Proceedings of the workshop Electro-encephalography in transport operations, Cologne 15-17 sept. 1985, 1-13.

Baker, C. H., 1959, 'Attention to visual displays during a vigilance task : II. Maintaining the level of vigilance', Brit. J. Psycho., **50**, 30-36.

Baker, C. H., 1960, 'Maintaining the level of vigilance by means of artificial signals', J. Appl. Psycho., **44**, 336-338.

Beatty, J., Ahern, S. K., Katz, R., 1977, 'Sleep deprivation and the vigilance of anesthesiologists during simulated surgery', Vigilance, New York : Plenum Press, ch. 23, 511-535.

Berch, D. B., Kanter, D. R., 1984, 'Individual differences', in J. S. Warm (ed) Sustained attention in human performance, . Chichester : John Willey & Sons, ch. 5, 143-170.

Broadbendt, D.E., 1958, Perception and communication, New York : Pergamon Press.

Broadbendt, D. E., Gregory, M., 1963, 'Vigilance considered as a statistical decision', Brit. J. Psycho., **54**, 309-323.

Broadbendt, D.E., 1971, Decision and stress, New York : Academic Press.

Coblentz, A. M., Mollard, R., Proux, S., Sauvignon, M., 1985, 'Variations of vigilance and human performance : a circadian approach, in : Proceedings of the workshop Electro-encephalography in transport operations, Cologne 15-17 sept.1985, 94-103.

Coblentz, A. M., 1986, Effets de l'automatisation sur la performance de l'opérateur humain.Comptes rendus des journées d'études CEE, 27-28 Octobre 1986, Paris : L.A.A. Université R. Descartes.

Craig, A., 1984, 'Human engeenering : The control of vigilance' in : J. S. Warm (ed) Sustained attention in human performance, Chichester : John Willey & Sons, ch. 8, 247-282.

Gale, A., Jones, D., Smallbone, A., 1974, 'EEG and short term memory', Nature, **248**, 439-440.

Gale, A., Davies, R., Smallbone, A., 1977, 'EEG correlates of signal rate, time in task and individual differences in reaction time during a five-stage sustained attention task', Ergonomics, **20**, 363-376.

Jerison, H. J., Pickett, R. M., 1964, 'Vigilance : The importance of the elicited observing rate', Science, **143**, 970-971.

Lehtonen, J. B.., Lehtonen, I., 1972, 'Alpha rhythm and uniform visual field in man ', Electroencephal. Clin. Neurophysio., **32**, 139-147.

Lille, F., Chéliout, 1981, 'Variations in diurnal and nocturnal waking state in air traffic controllers', Eur. J. Appl. physio., **49**, 319-328.

Lucaccini, L. F., Freedy, A., Lyman, J., 1968, 'Motivational factors in vigilance : The effects of instructions on performance in a complex vigilance task', Perceptual and Motor Skills, **26**, 783-786.

Mackie, R. R., O'Hanlon, J. F., 1977, Vigilance, New York : Plenum Press, ch. 25, 537-558.

Macworth, J. F., 1968, 'The effect of signal rate on performance in two kinds of vigilance tasks', Human Factors, **10**, 11-18.

Macworth, J. F., 1969, Vigilance and habituation, Baltimore : Penguin Books.

Macworth, J. F., 1970, Vigilance and attention, Baltimore : Penguin Books.

Macworth, N. H., 1948, 'The breakdown of vigilance during prolonged visual search', Quat. J. Exp. Psycho., **1**, 6-21.

Macworth, N. H., 1950, Researches on the measurement of human performance, M.R.C. Special report series, **68**, London : H.M.S.O.

Maningat, J., Benett, C., 'Video games and human performance' in : Proceedings of the human factors society, 30 th Annual Meeting, 385-389.

Mollard, R., Ignazi, G., 1986, 'Automatisation et performance : besoins et perspectives' in : les Comptes rendus des journées d'études CEE sur les effets de l'automatisation sur la performance de l'opérateur humain, Paris, 27-28 Octobre 1986, 3-16.

Mulholland, T., Runnals, S., 1962, 'Increased occurence of EEG alpha during increased attention', J. Psycholo., **54**, 317-330.

Nachreiner, A., 1977, 'Experiments on the validity of vigilance experiments', Vigilance, New York : Plenum Press, ch. 32, 361-421.

Nicholson, A. N., Stone, B.M., Borland, R. G., Spencer, M.B., 1984, 'Adaptation to irregularity of rest and activity', Aviat. Space Environ. Med., **55**, 102-112.

Nicholson, A. N., Pascoe, P. A., Roehrs, T., Roth, T., Spencer, M. B., Stone, B.M., Zorick, F., 1985, 'Sustained performance with short evening and morning sleeps', Aviat. Space Environ. Med., **56**, 105-114.

O'Hanlon, J. F., Beatty, J., 1977, 'Concurrence of Electro-encephalographic and performance changes during a simulated radar watch and some implications for the arousal theory of vigilance', Vigilance, New York : Plenum Press, ch.11, 189-201.

Parasuraman, R., 1979, 'Memory load and event rate control sensitivity decrements in sustained attention', Science, **205**, 924-927.

Peter, J. H., Fuchs, E., Langanke, P., Meinzer, K., Pfaff, U., 1983, 'The SIFA train function safety circuit. I. Vigilance and operational practice in psycho-physiological analysis', Int. Arch. Occup. Environ. Health, **52**, 329-339.

Smith, R. L., Lucaccini, L. F., 1969, 'Vigilance research : its application to industrial problems,' Human Factors, **11**, 149-156.

Wegmann, H.M., Gundel, A., Naumann M., Samel, A., Schwartz E., Vejvoda, M., 1986, 'Sleep, sleepiness, and circadian rhythmicity in aircrews operating on transatlantic routes', Aviation, Space and Environmental Medicine, **57** (12 suppl.), 53-64.

Wiener, E. L., 1987, 'Application of vigilance research : rare, medium, or well done ?', Human factors, **29**, 725-736.

Wilkinson, R. T., 1963, 'Interaction of noise with knoledge of results and sleep deprivation', J. of Exp. Psycholo., **66**, 332-337.

Wilkinson, R. T., 1969, 'Some factors influencing the effect of environmental stressors upon performance', Psychol. Bull., **72**, 260-272

EXPERIMENTAL APPROACHES TO VIGILANCE PROCESSES IN MONOTONOUS SITUATIONS: ONE-WEEK RETEST RELIABILITY OF PSYCHIC FATIGUE DATA

J. P. Leonard
Boehringer Ingelheim KG
Labor für Experimentelle
Pharmakopsychologie und EEG
Abt. Medizin / HPZ
D-6507 Ingelheim
F.R.G.

ABSTRACT. In earlier laboratory studies we were able to prove that habituation to irrelevant monotonous acoustic stimulation accelerated psychic fatigue. Psychic fatigue was assessed by EEG sleep onset times and habituation times of SRRs. In the study presented here, one-week retest reliability of several EEG and SRR criteria as well as blood pressure and psychometric variables was estimated. EEG sleep stage 1 was found to be the most suitable criterion of psychic fatigue, showing sufficient relability and no first-day effect. Blood pressure and psychometric variables had high reliabilities but failed to be sensitive measures of recovery during a 40-minute nap.

1. INTRODUCTION

Studying psychic fatigue in automatized systems the problem arises how to measure psychic fatigue, adequateley. Psychic fatigue leads to a change in structure of the task mastering and to psychological and physiological mechanisms which (try to) compensate the decrement of vigilance: If somebody gets tired during work he will not let himself simply fall asleep but he will try to counteract psychic fatigue. This may, for instance, lead to increased effort and tension, increased desire for motion, and compensatory increase in heart rate. Only when undemanding work lasts for a longer period psychological and physiological signs of deactivation will occur.

In laboratory experiments some of these problems may be excluded or controlled. Results are easier to interprete as vigilance regulation is based on a lower degree of complexity as in field experiments. Furthermore, in experi-

A. Coblentz (ed.), Vigilance and Performance in Automatized Systems/ Vigilance et Performance de l'Homme dans les Systèmes Automatisés, 55–66.

mental studies it is intended to exclude compensatory mechanisms. Thus, psychic fatigue should appear much earlier in the lab than in the free field which makes investigations more economical.

Our studies and considerations are concerned with monotony as a factor which causes psychic fatigue. States of monotony are undemanding states which are specified by the work. The undemanding state results from the fact that only few and unqualified demands of activity exist.

Expressed in more operationally defined language, monotonous situations are characterized by the deficiency of stimulation and/or its lower variability. The term "stimulus" used here has a variety of meanings: stimuli can be specific according to modality, quality, and quantity. The same physical stimuli may have different psychological meanings. Even elementary movements and complex activities possess stimulative, i.e. enteroceptive, qualities.

In our laboratory the effect of monotonous acoustic stimulation on sleep onset was investigated (a condensed review is given by Leonard, 1988). EEG sleep-onset time which indicates unambigously psychic fatigue, was our target variable. Monotonous acoustic stimulation is a psychophysiological paradigm which can easily be varied.

In the literature on psychophysiology this paradigm is commonly known as habituation procedure. Habituation processes may be indicated by Skin Resistance Responses (SRRs), for instance. By means of various experiments we were able to show that monotonous acoustic stimulation had fatigue promoting effects compared with quietness (cf. Leonard, 1988). Psychic fatigue was more intense during 500 cps tones of lower intensity, shorter interstimulus intervals, and irrelevant meaning. Subjects with a "weak" C.N.S. according to Strelau's Temperament Inventory (Strelau, 1983) rapidly developed psychic fatigue under monotonous acoustic stimulation but they rather remained awake during periods of quiet. Subjects with a "strong" C.N.S. showed contrary effects of monotonous acoustic stimulation on psychic fatigue, i.e. they fell asleep during periods of quiet and remained awake during stimulation.

High correlations between sleep onset time and speed of SRR-habituation would be able to prove how close the relationship is between psychic fatigue and information processing. Correlations between responding to stimulation and development of psychic fatigue were between .50 and .76 (all $N > 30$, Bohlin, 1971, Leonard, 1986). Thus, they were significant, in part impressive high, and substantial.

The hypothesis of our earlier experiments was based on the assumption that monotonous stimulation could actively accelerate psychic fatigue. The outcomes seemed to confirm the hypothesis and are in accordance with Pavlov's (1953) and Moruzzi's (1960) theory and results, which were discussed briefly in Leonard (1986).

2. PROBLEM

The study described below was performed to assess the reliability of physiological and psychometric variables which were used in earlier experiments on psychic fatigue. One-week retest reliability of different parameters was tested. Dependent variables were collected during 40-minute daytime sleep periods under monotonous acoustic stimulation. In particular we were interested in the following questions:

1. Which are the most reliable criteria of sleep onset and SRR habituation?
2. Which parameters are sensitive to situational influences and show a "first-day effect"?
3. May the correlations between sleep onset time and SRR habituation be confirmed with a tone stimulus of 10 sec. duration?
4. How reliable are blood pressure and psychometric variables in assessing deactivational processes?
5. Which psychometric parameters may indicate recovery functions during resting and reliable pre-post differences?

Suitable criteria should show high reliability and no first-day effect. By identifying variables which are sensitive to situational influences like being familiar with the laboratory or not, error variances in further studies may be diminished by using variables without first-day effects.

For several reasons it was not intended to record subjects under control conditions, i.e. quietness. Firstly, sufficient information about spontaneous daytime sleep during a 40-minute period has been collected in earlier studies. The present study was exactly paralleled to these studies. Secondly, during quietness no SRR habituation to tones is measurable. Thus, information concerning the habituation process is not available. At last, a third of the subjects of the study were expected to be non-nappers. Thus, we were to record about 50 subjects in order to get a sample of about 30 sleepers for valid interpretations of correlational statistics. Studying two groups of that size was rejected for economical reasons.

3. METHODS

One-week retest reliability of physiological and psycho-

TABLE 1

Times of day when experiments were performed.

start of exps	start of nap period
11.00	12.30
12.15	13.45
14.00	15.30
15.15	16.45

TABLE 2

Procedure and time course of experiments.

Psychometric tests (pre)	45'
Perdue Peg Board Short Term Memory Test (digit span forward / backward) Concentration Test d2 (Brickenkamp, 1981) Mood Check Lists (not reported here)	
Polygraphy	90'
Attaching electrodes for recording EOG, EEG, SRR and cuff for BP	
Instructions: "Some stimuli will be presented to you. We shall record your physical responses to these stimuli. You have control neither of the sequence of stimuli nor of the physical responses. Just rest quietly on your back."	
Adaptation period, 7' Foreperiod (in parallel to Leonard, 1985, 1986), 13' Nap period, 40'	
Detaching electrodes	
Psychometric tests (post)	45'
see above	

metric parameters was tested. Experiments took place four times a day (see table 1) and four days a week (Monday to Thursday) and in 3 consecutive blocks of 2 weeks. In 3 cases we recorded incomplete psychological data (which are not reported here). Thus, a total of 51 instead of the planned 48 paid male students participated in the experiment. None of them studied psychology. They were healthy, had no sleep disturbances, and did not usually nap. Age range was from 19 to 31 years.

4. PROCEDURE

Procedure was performed in parallel to earlier experiments (Daum et al. 1988; Leonard, 1985, 1986). Before and after 40-minute daytime sleep periods psychometric variables were assessed (see table 2).

Treatment during 40-minute naps was identical for each subject. Monotonous acoustic stimulation was realized by administration of 500 cps sine wave tones with a 10 sec. duration and an intensity of 38 dB(A), i.e. +10 dB(A) above background noise level of the camera silens. Interstimulus intervals varied between 30 and 60 sec., with a mean of 45 sec. The interstimulus intervals were randomized but identical for each administration.

Monotonous acoustic stimulation used here slightly differed from the earlier paradigms with respect to duration of the single tone. With a given interstimulus interval of about 45 sec. the longest tone used before lasted 5 sec. (Leonard, 1985). Most experiments run with a tone of 2 sec. duration (Daum et al. 1988; Leonard, 1986). The consideration for lengthening the tone was our interest in discovering whether sleep-promoting processes vary with the total amount of stimulation.

Sleep stages were recorded according to Rechtschaffen and Kales (1968). The 40-minute naps were analysed in 30-sec. epochs. Epochs which were covered by at least 15 sec. of a specific EEG activity were recorded as the corresponding sleep stage. Analyses were performed by two judgers, independently. In cases of non-accordance (8 out of 102

TABLE 3 (see next pages)

Means and SDs of the absolute values of nap 1 and nap 2 as well as of the differences between nap 1 and nap 2 (first-day effect) and one-week retest reliability of physiological and psychometric variables (* = $p \leq .05$; ** = $p \leq .001$; 2-tail prob.).

Table 3

Variable (unit)		nap 1 M	nap 1 SD	nap 1 n	nap 2 M	nap 2 SD	nap 2 n	nap 2 - nap 1 M	nap 2 - nap 1 SD	nap 2 - nap 1 n	nap 2 - nap 1 first day effect paired t test t	df	retest reliability Pearson's r
Blood Pressure (mmHg)													
Initial Value	SYS BP	112.80	11.78	51	108.16	11.57	49	-4.16	9.38	49	-3.11**	48	.66***
	DIA BP	66.85	12.13	51	62.61	10.70	49	-4.10	12.66	49	-2.27*	48	.40**
	BP AMPLITUDE	45.95	15.18	49	45.55	13.65	49	-.06	15.38	49	-.03	48	.43**
Before napping	SYS BP	109.63	13.73	51	104.48	11.80	50	-4.10	10.92	50	-2.66*	49	.57***
	DIA BP	65.80	12.35	51	60.68	12.41	50	-4.78	14.38	50	-2.35*	49	.32*
	BP AMPLITUDE	43.82	13.76	51	43.80	15.49	50	.68	15.78	50	.30	49	.40**
After napping	SYS BP	108.88	12.05	51	105.50	13.67	50	-2.50	10.81	50	-1.63	49	.63***
	DIA BP	67.10	9.22	51	65.62	12.29	50	-1.28	9.47	50	-.93	49	.63***
	BP AMPLITUDE	41.78	12.25	51	39.88	13.89	50	-1.22	12.46	50	-.69	49	.53***
Psychometric variables													
Perdue peg board (correct pins)	pre	31.37	4.71	51	33.10	4.36	50	1.88	4.35	50	3.06**	49	.53***
	post	33.25	4.62	51	34.74	4.98	50	1.62	3.22	50	3.54***	49	.77***
	post - pre	1.88	3.55	51	1.64	3.82	50	-.26	5.23	50	-.35	49	.00
Digit span forward (digits)	pre	8.49	2.12	51	8.96	2.37	50	.50	1.42	50	2.49*	49	.81***
	post	9.06	1.88	51	9.88	2.00	50	.84	1.58	50	3.75***	49	.67***
	post - pre	.57	1.45	51	.92	1.44	50	.34	2.02	50	1.19	49	.03
Digit span backward (digits)	pre	7.69	1.99	51	7.64	2.49	50	-.08	2.05	50	-.28	49	.60***
	post	8.51	1.95	51	8.48	2.24	50	-.06	1.56	50	-.27	49	.73***
	post - pre	.82	1.14	51	.84	1.78	50	.02	2.13	50	.07	49	-.01
Concentration test d 2 (% of errors)	pre	2.86	2.45	51	1.17	1.37	50	-1.66	1.84	50	-6.39***	49	.68***
	post	1.51	1.82	51	.83	1.09	50	-.70	1.10	50	-4.54***	49	.84***
	post - pre	-1.35	1.43	51	-.35	.84	50	.96	1.56	50	4.34***	49	.10

Table 3, cont.

Variable (unit)	nap 1 M	nap 1 SD	nap 1 n	nap 2 M	nap 2 SD	nap 2 n	nap 2 - nap 1 M	nap 2 - nap 1 SD	nap 2 - nap 1 n	nap 2 - nap 1 first day effect paired t test t	df	retest reliability Pearson's r
Sleep onset time (No of epochs)												
First stage 1 epoch	24.39	20.23	33	19.60	19.30	35	-3.48	15.11	29	-1.24	28	.63***
First 2 succ. stage 1 epochs	25.78	20.89	27	20.39	16.22	31	-4.92	15.63	24	-1.54	23	.63***
First 3 succ. stage 1 epochs	27.21	17.35	24	22.20	18.51	30	-8.10	16.90	21	-2.20*	20	.36
First 4 succ. stage 1 epochs	28.87	18.13	23	20.07	15.20	27	-12.28	12.08	18	-4.31***	17	.62**
First 5 succ. stage 1 epochs	30.00	18.61	23	18.15	11.69	26	-13.00	12.07	18	-4.57***	17	.59**
First 6 succ. stage 1 epochs	31.57	18.87	23	18.88	12.55	26	-14.28	14.41	18	-4.20***	17	.46
First stage 2 epoch	39.69	15.74	16	29.31	16.80	26	-12.20	22.92	15	-2.06	14	.13
First 2 succ. stage 2 epochs	39.69	15.74	16	29.31	16.80	26	-12.20	22.92	15	-2.06	14	.13
First 3 succ. stage 2 epochs	39.40	16.25	15	28.72	16.88	25	-13.07	23.53	14	-2.08	13	.11
First 4 succ. stage 2 epochs	40.00	17.03	13	28.96	16.74	25	-12.42	25.16	12	-1.71	11	.11
First 5 succ. stage 2 epochs	41.15	16.12	13	28.96	16.74	25	-13.67	25.77	12	-1.84	11	.02
First 6 succ. stage 2 epochs	41.15	16.12	13	29.13	16.79	24	-13.67	25.77	12	-1.84	11	.02
First appearance of theta waves	14.05	10.37	40	11.27	13.40	44	-5.64	9.94	39	-3.55***	38	.41*
Time until SRR habituation to tones (minutes)												
First Non-Response	1.53	1.92	47	.70	1.02	47	-.70	1.76	46	-3.35**	45	.43**
First 2 Non-Responses	2.40	3.27	47	1.13	2.01	47	-1.24	2.73	46	-3.07**	45	.55***
First 3 Non-Responses	3.00	4.29	47	1.74	3.17	47	-1.09	3.86	46	-1.91	45	.47***
First 4 Non-Responses	3.00	4.29	47	2.47	4.42	47	-.35	3.96	46	-.60	45	.58***
First 5 Non-Responses	3.40	4.65	47	2.85	5.12	47	-.28	4.24	46	-.45	45	.61***
Time until absence of spontaneous SRRs (No of epochs)												
... for 2 succ. epochs	3.13	3.17	47	2.79	4.28	47	-.70	4.15	45	-1.14	45	.16
... for 4 succ. epochs	4.76	6.00	45	3.04	4.67	45	-1.39	5.24	44	-1.76	43	.50**
... for 6 succ. epochs	7.14	11.95	43	4.84	11.10	44	-2.10	11.76	42	-1.15	41	.49**
... for 10 succ. epochs	12.05	16.51	40	8.56	17.41	41	-5.03	19.23	37	-1.59	36	.18

assessments) agreement was reached after discussion. SRRs were only counted if they had an amplitude of 500 Ohm or higher and latencies from 1.0 to 5.0 seconds. All parameters were recorded on a R 511 Beckman polygraph with a paper speed of 10 sec.

5. RESULTS

All numeric results are listed in table 3.

5.1. EEG and SRR criteria of psychic fatigue.

The first appearance of sleep stage 1 is the most reliable criterion of psychic fatigue in the EEG (r = .63). It is also a very sensitive measure, since 33 out of 51 subjects reached sleep stage 1 for at least 15 sec. during the first nap. Sleep onset lasted about 12.2 min. during nap 1 und 9.8 min. during nap 2. The difference between sleep-onset times of both naps is not significant. Thus, no first-day effect related to this EEG criterion was observed.

The most reliable SRR habituation criterion was the time from starting the record until the last response to a tone before 5 successive non-responses followed (r = .61). The criterion was reached by 47 out of 51 subjects during each nap. Four subjects did not habituate at all. No first-day effect was observed.

5.2. Correlation between sleep onset time and SRR habituation.

Following the criteria used in earlier studies, correlations were calculated between the time until 2 successive stage 1 epochs occurred and the time until 3 successive SRR responses failed to appear. During nap 1 correlation was r_1.36 (n_1 = 27, p_1 > .05) and during nap 2 it was r_2 = .10 (n_2 = 30, n.s.). Thus, both correlations were much lower than in earlier studies and did not correspond with each other. They failed to confirm previous correlations.

5.3. Blood pressure and psychometric variables.

Blood pressure before napping and after a 40-minute nap was reliably measured (systolic and diastolic BP with an r = .63). The reliability of BP amplitude was sufficient high (r = .53). The first-day effect was not significant.

On the other hand, blood pressure before and after

napping showed no influence of the napping period on blood pressure.

The pre- and post-test values of the perdue peg board test, both short term memory tests, and the concentration test had sufficient or high reliabilities. Looking at the nap effect (post - pre), correlations of the psychometric measures were about zero. This could mean that the psychometric methods used in this study were not able to indicate recovery functions during resting when measured by post-pre changes. Apart from the rather difficult task of digit span backward, all psychometric tests demonstrated well known training effects. First-day effects were significant.

6. DISCUSSION

Measuring psychic fatigue by EEG sleep onset time, the rather weak criterion "appearance of the first stage 1-epoch" seemed to be the most suitable. It showed high reliability and no adaptative effects between nap 1 and nap 2. Thirty-three out of 51 subjects were able to reach the criterion during nap 1. Increasing the sleep-onset criteria from one to six successive stage 1 epochs resulted in a significant prolongation of sleep-onset times in nap 1. Using stage 2, more than 2/3 of the subjects failed to reach the sleep onset criterion during nap 1. In contrast, about half of the subjects reached stage 2 sleep during the second napping period. Reliabilities decreased with increasing sleep onset criteria.

Many subjects (40/51 in nap 1, 44/51 in nap 2) had theta waves after a relatively short time (7 min. in nap 1, 5.6 min. in nap 2). However, a strong day effect and a low reliability do not recommend theta as a criterion of psychic fatigue.

With regard to SRR habituation, the strongest criterion seemed to be the best one. Five non-responses were reliably reached by 47 subjects within about 3.4 min. (nap 1) and 2.9 min. (nap 2). A first-day effect could not be observed.

Habituation times clearly preceded psychic fatigue assessed by EEG sleep stage 1. The difference was about 8.8 min. in nap 1 and 7.0 min. in nap 2. Neither evoked nor spontaneous SRRs were observed after subjects fell asleep. The SR recordings were completely silent. Thus, SRRs - which are solely mediated by n. sympathicus and acetylcholine - disappeared during sleep onset. After spontaneous awakenings SR responses occurred again. However, they did not only recur after awakenings but also after stage 2 sleep. Although we do not go into further discussion of these observations, it is likely that the development of psychic fatigue, i.e. the transition from a state of awake

EEG to a state of consolidated sleep, could physiologically be accompanied by inhibition of sympathetic activity. According to Pavlov's theory this inhibition process should probably emanate from areas in the cortex in which acoustic perception is represented. The inhibition of acoustic centres should spread over the cortex and then attenuate sympathetic activity in the a.n.s. via centrifugal pathways. This dampening of sympathetic activity during development of sleep seems to be a transient one which ends with awakening from stage 1 sleep or with development of stage 2 sleep. In sleep stage 2 both sympathetic activity and sleep regulatory mechanisms are assumed to come under different control mechanisms, with the beginning of the EEG compared to the awake state.

A switching of the regulation of awareness seems to occur, whereby an involuntarily developing deactivational process is triggered. Such a "swich" of awareness is frequently experienced. Many of us remember a sudden, uncontrollable drop of awareness with rising psychic fatigue. For instance, when our concentration is fixed on a speaker in whose paper we are interested, suddenly, eyelids close or the head sinks without warning.

Correlation between sleep onset time (first 2 stage 1 epochs) and SRR habituation (2 non-responses) was rather low and not repeatable one week later ($\underline{r}_1$ = .36, $\underline{r}_2$ = .10). Calculations between all further criteria did not result in different coefficients. Thus, different criteria did not explain why correlations were lower than in our earlier studies. With regard to the sleep onset times, it was surprising that it took subjects during this monotonous acoustic stimulation test slightly longer to fall asleep than subjects from earlier studies with or without stimulation. During this study comparable sleep onset times (two successive stage 2 epochs) lasted about 12.9 min., whereas in earlier studies with a paralleled paradigm, control subjects and subjects with irrelevant monotonous acoustic stimulation showed means of 10 min. or lower.

Thus, afterwards it seemed likely to us that prolongation of the tones from 2 sec. to 10 sec. duration was a fatigue-delaying and not a fatigue-promoting variation. We therefore took a detailed look at the original recordings and found a possible explanation for the results: SR responses were not only evoked by the onset of tones but also by their offsets. Tone-offset responses in the SRRs were always accompanied by arousal reactions in the EEG. Thus, offset responses may have delayed the process of falling asleep.

While the SRR habituation to tone onset developed rapidly and systematically, tone-offset responses appeared unpredictably. The explanation proposed for this is that the onset of tones probably worked as a subliminal warning stimulus even when it was no longer consciously perceived by the subjects. However, such a stage of vigilance, during which subliminal perceptions are still processed, should pass fairly rapidly. That could explain the unsystematic response to tone offsets and thereby the low correlations between SRR habituation and sleep onset times.

The reliability measures of blood pressure and psychometric variables were sufficient high. After a 60-minute resting period even the one-week retest reliability of blood pressure amplitude was r = .53 and the diastolic value reached the same stability as the systolic value. The significant first-day effects in the initial values (at the start of the foreperiod) and before napping (about 15 min. later) indicated a remarkable sensitivity of blood pressure to situational aspects. Recovery functions during rest were not observed either in the blood pressure or in the psychometric data. This was easily explained by the fact that subjects were not tired or exhausted by tests before the resting period. The reliability of the psychometric variables was between r = .53 and r = .84. Post-values were higher than pre-values. A strong training effect was seen for perdue peg board, digit span forward and concentration test d 2. Results confirmed the necessity of the common practice of ensuring that subjects are well trained in the psychometric tasks before experiments. Because resting was not accompanied by recovery post-pre differences were low. Standard deviations of the pre- and post-values were also low. The reliabilities of the different scores were about zero. Thus, post-pre changes did not show interpretable differences between subjects (Petermann, 1978).

7. CONCLUSIONS

The most suitable criterion of sleep onset was the first appearence of EEG sleep stage 1. The time until SRR habituation, marked by the first 5 non-responses appeared to be the best SRR criterion in assessing psychic fatigue. The drop of awareness during "falling" asleep seemed to correspond with a switch in regulating sympathetic activity, indicated by a transient silence of the SRR recording. Correlations between sleep onset time and SRR habituation were low and not consistent. A reason for this may be offset responses to the 10 sec. tone. Blood pressure and psychometric variables were reliable measures but were not sensitive to resting. The well known training effect in the psychometric tests could also be confirmed in this study.

8. REFERENCES

BOHLIN, G. (1971). Monotonous stimulation, sleep onset, and habituation of the orienting reaction. Electroencephalography and Clinical Neurophysiology, 31, 593-601.

BRICKENKAMP, R. (1981). Test d 2. Aufmerksamkeits-Belastungs-Test. Göttingen, Hogrefe.

DAUM I., LEONARD J.P. & HEHL F.-J. (1988, in press). Development of sleep during monotonous stimulation as related to individual differences. Pavlovian Journal of Higher Nervous Activity.

LEONARD, J.P. (1985). Experimentelle Schlafinduktion: Kontrollierte Einschlafbeschleunigung und Einschlafverzögerung unter monotoner akustischer Stimulation. Zeitschrift für experimentelle und angewandte Psychologie, 32 (1), 120-134.

LEONARD, J.P. (1986). Experimentelle Schlafinduktion: Physikalische und kognitiv-emotionale Variationen psychischer Ermüdung im Tageskurzschlaf. Zeitschrift für experimentelle und angewandte Psychologie, 33 (4), 636-655.

LEONARD, J.P. (1988). Experimental models of vigilance regulation: Accelerated and decelerated sleep onset under the control of monotonous stimulation. In J.P. Leonard (Ed.), Vigilance: Methods, Models, and Regulation, (p. 119-128). Frankfurt/M., P. Lang.

MORUZZI, G. (1960). Synchronising influences of the brain stem and the inhibitory mechanisms underlying the production of sleep by sensory stimulation. Electroencephalography and Clinical Neurophysiology (Supplement), 13, 231-256.

PAVLOV, I.P. (1953). Sämtliche Werke. Band III/1, III/2. Berlin, Adademie Verlag.

PETERMANN, F. (1978). Veränderungsmessung. Stuttgart, Kohlhammer.

RECHTSCHAFFEN, A. & KALES, A. (1968). A manual of standardized terminology, techniques and scoring system for sleep stages of human subjects. Los Angeles, Brain Information Service / Brain Research Institute, UCLA.

STRELAU, J. (1983). Temperament Personality Inventory. London, Academic Press.

LONG-RANGE OPERATIONS IN THE GLASS COCKPIT: VIGILANCE, BOREDOM, AND SLEEPLESS NIGHTS

R. Curtis Graeber
Aerospace Human Factors Research Division
NASA-Ames Research Center
Moffett Field, California 94035
U.S.A.

ABSTRACT. The introduction of the two-person glass cockpit to long-haul flight operations raises some critical concerns about flight safety. While the pilot is further removed from the operation of his airplane, his role as the primary decisionmaker is more apparent. Man alone can provide the judgment, flexibility, and originality to solve problems and act decisively. While advanced automation can help by providing sophisticated data on aircraft status, it may also increase boredom and reduce alertness. This is especially likely during long-haul flights where research has documented considerable sleep loss and degraded alertness in less automated three-crew aircraft. While anomaly detection software may offset the resultant loss in monitoring by the crew, the uniquely human qualities associated with successful decisionmaking and problem solving may be severely compromised. As cruise range is extended well beyond present limits strategies must be developed to manage crew alertness. One approach may be to capitalize on the powerful computer systems and color displays already onboard.

1. Introduction

Within the year the first true long-haul "glass" cockpit airliner will be introduced to commercial passenger service. In anticipation of this event, an intense debate has developed about the performance problems associated with operating highly automated aircraft on extended international routes. The major concern has been the potentially detrimental effects of boredom and fatigue on two-person crews confronting a bank of computer generated displays Design improvements resulting in improved lift capacity and lower fuel requirements will enable such aircraft to fly further than previously possible. Single flights may extend up to 16 hours, and crews will be exposed to time zone shifts even more dramatic than those experienced today. With the rapidly increasing demand for nonstop air travel, there is little reason to doubt that such flying will soon become commonplace especially in such widespread regions as the Pacific Rim.

Given the highly dynamic nature of aviation and its catastrophic intolerance for human error, the issue of vigilance in automatised cockpits must be addressed. In doing so, this paper will first attempt to survey the subtle changes in pilot duties that have accompanied increased flight deck automation. Then, in order to provide some insight into the nature of the potential problem, specific findings will be described from two sources: (a) field studies on fatigue and sleep loss in current long-haul flight crews, and (b) airline incident and accident reports involving flight crew errors in international flight operations. Finally, the feasibility of developing strategies to manage pilot alertness will be explored with the aim of suggesting practical guidelines for operating such advanced aircraft in tomorrow's skies.

A. Coblentz (ed.), Vigilance and Performance in Automatized Systems/ Vigilance et Performance de l'Homme dans les Systèmes Automatisés, 67–76.

2. Advanced Transport Aircraft

2.1. AUTOMATING THE PILOT

Pilots traditionally enter their profession for the love of flying. It is the thrill of controlling one's first airplane that provides the initial motivation to continue through interminable ground school sessions and check rides. Eventually, the promise of financial gain may motivate the student to meet the requirements for becoming an air transport pilot. Yet, senior airline pilots still pride themselves on executing a smooth approach and landing after each flight. The undercurrent of cockpit competition reveals itself in the universal compliment for a "good" landing when everything goes right, and long-haul pilots continue to complain about their relative lack of landing opportunities.

The point of this emphasis on landing airplanes is to convey some sense of the importance that controlling the aircraft has for the vast majority of flight crews. While the installation of autopilots initially removed a large portion of this occupational reward, most pilots today still handfly the airplane until it reaches cruise altitude and then repeat the process through descent to landing. More recent advances in automation continue to encroach upon the control aspects of piloting. For example, within the past two decades aircraft automation has progressed from the operation of control surfaces to the total operation of systems controlling fuel distribution, engine thrust, navigation, and cabin pressurization. In today's glass cockpit the pilot is almost left with the feeling that his main function is to drive the airplane into take-off position and taxi it back after landing - - - and, of course, talk to the passengers during flight to assure them that everything is going smoothly. The projected flight deck implementation of datalink ground communication systems in the near future will further reinforce this feeling by eroding the need for verbal communications to the outside world.

Of course, there are strong arguments for implementing all of these developments. They result in the more efficient operation of aircraft systems and can provide a flood of data on current aircraft status whenever needed. They make flight planning easier, reduce workload during the critical phases of flight, and minimize the potential for human errors resulting from miscalculations, memory lapses, and miscommunications. Their economic benefits in terms of lower fuel costs, reduced crew requirements and more reliable airplane performance are undeniable. However, when combined with increased flight deck comfort and long-haul operations, advanced automation produces an environment ripe for performance atrophy. The major reason is that crew workload has been reduced to the point that monitoring has become the only remaining human activity for more than 90% of any routine flight. One is left with the nagging impression that the design of the cockpit is being led by the technology available and not by the needs of the crew who occupy it.

2.2. DECISIONMAKING ON THE FLIGHT DECK

Another way to describe this evolution in cockpit technology is to view it as a shift in the pilot's role from that of a control manipulator to that of a systems manager. While the computers perform the repetitive and mundane tasks, the pilot retains his command responsibility for the overall operation and safety of the aircraft. Thus, while this role has always been the essence of piloting, it has now been stripped of almost all those ancillary duties more associated with flying skill than intelligence. Despite all the technological power around him, only the pilot is able to use foresight, judgment, and innovation to integrate information and make flexible decisions in this very dynamic and sometimes unpredictable environment. While recent developments in artificial intelligence enable computers to sort through decision alternatives and recommend choices, they are limited in their applicability and validity by the inflexible boundaries of their databases and algorithms. Conversely, the experienced airline pilot possesses a unique aviation background against which to judge computer generated alternatives as well as those alternatives which are less obvious to an automated decision aid. Unfortunately, it is precisely the higher level human cognitive skills which are critical for decisionmaking that are most vulnerable to the combined impact of sleep loss, circadian desynchronization, and boredom. While systems monitoring will also suffer, it represents a relatively smaller part of the overall concern since onboard computers are already able to detect and announce many systems anomalies.

There is one other development associated with the glass cockpit that may detrimentally affect decisionmaking, that is, the reduction in crew size by elimination of the flight engineer position. While there is nothing sacred about a three-person flight crew, there are reasonable arguments to recommend the value of the third individual especially in long-haul operations. Most justification for elimination of the flight engineer is based on the ability of automated systems to carry out his usual functions and to prevent the two pilots from experiencing excessive and unsafe workloads during a flight emergency. While these arguments are valid for long-haul aircraft as well as for shorter range aircraft, there is little doubt that three people are more stimulating than two and that a third crewmember can increase the collective aeronautical intelligence of the cockpit team by 50%. The concern here is low workload and crewmembers whose mental efficiency is likely to be compromised especially at night after successive time zone crossings. While it is questionable whether such concerns are sufficient to justify the cost of maintaining the flight engineer position, the decrease in crew complement must be considered when evaluating the risk of crew performance deficits in these advanced aircraft. In this regard, the well known effects of sleep loss and circadian desynchronization on mood and interpersonal relations are another vulnerable avenue whereby decisionmaking could be negatively impacted. There is already some preliminary data suggesting that crews in advanced cockpits communicate less often with each other than do those in more traditional two-man cockpits and that this communication is even less frequent at night (Costley & Johnson, 1987). The importance of communication in the cockpit has recently received considerable attention in the industry because poor crew coordination has been identified as one of the most prominent causes of aviation accidents and incidents (Lauber, 1984). The importance of this factor and others will become more obvious in the discusion of the operational record, but it will be helpful to first describe some of the recent findings concerning the effects of transoceanic flying on international crews.

3. Sleep Loss and Circadian Factors in Long-Haul Flight Crews

In order to obtain some objective information about the fatigue associated with long-haul transmeridian flying, our laboratory has been engaged in a series of studies designed to document sleep patterns, circadian desynchrony, and changes in subjective state in B747 pilots and flight engineers. While the results are too extensive to describe here in detail, there are several major findings which help clarify the problems facing such crews. More complete descriptions are available in several other publications (Graeber, 1986, 1988; Graeber, Connell, Schreiber, Sasaki, & Kurosaki, 1988).

3.1. LAYOVER SLEEP

The first set of findings comes from multiple studies done in collaboration with the

the DFVLR Institute of Aerospace Medicine, the RAF Institute of Aviation Medicine at Farnborough, the Jikei University School of Medicine, and the Stanford University Sleep Disorders Center. Using San Francisco as a hub, B747 crews volunteered to stay at a sleep laboratory during the first outbound layover after operating across 7-9 time zones. After providing baseline sleep data in a laboratory near their home domicile, four crews from each of three airlines were studied after flying westward, and four more crews from two carriers were studied after flying eastward. While at the destination laboratory, the volunteers were asked to follow their usual daily routine for that destination and were free to eat, sleep, or exercise whenever they chose. All sleep episodes were recorded polygraphically (EEG, EOG, and EMG), and physiological daytime sleepiness tests (MSLTs) were administered on even GMT hours whenever a crewmember was awake.

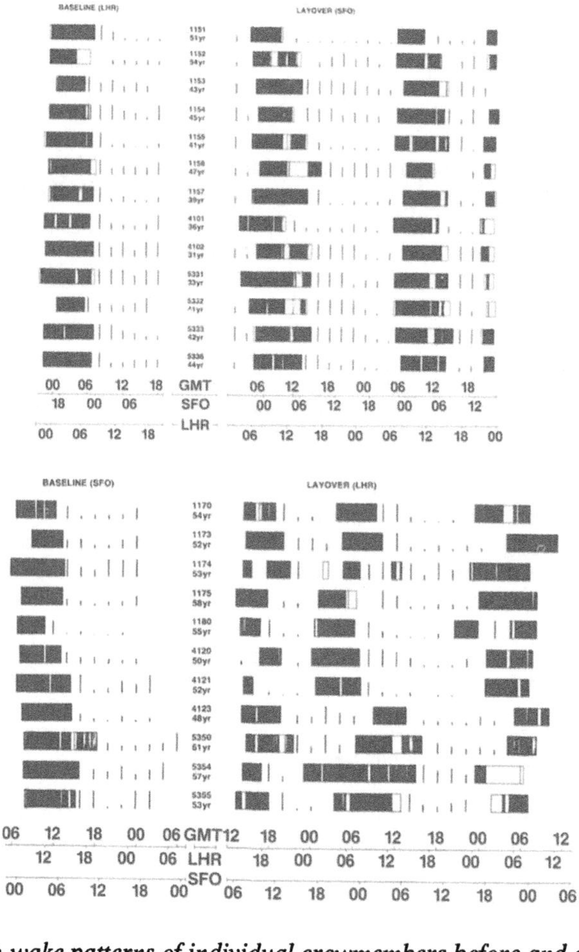

Figure 1. Sleep-wake patterns of individual crewmembers before and after westward (upper, from Nicholson et. al., 1986) or eastward (lower) flights across eight time zones. Dark bars indicate sleep, white bars indicate awakenings. Vertical lines indicate sleep latency scores ranging up to 20 min.

The findings were remarkably consistent despite the cultural disparity among the crew samples. On the average, most crewmembers were able to obtain enough sleep during the layover either by sleeping efficiently at selected times or by sleeping less efficiently but staying in bed longer than usual. One very clear result emerged regarding the direction of flight. Sleep quality decreased more after eastward flights than after westward flights. This directional difference can be seen by comparing the patterns of sleep for two sets of crews shown in Figure 1. Sleep was much more variable and fragmented after eastward night flights than after westward day flights over an equivalent number of time zones. This directional difference is highly consistent with the fact that the body clock's natural period is longer than 24 hours. While westward flights lengthen the day, eastward flights shorten it and hence should result in greater circadian induced sleep disruption. Research on passengers has shown that such directional differences in circadian adjustment occur regardless of the time of flight, day or night (Klein & Wegmann, 1980)

While there were dramatic sleep losses in some individuals, sleep quality decreased only slightly in most cases. This was reflected in the average total sleep times for each airline group and in the crewmembers' subjective ratings of sleep quality. In the case of westward flights, some of these means exceeded the comparable baseline values. There was, however, a wide range of individual differences in sleep quality and efficiency (i.e., percent of time-in-bed when the subject was asleep) among the crews. For example, in the U.S. group layover sleep efficiency varied from 54.6% of 9.8 hrs. to 90.0% of 16.2 hrs. The number of spontaneous awakenings during a single layover sleep period ranged from fewer than 10 to more than 50.

One individual factor was found to strongly affect sleep and sleepiness. This factor was age. An analysis of the baseline sleep of all 56 volunteers revealed that age was significantly correlated with an increased number of awakenings, a higher percentage of light drowsy (i.e., restless) sleep, a lower percentage of deep (slow-wave) sleep, and lower sleep efficiency. This was particularly true for those crewmembers over 50 yrs. of age.

3.2. SLEEPINESS

The average daytime sleepiness scores also showed no substantial increase from baseline conditions. They continued to exhibit a consistent 12-hour rhythmic pattern which remained synchronized to home domicile time despite a 48-hr. stay in the destination time zone. For example, the daytime sleepiness of the German crewmembers was greatest in both San Francisco and Cologne during the late afternoon at home (15:00-17:00 GMT+1). This stability of the sleepiness rhythm has practical implications for anticipating when sleepiness will occur on the flight deck and for planning when you are most likely to nap successfully during layovers or before flights.

The sleepiness data led to another finding that has important operational implications. Contrary to popular belief, we found that crewmembers were very poor predictors of how sleepy they really were. An analysis of all 56 subjects from the four airlines revealed no correlation between the objective physiological sleepiness scores and the subjects' own subjective ratings of sleepiness given just beforehand. This finding should serve as a warning to operating crewmembers and their managers about the validity of claims regarding the relative lack of sleepiness on the flight deck. The soporific nature of the long-haul cockpit, especially at night, makes it an almost ideal setting to expose the crew's underlying level of physiological sleepiness.

We have seen manifestations of this vulnerability in our research on multiple-segment long-haul trip patterns. While the collaborative effort just described is the first study to objectively measure crew sleep and sleepiness during layovers, it has its limitations. Single out-and-back trips are relatively simple for flight crews and not representative of most long-haul commercial flight operations. Consequently, observers from our laboratory have been accompanying international flight crews in the cockpit during the course of trip patterns lasting up to nine days. During this time, as well as for several days before and after, the crewmembers continuously wear a solid-state biomedical monitor which records heartrate, non-dominant wrist activity, and rectal temperature every two minutes. The observer, who is at least a private pilot, records all operational events during flight and provides periodic feedback to the volunteers by displaying timeplots of their data on a small video display unit. The subjects maintain a daily logbook which chronicles the time and quality of all sleeps, daily activities (e.g., meals, caffeine and alcohol consumption, exercise, showers, etc.), and subjective mood and fatigue every two hours. These studies have included 13 passenger crews from N. American carriers as well as 11 military cargo crews providing a total of 88 subjects and 155 flights (850 flight hours).

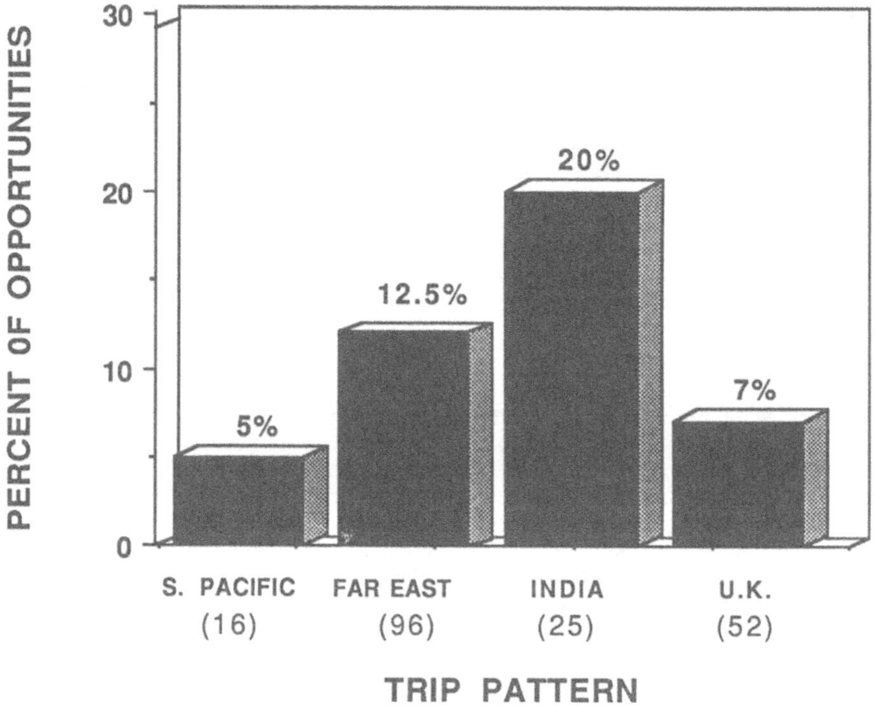

Figure 2. Incidence of cockpit napping during cruise on long-haul trip patterns. Opportunities (given in parentheses) = No. of flights x No. of subjects.

One of the most revealing findings of this extended study came from the records of crewmembers napping in their seats during cruise. The incidence of this behavior on commercial flights is shown in Figure 2 as the percentage of opportunities for cockpit napping (i.e., denominator = No. of subjects x No. of flights). Only naps lasting at least 20 minutes are included. The trips varied in the total number of flights, number of night flights, and total number of time zones crossed. The trip to India was an 7-day eastward trip originating from the East coast and transiting through Europe and the Mideast (16 time zones crossed, 2 night + 4 day flights). The U.K. trip was an 8-day pattern requiring 6 transatlantic shuttles between the Pacific time zone and the U.K. (48 time zones, 3 night + 3 day flights). The Far East trip was a 9-day westward trip departing from the Pacific time zone and stopping at three Far East destinations and Hawaii (28 time zones crossed, 3 night + 6 day flights). The South Pacific trip was primarily a North-South 5-day trip departing from the Pacific time zone and transiting through Hawaii but requiring all-night flights to and from the South Pacific destination (8 time zones, 2 night + 1 evening + 1 day flight). Inflight data from the military crews has been omitted from the analysis because crewmembers are allowed, and sometimes encouraged, to leave their seats and go back to a bunk for sleep on a one-at-a-time basis.

All the naps shown in Fig. 2 were unplanned. When a crewmember did nap, he occasionally told the others that he needed a brief rest, but, more often he just fell asleep. Cockpit napping usually occurred during local night, but there were also occurrences during daytime flights. Typically, the other members of the crew would continue to perform their duties and not say anything. No more than one person was asleep at any time, and naps never lasted more than one hour. The only planned naps were taken during the first and last U.K. flights. The crews on these two segments were augmented by an extra pilot and flight engineer, so that these naps occurred in the cabin during time off.

It is unlikely that these percentages reflect the actual incidence of cockpit napping during long-haul passenger flights. In fact, they probably underestimate the extent for three reasons. First, while the observer's presence did not prevent subjects from napping, it most likely discouraged some napping in that it is not approved by management. Second, some of the non-transoceanic flights were relatively short (1-3 hrs), and, though occurring at night, they provided little cruise time available for sleep. Finally, the lowest incidence of napping (5%) on a trip may be attributable to the extra presence of management pilots during two of the four all-night flights. While the relatively high age (45 to 60 years) of the crewmembers may be a contributing factor, these data confirm the general belief throughout the industry that long-haul crewmembers do occasionally nap during cruise, especially at night (e.g., Feedback, 1984). Moreover, the findings are symptomatic of the degree of sleepiness that can develop on the flight deck and suggest that the vigilance of individual pilots may be impaired at least during cruise; however, it should be realized that napping may have a beneficial effect on the overall vigilance level of the crew. These conclusions are also consistent with the confidential reports on napping received from flight crews in both the U.S. and Europe. Similar reports of fatigue related crew errors will be discussed later in a review of the operational record.

3.3. SLEEP LOSS

Reduced alertness and sleepiness on the long-haul flight deck can result from two different sources: (a) the external circadian desynchronization arising from duty at a time of day when sleep is the norm, and (b) the cumulative sleep loss that can develop from the inability to sleep after crossing time zones. While a pilot may obtain his usual amount of sleep during a layover, he may still succumb to the influence of the body clock at the "wrong" time of day, or he may be flying at the "right" body time but suffer from prior sleep loss. The latter has been the focus of a follow-up set of collaborative polysomnographic sleep studies on crews flying polar routes.

The procedures used in these studies paralleled those used during the first set of layover studies except that portable EEG recorders were used to record sleep at each layover site. Crews stayed in their usual hotels instead of sleep laboratories, and no sleep latency tests were administered. The basic protocol consisted of four crews from each airline flying trips between Japan and Europe with a layover at Anchorage in each direction. Again there were directional differences, with initial eastward night flights resulting in a more dramatic cumulative impact on sleep consolidation and resynchronization throughout the trip and during recovery at home. Data from the eastward-westward crews also revealed substantial individual differences in the amount of cumulative sleep loss experienced by the end of the trip. While the average sleep loss was 1.5 hours per night, a few subjects actually gained sleep while some lost over 20 hours during the 6 days away from home (Sasaki, Kurosaki, Takahashi, Mori, Spinweber, and Graeber, 1987).

Such wide-ranging individual differences have been consistently observed in all our studies of flight crews engaged in international long-haul operations. While longer trip patterns tend to magnify these differences, they can be seen very early in a trip, especially if the initial flight is eastward. A re-examination of Fig. 1 readily demonstrates this observation. The implication of such observations for the automatised cockpit is that manufacturers and carriers alike must assume that at least one crewmember is very likely to be in a low state of alertness during a substantial portion of any flight. Operational procedures and design guidelines must allow for the resultant degradation of performance, as should regulatory limits on flight and duty time. Some insight into the nature of this degradation can be gained from examining the operational record.

4. The Operational Record

In discussing the incidents and accidents associated with modern long-haul passenger jets one must not loose sight of the incredible safety record associated with this form of transportation. Every day thousands of airplanes take off and land successfully without any serious problems. It is in the spirit of maintaining and improving this success for the next generation of aircraft that an examination of some reports is appropriate.

In a recent review of the worldwide jet fleet accident record, Capt. Caesar (1987) continues to report that the loss ratio (based on number of sectors flown) for long-haul operations is consistently almost three times higher than that for the combined short- and medium-range fleets. Similar statistics describe the record for the past 20 years. While numerous other adverse circumstances (e.g., language problems, more difficult handling characteristics, unfamiliar and poorly equipped airports) undoubtedly contribute to this difference, there is little doubt that one significant factor is the impact of time zone changes and long flights on crew performance.

An examination of confidential reports submitted to NASA's Aviation Safety Reporting System (ASRS) from long-haul flight crews reveals that most reports are concerned with errors associated with the take-off or landing phases of flight and not cruise. Conversely, a considerable number of reports submitted to the United Kingdom's Confidential Human Factors Incident Reporting Program (CHIRP) relate performance deficits during cruise (see Feedback, 1984). This discrepancy most likely does not reflect a difference in the frequency of such events but rather the apparent tendency for U.S. pilots to report to ASRS primarily when regulatory immunity is sought. The types of errors reported to the system include failure to obtain clearance for landing, landing on the wrong runway, altitude busts, taxiway incursions, heading deviations on approach, and gross track deviations in cruise. One of the more interesting reports consisted of a 900 ft. altitude deviation during the entire flight across the Pacific due to incorrect altimeter settings by both pilots after departure.

While a low barometric pressure in Anchorage facilitated the error, the crew cited an early departure and inability to sleep the night before due to circadian desynchronization. Yet no one caught the mistake during the entire flight until Tokyo radar noticed the deviation. A similar situation occurred in a fuel error. In this case, an incorrect computer flight plan resulted in too little fuel being put on board because it showed the aircraft 100,000 lbs. lighter than it actually was. Despite several opportunities to catch the error before departing, no one noticed it until after the second in-cruise fuel reading. Again the crew cited muliple trans-pacific operations and cumulative sleep loss.

There are two recent accidents that also reveal the potential for serious operational errors, especially when automated systems are involved. The first of these involved a DC-10 which ran off the end of the runway at John F. Kennedy airport due to a failed autothrottle. The accident report cited the crew's apparent total lack of awareness of the resulting excessive airspeed during the approach (NTSB, 1984). The crew had just completed a transatlantic flight from Europe. The second accident was more dramatic and involved a B747-SP flying from Taipei to Los Angeles. At 41,000 ft. the aircraft began to experience loss of power in its Number 4 (right outboard) engine. The crew left the autopilot "On" while focusing on restoring engine power. Because of insufficient rudder authority, the autopilot could not correct for the rightward yaw, and the plane rolled to the right with the nose eventually falling through the horizon. The plane fell 30,000 ft. before it was recovered and diverted to San Francisco with major stuctural damage. According to the report, the crew failed to recognize the rolling of the aircraft and its approach to extreme attitude until it was too late. In its probable cause statement, the Safety Board cited the captain's preoccupation with the inflight malfunction and his failure to monitor the flight instruments properly (NTSB, 1986).

5. Facing the Challenge

Incidents such at those described above will continue to occur. Although the changing nature of automation will eliminate some errors, such as incorrect waypoint entries, others are bound to appear. As researchers, we have an obligation to translate our knowledge about human performance into real-world dividends for safety and, hopefully, efficiency as well. However, there is now a growing awareness among human factors experts that many of the classic vigilance studies involved tasks and subjects that bear little resemblance to actual airline operations (Weiner, 1987). Consequently, more effort will have to be directed to examining vigilance in real-world situations or carefully designed full mission simulations. In the meantime, both regulatory authorities and flight managers will have to realize that past efforts to police and control pilot fatigue have met with limited success and will continue to do so given the human limits of flight crews. By placing tired pilots in automated two-person cockpits, we run the risk of increasing the boredom and complacency already characteristic of long-haul operations in three-person cockpits.

One approach may be for manufacturers to capitalize on the power of on-board computers to help minimize the influence of fatigue and sleepiness on the flight deck. The ability of these machines to generate high levels of attention and to monitor operator responses should be applied in a positive manner toward this end. They should be used not only to monitor whether pilots are awake, but also, to provide feedback to crewmembers about their alertness and to provide stimulation to overcome transient boredom and sleepiness while on duty. Thus, they should be used to assist the crew not only in managing information but also alertness on the flight deck. Likewise, on-board sleeping facilities should be carefully designed in light of the latest scientific knowledge to assure appropriate rest, and the rostering of augmented crews should be carefully worked out in advance to assure the optimum level of alertness at all times, but especially during descent and landing.

References

Caesar, H. 'Safety statistics and their operational consequences.' *Proc. 9th Orient Airlines Assoc., Flight Safety Seminar*, Manilla, Philippines, May, 1987.

Costley, J., and D. Johnson. 'A comparison of cockpit communication, B737-B757.' Interaction Trainers International, Ltd., Huntingdon, Cambs., U.K., 1987.

Feedback. Farnborough, U.K.: RAF Institute of Aviation Medicine, 1984, No. **5**, 6.

Graeber, R.C. (Ed.) Sleep and wakefulness in international aircrews. *Aviation, Space & Environmental Medicine*, 1986, **57** (Supplement), B1-B64.

Graeber, R.C. 'Aircrew fatigue and circadian rhythmicity.' In E.Wiener and D.C. Nagel (Eds.), *Human Factors in Aviation*. New York:Academic, 1988, in press.

Graeber, R.C., L.J. Connell, H.G. Schreiber, M. Sasaki, and Y. Kurosaki. 'Aircrew sleep and fatigue in long-haul flight operations.' *Proc. of 40th Flight Safety Foundation International Air Safety Seminar*. Arlington: Flight Safety Foundation, 1988, in press.

Klein, K.E., and H.M. Wegmann. *Significance of circadian rhythms in aerospace operations*. AGARDograph No. 247. NATO AGARD: Neuilly sur Seine, 1980.

Lauber, J.K. 'Resource management in the cockpit.' *Air Line Pilot*, 1984, 20-24, 38.

National Transportation Safety Board. *Scandanavia Airlines System DC-10-30, John F. Kennedy Airport, Jamaica, New York, February 28, 1984* (Report NTSB-AAR-84-15). Washington, D.C., 1984.

National Transportation Safety Board. *China Airlines Boeing 747-SP, N4425V, 300 nautical miles northwest of San Francisco, California, February 19, 1985* (Report NTSB/AAR-86-03). Washington, D.C., 1986.

Nicholson, A.N., P.A. Pascoe, M.B. Spencer, B.M. Stone, and R.L. Green. 'Nocturnal sleep and daytime alertness of aircrew after transmeridian flights.' *Aviation, Space & Environmental Medicine*, 1986, **57** (Suppl.), B43-B52.

Sasaki, M., Y. Kurosaki, T. Takahashi, A. Mori, C.L. Spinweber, and R.C. Graeber. 'Flight crew sleep after multiple layover polar flights.' *Sleep Research*, 1987, **16**, 637.

Wiener, E.L. 'Application of vigilance research: Rare, medium, or well done?' *Human Factors*, 1987, **29**, 725-736.

PERFORMANCE ASSESSMENT TECHNIQUES IN ADVANCED COCKPIT SYSTEMS

Clark A. Shingledecker
NTI, Incorporated
Dayton, Ohio
USA

ABSTRACT. This paper describes a group of available physiological and behavioral assessment methods that show promise as meaningful indicators of aircrew performance capability for future aircraft systems. While many of these techniques may be profitably used as aids to the cockpit design process, this paper focuses on their potential use as operational indices of pilot status in future systems which will incorporate revolutionary types of displays and advanced levels of automation. Preliminary results are presented which demonstrate a correlational methodology designed to evaluate the predictive capabilities of candidate biocybernetic measures.

1. INTRODUCTION

Within the next two decades, advances in engineering and computer science will be responsible for a variety of dramatic changes in military and commercial aircraft. Current improvements in avionics, airframe construction, propulsion and control systems have already produced significant changes in the remote sensing abilities, appearance and performance envelopes of aircraft that were not considered possible only a few years ago. Even further benefits can be expected in the aircraft of the near future because of the maturing fields of display technology and artificial intelligence.

While each of these enhancements will undoubtedly improve the potential effectiveness of future aircraft, they will also produce fundamental changes in the roles and importance of critical flight crew members and in the performance requirements that will be imposed upon them. As a consequence, serious consideration must be given to methods and approaches which can be used to insure optimal human performance in future cockpit systems.

A. Coblentz (ed.), Vigilance and Performance in Automatized Systems/ Vigilance et Performance de l'Homme dans les Systèmes Automatisés, 77–85.

Several factors contribute to growing concerns over the possibility that aircrew performance will be a major limiting factor in future systems. In present aircraft, the use of sophisticated flight computers is rapidly relieving aircrew of labor-intensive duties, and shifting their tasks to those of system monitors and supervisors. Future aircraft designs are continuing this trend so that even higher levels of system automation will be employed. Since such automation can be expected to result in crew size reduction, the advanced capabilities of future airborne systems will often be the sole responsibility of a single pilot whose performance can easily be the most important determinant of the outcome of a major battle or the safety of hundreds of passengers.

The problem of reduced crew redundancy in future aircraft will be compounded by the way in which automation affects the nature of the tasks performed by the pilot. Removed from the active control loop, the pilot is likely to be more susceptible to vigilance decrements. Furthermore, the types of passive monitoring activities that will be required of the future pilot have been shown to reduce situation awareness and human abilities to effectively diagnose and solve system problems when they are detected (Kessel and Wickens, 1982).

At the other extreme, enhanced sensing and display technologies as well as expanded computer systems will present the possibility that the future pilot will be subject to excessive mental workload under certain critical conditions. While "natural" display codes and intelligent formats will alleviate this problem to some extent, it is not entirely certain that these design technologies will be sufficient to overcome the need for processing increased amounts of data in shorter periods of time.

A final group of variables which will impact future pilot performance capabilities are not products of future technology, but are stressors whose effects may be magnified by mental underload or overload. Included among the most obvious physical and psychological threats to future aircrew performance capacities are fatigue and sleep loss in extended operations, the use of prescribed or illegal drugs, and in combat aircrews, exposure to chemical, biological and nuclear agents.

Taken together, the rising criticality of the performance exhibited by key crew members, changing task demands, and the incapacitating potential of operational stressors suggest that specific, active subsystems will be needed in future aircraft to guard against catastrophic failures due to human error.

A technologically feasible approach to preventing such error would be to monitor for decrements in the performance capabilities of the human operator, and to take appropriate corrective action. At the simplest level, biocybernetic intervention could involve the evaluation of performance capabilities prior to a flight in order to select those personnel who exhibit an optimal fitness for duty. In a more advanced application made possible by the hardware and software advances anticipated for the future cockpit, performance capabilities could be monitored on a moment-to-moment basis during flight. Thus, impending performance decrements could be detected automatically, and the information used to alert the pilot, inform command personnel or even initiate computer control of the system. Employment of sophisticated algorithms designed to track the current intent of the pilot could permit even more finely tuned utilization of the monitored state of the operator.

Although a number of technical advances must be made to produce practical biocybernetic monitoring systems for routine operational application, sufficient computing and sensing technology is currently in place to judge the overall concept as technically feasible. In addition, a number of existing human metrics are adequately understood to suggest their inclusion on a candidate list of monitoring techniques. The following section of this paper briefly reviews these assessment methods in view of a multiple resource model of performance capabilities.

2. CANDIDATE MEASURES

A majority of the models that have been forwarded to explain human performance capabilities concur that unitary factor approaches are inappropriate. Thus, it appears that any attempt to accurately predict decrements in human performance must account for the possibility that two or more mechanisms may underlie performance capability, and that they may be differentially affected by task demands, stressors and operator states. A corollary to this position is that multiple measures will be required to predict performance failures caused by impairment of the separate mental elements.

A typical multi-element form of human performance model is one which incorporates both specific information processing structures and a more general, energistic component. In such a model, performance may be degraded because of impairment of one or more information processing mechanisms (e.g. perceptual, central and motor processing stages), or because of a non-optimal state of activation. The following candidate assessment techniques were selected primarily on the basis of evidence suggesting their sensitivity to one or more of these components.

2.1. Physiological Predictors

Physiological measures of central and peripheral nervous system activity are inherently suited to the task of providing biocybernetic feedback about the cognitive status of the human operator. Although instrumentation and artifact problems can impose limitations on their use, many of these measures can be obtained non-invasively and without interfering with the ongoing behavior of the operator (see O'Donnell, 1979 for an extensive review of these methods). Furthermore, a reasonable body of evidence suggests that four of these electrophysiological indices may be useful for real-time assessment of pilot status.

Voltage variations recorded from scalp sites are known correlates of the state of activation of a subject. Frequency analysis of the electroencephalographic (EEG) signal demonstrates dynamic changes such that low frequency, high amplitude waves tend to dominate the spectrum in subjects who are fatigued, while high frequency activity contributes more power in alert subjects.

Although such EEG measures are not directly useful for assessing operator workload levels, they appear to provide one possible method for assessing the general arousal state of pilots in automated systems. Measures of the status of certain cognitive processing structures appear to be available from event related potentials (EP) extracted from the EEG signal. These averaged brain responses to discrete stimuli can be elicited by a variety of cockpit events, and contain identifiable waveform components whose latencies and amplitudes have been related to perceptual efficiency and central processing capability. In addition, the P300 wave of the EP has been used as an indicator of the level of processing resource investment required by a task.

A third physiological metric which has strong potential for use in biocybernetic monitoring of pilots is heart rate variability (HRV). The typical result of studies in which the beat-to-beat interval variation has been assessed during work periods is that HRV decreases as workload increases. Although considerable controversy has arisen over appropriate methods for assessing HRV, recent results obtained with techniques that assess only highly sensitive components of sinus arrhythmia have proven most useful (e.g. spectral analysis and vagal tone metrics).

A final candidate metric which appears promising is the evaluation of involuntary eyeblink behavior recorded from the electroculographic (EOG) signal. Increases in blink rate have been shown to be related to performance decrements which occur as a function of time on task. In addition, increases in eyelid closure times and blink duration appear to signal similar deficits.

2.2. Behavioral Predictors

Although not as obviously relevant to the problem of biocybernetic monitoring, several behavioral measures also offer opportunities for predicting the performance capabilities of future pilots. In particular, recent developments of standardized batteries of perceptual, cognitive and motor tasks have provided ideal methods for evaluating mental status prior to a mission (e.g. Shingledecker, 1984). While such an application of these test batteries has not been instituted, data bases are rapidly accumulating on the effects of drugs and stressors for standardized tasks. Thus, in the near future it should be possible to develop normative criteria for assessing pre-mission mental fitness for duty.

A second behavioral technique that might be employed as a biocybernetic method would involve the use of embedded secondary tasks for assessing the information processing capacity of the pilot in a real-time fashion. The embedded task technique was developed as a non-intrusive secondary task workload measure for high fidelity cockpit simulation research (Shingledecker and Crabtree, 1982). Using this method, naturally occurring, mission-relevant secondary tasks are extracted from the normal repertoire of pilot duties and used as probe tasks to determine the processing demands of other concurrent duties. The advantage of this technique is that tasks are selected which are lower in priority than those whose workload is to be assessed. Thus, the secondary probe task (e.g. responding to noncritical radio messages) is unlikely to intrude on primary performance, but provides an index of the capability of the pilot to assume additional workload. While this method has not been evaluated as an approach to automatically evaluating the performance status of operational pilots, it presents an intuitively appealing and simply instrumented technique that may prove useful in future systems.

3. EVALUATION OF CANDIDATE METHODS

Although the physiological and behavioral techniques outlined above are promising candidates for pilot monitoring in future cockpits, evaluative research is needed to assess their sensitivity and reliability for this purpose. Unfortunately, traditional experimental methodologies that have been used to investigate measures of performance capability are inappropriate for such research. Rather than assessing a measure's capability to reflect the presumed impact of an intervening variable (e.g. fatigue, boredom), a method is needed for this research which can be used to identify measures that predict variations in human performance. One approach that can be used to achieve this goal is a correlational method in which a relationship is sought between a predictor metric and a criterion performance index.

Our organization has employed this type of correlational approach in preliminary research designed to identify useful predictor metrics for biocybernetic applications. In this methodology, candidate performance predictor metrics are correlated with simultaneous and temporally succeeding measures of performance on a simulated systems operation task. Performance variability is induced by exposing subjects to the combined stressors of sleep loss and continuous performance. It should be noted that the intent of imposing these stressors is not to produce some predicted pattern of decrement due to fatigue. Instead, the technique is simply designed to capitalize on the performance variation likely to be induced under these conditions in order to examine correlations across a broad range of within-subject variability.

3.1. Method

A limited experimental implementation of the methodology has been completed in which two subjects performed the complex time sharing task continuously for eight hours following twelve preceding hours of sleep deprivation. The task was designed to simulate a generic aircraft operation activity and contained two primary components which were performed with equal priority. The first of these components was a single axis, unstable tracking task. The second was a visual monitoring task which required the subjects to monitor four continuously varying tape displays. Signals requiring a manual detection response were biases in the normally random movement of the display indicators to the upper or lower sections of the display. Signals appeared at a rate of four to five each minute.

Candidate predictor measures were selected to match the information processing demands of the simultaneously performed tasks. In order to assess activation level, four frequency bands of the EEG spectrum were identified for power spectrum analysis. Additional measures of alertness were eyeblink rate, duration, amplitude and eyelid descent time. Interpolated behavioral measures were used to assess the perceptual and response capabilities of the subjects.

The memory search paradigm (Sternberg, 1969) and the interval production tapping task (Michon, 1966) were chosen for this purpose.

3.2 Results

Data were collected on the criterion tracking/monitoring tasks and on the physiological metrics in five minute intervals. The interpolated behavioral measures were collected during a break preceding each 50 minute performance period. To permit comparisons among the candidate predictors, only advance predictor correlations for the eight one-hour performance periods are discussed here.

These correlations are shown in Table I. Although the results are based on only two subjects, a number of tentative observations can be made about the potential predictive capacity of the candidate metrics.

TABLE I.

Performance Prediction Correlations

EEG Proportional Power

		Delta	Theta	Alpha	Beta
Tracking Error	S1	.14	.29	-.17	-.21
	S2	.93	-.88	-.94	-.93
Missed Signals	S1	.95	-.64	-.86	-.97
	S2	.17	-.07	-.26	-.29

EOG Eyeblink Parameters

		Interval	Amplitude	Duration	Descent
Tracking Error	S1	-.14	.22	.37	0
	S2	-.76	.10	.99	.94
Missed Signals	S1	-.84	.62	.80	.81
	S2	-.22	-.02	.04	-.20

Interpolated Behavioral Tests

		IPT Duration	IPT Duration	Sternberg RT
Tracking Error	S1	-.28	.22	.01
	S2	-.66	.69	.35
Missed Signals	S1	-.76	.53	.55
	S2	-.30	-.05	-.07

A strong relationship was obtained between performance and the proportion of the total EEG power in each of the four bands. The pattern of correlation across the bands is a general shift in power, such that poorer performance occurred when the relative power in the low frequency (delta, 1-3 Hz) band increased, and relative power in higher frequency bands decreased (4-30 Hz). Similarly positive predictive relationships were obtained for the measures of eyeblink behavior. Increases in tracking error as well as poorer signal detection were predicted by larger amplitude blinks, higher blink rates, longer eyelid descent times and longer closure durations.

A more variable set of relationships was obtained between interpolated behavioral task measures and criterion performance. In general, criterion task decrements were associated with a decrease in the interval between finger taps and an increase in the variability of intertap intervals. Longer memory search task reaction times were also predictive of poorer criterion performance.

Although the results summarized above are generally descriptive of the average correlations obtained, inspection of Table I. reveals marked individual differences between the two subjects. Specifically, for Subject 1, correlation coefficients were consistently larger for the monitoring index than the tracking measure. In contrast, the predictor measures were more strongly associated with tracking error than monitoring misses for Subject 2.

These findings are congruent with the hourly mean performance scores obtained by the two subjects. In each case, the criterion tasks which displayed the stronger correlations with the predictor metrics were those which showed the largest decrements during the eight hour testing period. This result indicates that the predictive power of the candidate measures may be dependent on the resource allocation policy of the performer. For the subjects tested in this study, the predictive measures most strongly anticipated the component of criterion task performance that appeared to receive the least effort expenditure.

4. CONCLUSIONS

Both human performance theory and the type of result reported here suggest that the development of useful techniques for monitoring the performance capabilities of pilots will require the development of a family of technologies from which appropriate measures can be selected to match the individual and the operating environment. In addition, it is highly likely that enhanced prediction will depend upon the development of combinatory methods which reflect the output of several simultaneously measured parameters.

Regardless of the way in which these measures are combined, at least three factors will have to be considered when selecting them for specific applications. As suggested by multifactor models of human performance, the structure of tasks and the types of stressors to which aircrew will be exposed will have to be taken into account. In addition,the results presented here indicate that task priorities may have to assessed in order to determine the specific aspects of performance that will be predicted by monitoring parameters.

A second factor which will determine the nature of appropriate measures is the temporal prediction requirement of the operational scenario. The full results of the demonstration study discussed here indicated that different metrics varied in terms of the time period for which significant predictions were obtained.

Finally, selection of measures will also be affected by the limits and practicalities of the operational environment. Intrusiveness, equipment size and weight, and aircrew acceptance will continue to be important issues in the implementation of advanced human factors methods for the cockpits of future aircraft.

5. REFERENCES

Kessel, C.J. and Wickens, C.D. 'The transfer of failure-detection skills between monitoring and controlling dynamic situations'. Human Factors, 1982, 24, 49-60.

Michon, J.A. 'Tapping regularity as a measure of perceptual motor load'. Ergonomics, 1966, 9(5), 401-412.

O'Donnell, R.D. Contributions of Psychophysiological Techniques to Aircraft Design and Other Operational Problems. NATO AGARDograph No. 244, 1979.

Shingledecker, C.A. A Task Battery for Applied Human Performance Research. Air Force Aerospace Medical Research Laboratory, AFAMRL-TR-84-071, 1984.

Sternberg, S. 'Memory scanning: mental processes revealed by reaction time experiments'. American Scientist, 1969, 57 421-457.

PERFORMANCE ASSESSMENT REQUIREMENTS FOR FUTURE COCKPIT SYSTEMS

Robert D. O'Donnell
NTI, Incorporated
Dayton, Ohio
USA

ABSTRACT. Advanced cockpit technologies which may drive automation requirements in the future are described, and future performance assessment metrics for such systems are explored. Two major changes in assessment methodology are discussed. The first involves a radical alteration in the direction of information flow - from the human to the system - and the second involves an increase in the need to measure complex hypothetical constructs such as situation awareness, workload, and pilot intent rather than measuring simple behavioral, subjective, or physiological states. Since traditional metrics will prove inadequate for assessing complex systems of the future, combinatorial approaches will assume greater importance, and some requirements and suggested approaches for such integrated measures are presented.

1. INTRODUCTION

In the earliest single-seat aircraft, the number of cockpit controls and displays was so low as to be completely manageable by an individual. By World War II, however, this simplicity was already beginning to disappear. The number of instruments, control devices, and switches had begun to rise dramatically. Cockpit complexity has continued to increase in virtually a logarithmic fashion until, in modern fighter aircraft, the cockpit suite presents an overwhelming array of dials, instruments, and activation devices.

Attempts to alleviate this situation in the context of the traditional cockpit have been less successful than would have been hoped. In the area of manual control, automation has had some success in reducing the complexity and sensitivity of the flying task. Unfortunately, automation of control tasks has frequently had the effect of increasing the operator's cognitive and monitoring responsibilities.

A. Coblentz (ed.), Vigilance and Performance in Automatized Systems/ Vigilance et Performance de l'Homme dans les Systèmes Automatisés, 87–97.

In these cognitive areas, automation has been much more difficult to achieve. While programs such as the "Pilot Associate" in the U. S. may help, they still place the pilot in the position of managing an impossibly large amount of processes and systems, and inevitably raise questions concerning the limits of human capabilities to take in, process, and act upon data.

Such considerations have led many individuals to suggest that the human will necessarily be forced to yield more and more actual control to the system, not only in the data processing and manual control sense, but even in the complex decision-making sense. To many, this is an unpalatable thought. The human operator wants, beyond all else, to feel in control. While some loss of manual control and simple data processing has been reluctantly tolerated, requiring the pilot to share actual responsibility with the system could cause a kind of psychological "detachment" from the task which would be unacceptable to the present type of system operators.

One possible solution to providing a true integration of the system and the operator is rooted in the well-established technology of helmet-mounted displays. Such displays, already being utilized in systems in many NATO Countries, provide heads-up display of system information to the operator at optical infinity, utilizing a series of television projections and images in the helmet visor. In its present form, this type of system presents a relatively austere set of data for the operator. However, it is not difficult to conceive of this technology providing an entire world view in virtual space - a "virtual world" - for the pilot. Such a "super cockpit" concept is currently under development at the Armstrong Aerospace Medical Research Laboratories at Wright-Patterson Air Force Base in the United States under the direction of Dr. Thomas Furness.

It is argued here that implementation of this "virtual world cockpit" will foster in the human a psychological bonding with the system itself, with implications far beyond obvious changes in control/display requirements. This bonding can drastically change the nature of traditional automation questions. Consideration of what to automate, what kinds of over-ride capabilities to provide, and how to assess performance will now require different kinds of data input from the performance assessment specialist. These differences are explored below.

2. THE SUPER COCKPIT OF THE FUTURE

The view of the world which is now given to the operator in traditional systems has been compared to looking at a scene through a series of narrow cones or straws. The instruments, in effect, provide the operator with a "two dimensional" view of the world, made up of several discrete pieces of data. From these, the pilot must construct the overall world status by integrating visual, auditory, vestibular, proprioceptive, and other kinds of data into a composite mental representation.

The creation of the virtual world capability, with its associated sensors, computer image generation, and panoramic view, radically changes this constricted view of the world. The operator can now be given a "total picture" of the environment in which the aircraft is operating. The environment surrounding the aircraft can be sensed and stored in on-board computers (updated by data-linked information from the ground) and can be made available to the pilot in fully processed form. This processing could theoretically be made to appear as "real" as the outside world itself. It could be panoramic - responding to the line of sight of the pilot - so that an entire 360 degrees is available for viewing in all dimensions. This would enable the pilot to "look" through the cockpit floor. In short, the world would be no longer a two dimensional construction carried out mentally by the pilot, but a full three-dimensional synthesis provided to the pilot by the system. From a sensory/perceptual input and integration viewpoint, this type of presentation greatly expands the amount of information that can be presented to the operator in condensed form. Even a "God's-eye" view of the situation inside the virtual space of the cockpit can be given at the same time that the outside view is being presented. This "situation dome" would portray all sensed objects in an aerial perspective, providing the operator with a situational picture in three dimensions. Decisions based on the "peephole" view of the world will be replaced by an integrated, holistic view.

With respect to other kinds of sensory information, similar advances are possible. Auditory localization as a source of data input has been considered feasible for some time. Adding auditory realism to the visual scene in the virtual world can further stimulate integration of the pilot into the scene. The problem of tactile feedback to the operator is somewhat more difficult, but still solvable. Vibrotactile stimulation at specific frequencies causes sensations of pressure. It is possible to construct such vibrotactile stimulators inside flight gloves such that a sensation of pressure is created in the fingertip without any "real" pressure. Other such "artificial" forms of pressure, movement, and even orientation sensations can be conceptualized. Together, these capabilities to create "artificial" sensations, and the ability to integrate them or dissociate them at will mean that the virtual world can become as "real" to the operator as the so-called "real" world. An environment could be created in the cockpit which mimics, complements, supplements, or even totally contradicts the outside environment. To the extent that the pilot becomes part of the virtual world, this world <u>becomes</u> the real world. It is this potential to capture to pilot's total perceptual world which provides the basis for believing that a synthesis between human and machine can be generated.

One of the biggest problems in cockpit design today is the rapidly shrinking amount of real estate available for displays and controls. To address this problem, alternative methods of system activation are already in existence, and these promise to be expanded and refined in the super cockpit of the future. For instance, magnetic sensing of head position is already a reality in certain systems. This capability, combined with eye position sensing, lets the system know precisely where the operator is looking. Aircraft equipment, such as guns, can thus be turned to a desired position by simply looking at the intended target. The same technology can be used for sensing the position of the operator's hands and fingers through magnetic strips located in each finger of the gloves. This would permit the system's computers to know, at all times, exactly where the operator was pointing.

The above capability could permit utilization of switches which exist only in virtual space. The pilot would see a set of (imaginary) switches in virtual space. By "pushing" on the desired switch, the system would be able to correlate the position of both the switch and the pilot's finger. It would therefore "know" that the switch had been pushed, and could activate the desired system. Further, through use of the vibrotactile feedback system described above, the pilot would "feel" the pressure of pushing the virtual space switch.

Naturally, such switches would not always be present in the pilot's field of view. They would have to be called up as required. Several techniques for this are feasible. Voice control, of course, is becoming much more accurate and flexible. A certain suite of switches (as well as associated displays) could be called up by simply saying certain key words (e.g., "landing", "fuel", "cruise", etc.). Alternatively, or in addition, eye position sensing can be used for this purpose. By looking at a certain position or word displayed in the field of view, and confirming it with a word or hand motion, the operator could totally change the display and control suite which is operative.

The overall effect of this type of control capability is that all realizations of the person/machine interface will be removed from the cockpit itself. In effect, the cockpit can become an enclosed box. Both the displays and controls will be located inside the helmet and apparel of the person. Again, from the pilot's psychological perspective, system control will appear to reside in his or her head rather than in the aircraft, (even though in fact "control" is now being completely shared between the person and the machine). This effect will cause a further bonding between the person and the machine.

As radical as the above changes are, the super cockpit concept goes far beyond sensory and motor concepts. The most radical change introduced by this concept will certainly be in the cognitive realm of the pilot. In view of the incredible wealth of information now available to both the pilot and the system, a capability exists which could truly revolutionize the nature of system/human interactions.

Throughout the history of such interaction, there has been a continual battle to provide more and better data to the person. One can visualize the increase in the number of control/display interfaces which has occurred as an attempt to provide more and more information about the world to the pilot. Automation, at least in the cognitive realm, can be seen as an attempt to reduce this need for more and more data by the operator. The foundation and guiding principle of this requirement, of course, has always been the decision-making task, which constituted the human's most characteristically human function.

The advent of the virtual cockpit can totally reverse the way information flows in the system. The Pilot Associate, as well as other decision-making adjuncts, introduce decision-making capabilities to the system which far exceed those of previous eras. For instance, the system may decide what information should be presented to the operator, and thus gain considerable power over any decision the human might eventually make. Similarly, the system may be able to determine that a certain action taken by the operator is not consonant with good system operation. The fact is that certain decisions may be better trusted to the machine than to the human.

In effect, then, the total and sole need to provide information from the machine to the operator has now been compromised. The goal of system design is no longer only to provide more and better quality data to the operator on the machine's status. It is also <u>to provide high quality data to the machine on the operator's status</u>. This represents a substantial turnaround in the direction of information flow from traditional system requirements. It means, in effect, that there will be an increased need in the future to monitor operator state, and to provide that data to the system in an on-line manner.

This establishes the assessment requirements for future systems. The need to monitor the operator continuously becomes a dominant assessment requirement. Assessment of basic sensory and motor capabilities is not seen as terribly difficult, or as involving methods which are radically different from traditional assessment methods. What will certainly be different is the requirement to monitor the cognitive state of the person. The nature of some of these differences, and their implications for performance assessment are discussed below.

3. AUTOMATION IN THE SUPER COCKPIT

The basic automation question to be faced in the super cockpit of the future will be a cognitive one - How can decision making tasks be optimized between the human and the machine? It is important to remember that these "decisions" are not only low-level decisions concerning things like optimal fuel consumption or even optimal route selection, but may involve very high-level decisions such as target selection, tactics of engagement, and even decisions on mission continuation or termination. However, one must question whether it is possible to provide the system with as much reliable information about the human as it is to provide the human with reliable information about the system. Clearly, the answer is no. For historical, scientific, and psychological reasons, the state of assessment methodology is not as advanced for the human as it is for the system. Thus, it is possible to see the area of human assessment as a major limiting factor in utilizing the full potential of the super cockpit capabilities presented above. Yet, the situation may not be as hopeless as it might first appear.

In the past, performance assessment methodology has concentrated on measuring single dimensions of human capability, and it has done this with considerable success (O'Donnell and Eggemeier, 1986). The problem, however, is that these unitary measures are not appropriate for the complex cognitive and performance requirements of an integrated system such as a super cockpit. <u>For appropriate performance assessment to occur in the context of automation decisions in the future, it will be necessary to define the complex constructs which are critical to system/human decision allocation.</u> In the following sections, a few of the known constructs are discussed in order to provide criteria and suggestive methodologies for performance assessment.

3.1 Workload

In considering automation in the super cockpit, workload forms a pivotal concern. Workload has been defined as a reserve capacity consisting of those resources which are still available for emergency situations once the primary task demands have been accommodated (O'Donnell & Eggemeier, 1986; O'Donnell, 1983). In this sense, workload is seen as resulting from a number of reasonably independent determinants. Further, the search for a single measure of workload - a "holy grail" - is doomed to failure precisely because of this multi-dimensional nature of the construct (Shingledecker, 1983). It is clear that if a workload metric is to evolve which will be useable in the practical, operational environment, several individual measures will have to be integrated into a composite assessment.

Thus, the measurement challenge for workload assessment is to develop integrated measures utilizing the relatively well developed and documented individual measures. In this area, there is a good body of theoretical work on which to base such integration (e.g., the multiple resource theory; Wickens, 1980). Future efforts, therefore, should be directed toward exploring how these resources combine under the demands of complex task performance. They should concentrate on creating an appropriate combinatorial formula for describing such combinations. Such a formula must take into account the fact that different kinds of assessment approaches (subjective, behavioral, and physiological) may measure independent, overlapping, identical, or composite resource structures. Combinatorial formulas may therefore become quite complex, containing many conditional statements. They should result, however, in an unambiguous statement of the "workload" (e.g., in terms of reserve capacity) of <u>this</u> individual performing <u>this</u> task at <u>this</u> time, and should do so in a form which can be used as an input for overall systems effectiveness decisions.

3.2 Situation Awareness

In the modern fighter cockpit, it is clear that the operator must maintain a constant awareness of the surrounding environment. Such situation awareness has been described as the single most important determinant of superior versus adequate performance. Somehow, if the super cockpit is to capitalize on the types of automation which are available, it must arrive at some reasonable assessment of the state of a pilot's situation awareness. This assessment then must serve as an input data source to aid the system's decisions.

Situation awareness is, like workload, essentially a construct which incorporates several interacting functions resulting in a final behavioral condition. In this case, the factors may include momentary state variables (such as workload, physiological condition, etc.), plus a longer-term process which includes learned and unlearned components. In any case, it has been suggested (Fracker - personal communication) that the net result of these process is that the operator "expects" certain things to happen based on an assessment of the current and future situation. This expectation is probabilistic and conditional. It is based on schema which the person has developed as the result of all the factors noted above.

This view suggests a number of possible ways in which the individual's situation awareness might be assessed. Certain physiological measures (such as the P300 of the evoked response) are appropriate for detecting a mismatch between an "expected" and a "surprise" event (Duncan-Johnson and Donchin, 1977). Behaviorally, one might ask the subject to use present awareness of the situation in order to predict where a target will appear next, or one might blank a display and require the subject to report on the contents of the display at random times. Subjective techniques, such as partial report of the status of the environment, might also be obtained. Unfortunately, none of these techniques are clearly appropriate for use in the operational environment. However, they could provide the foundation for measures which are appropriate.

3.3 Pilot Intent

With the advent of the system's ability to override the decisions of the pilot, it has become increasingly important to know in advance what is being contemplated by the pilot. This represents perhaps one of the most difficult constructs to conceptualize. Intent consists of a large number of contingent choices. In a very real sense, intent can not correlate perfectly with ultimate action. At any given moment, intent may be expressed as a probability based on many individual sub-probabilities.

Techniques for determining pilot intent in any operationally useable way have not been well explored. Such techniques will obviously be very complex. One possibility is to determine a set of "nominal" behaviors which would constitute the normal behavior (as well as a range of variability in that behavior) of the pilot when faced with certain preceding events. Within these normal behaviors (as well as the preceding events), certain "markers" could be identified (e.g., control movements, physiological states, etc.) which correlate with the future "decision" or behavior of the individual. These markers would then constitute the predictor set for the future decision or intent of the operator (See Shingledecker - this volume). Again, it is clear that such a final metric will involve integration of several separate types of measure. In this case, the integration will be directed to a predictive criterion, and will result in a set of probabilities dealing with the final intent of the pilot.

4. THE NEED FOR INTEGRATED MEASUREMENT

The brief and selective description of constructs presented above argues that there is a severe need to develop a methodology for measurement integration. Obviously, no one would suggest that existing individual metrics are perfect, or that there is not room for considerable progress in generating new and creative testing approaches. However, by and large, the researcher or designer has a fairly large number of test options from which to choose. What is still missing, however, is a well developed methodology to integrate measures into the kind of construct assessment described above. Individual metrics, even if they were perfect for assessing an aspect of these constructs, would fall far short of providing the necessary data for carrying out such things as automation in the super cockpit. The need for multi-dimensional assessment has been accepted. The requirement for combinatorial techniques to integrate various measures is just beginning to be addressed.

Although no clear-cut dominant approach has yet emerged, certain general criteria for such integrated measures can be described. First of all, an integrative technique must be able to be applied in near-real-time if it is to be useable for automation decisions. Similarly, such techniques must not be so unique in their solution to specific problems that generation of the data base for decisions would be prohibitive. Thus, they must be reasonably general in form. A third criterion for these approaches is that they must be capable of accepting a range of data, from nominal data to ratio data, and also a range of types of data, from operator state (physiological) data, to subjective data, to system-status data. Further, such a technique must be able to accommodate specified levels of certainty for each kind of data, as well as other modulators of such data, such as criticality, etc. Conditional data, as well as specification of the probability of conditional conclusions, must be able to be accommodated.

Potential techniques for such integration involve algorithmic approaches, as well as so-called non-algorithmic approaches. In the first category are the pragmatic approaches such as those employed by Gevins (see this volume) and those which utilize n-dimensional space to relate many measures to a cluster of outcomes. The non-algorithmic approaches are principally focused on application of one or more types of neural net programs (Lippmann, 1987). To date, however, little or no attention has been paid to the need to utilize such combinatorial approaches for integration of whole system data based on well-defined constructs. There is a clear need for work in this direction.

With respect to the question of automation in the super cockpit, a possible approach would involve, first, exhaustively defining the kinds of constructs of interest to this type of futuristic development. Then, the measurement techniques which tap elements of the construct would be systematically identified. Finally, several approaches could be empirically tried to optimize classification of these several individual measurements against pre-defined criterion conditions. Ultimately, a hierarchical approach could be envisioned. At a low level would be the individual measurements of things like heart rate, brain waves, reaction times, error scores, etc.. These particular scores combine one or more times, using one or more combinatorial approaches, to yield an overall assessment of constructs such as workload, situation awareness, or pilot intent. This assessment then becomes a data point for the next level of assessment. At this level, these intermediate constructs are combined with similar assessments which have been carried out on system hardware status, environmental status, and other factors of importance to overall performance. Such a higher-level combination almost certainly will require a collective computational approach which will permit vast quantities of data to be processed rapidly.

The above approach and series of steps represent only one of many possible approaches. Whatever the final realization of the steps, it appears clear that the future of assessment technology rests with the development of techniques to <u>integrate</u> specific measures, rather than with breakthroughs in the development of new testing techniques. With respect to automation in future systems, development of such integration techniques appears critical, and will certainly provide the next level of evolution in this field.

5. REFERENCES

Duncan-Johnson, C. C. and Donchin, E., On quantifying surprise. The variation in event-related potentials with subjective probability, Psychophys, 14 (1977) 456-467.

Lippmann, R.P. An introduction to computing with neural nets. IEEE ASSP Magazine, April, 1987, 4-22.

O'Donnell, R. D., The USAF neuropsychological workload test battery: Concept and validation, Proceedings 338 of the NATO Advisory Group for Aerospace Research and Development, Paris (1983) 5/1-5/9.

O'Donnell, R. D. and Eggemeier, F. T., Workload assessment methodology, in: Boff, K. R., Kaufman, L. and Thomas, J. P. (eds.) Handbook of Perception and Human Performance Vol II (Wiley, New York 1986).

Shingledecker, C. A., Behavioral and subjective workload metrics for operational environments, Proceedings of the AGARD (AMP) symposium on sustained intensive air operations: Physiological and performance aspects, AGARD-CP-338 (1983) 6/1-6/10.

Wickens, C. D., The structure of attentional resources, in: Nickerson, R. (ed.) Attention and Performance VIII (Erlbaum, Hillsdale, N. J. 1980).

II

CHARGE DE TRAVAIL ET AUTOMATISATION

WORKLOAD AND AUTOMATION

EFFECTS OF BRIDGE AUTOMATION ON MARINERS' PERFORMANCE

Nadine Sablowski
c/o Shiphandling and Simulation Facility SUSAN
Elbchaussee 23
2000 Hamburg 50
F.R.G.

ABSTRACT. A maritime simulation study is presented which aims at defining limitations of the intended one-man-manning of ship's bridges at night-time. Seventy mates/masters run through the same scenario under different experimental conditions concerning bridge equipment, bridge manning, and time of day. Throughout the experimental run workload is assessed by means of primary and secondary task, physiological data (heart rate), and subjective ratings. Whereas workload variations within the scenario are reflected by several of the applied measurement techniques no significant effects of the different work conditions can be shown. Consequently, safety at sea is supposed to depend on improved training and qualification of the nautical officer rather than manning regulations or bridge equipment. The results also emphasize that there is a need for developing selection criteria for seagoing personnel concerning personality traits.

1. INTRODUCTION

European merchant marine today has to cope with severe financial problems. Therefore shipping agencies as well as legal authorities think about ways to save costs by e.g. increasing automation of vessels and reducing crew manning.

Within this context the study presented in this paper aims at defining limitations of the intended one-man-manning of ship's bridges at night-time by assessing workload experienced by nautical officers in different situations and work conditions. The research is carried out on a shiphandling simulator because of its numerous advantages compared with field studies: any kind of situation and environment can be simulated with the possibility of exactly reproducing experimental conditions and of recording all important data. Besides being cost- and time-saving it is a safe way to investigate even critical situations (1). Unfortunately very little is known yet about the validity of results obtained in simulation studies (2).

Workload assessment in a maritime context is difficult because of the

A. Coblentz (ed.), Vigilance and Performance in Automatized Systems/ Vigilance et Performance de l'Homme dans les Systèmes Automatisés, 101–110.

lack of qualification and procedural standards and the lack of marine-specific measurement techniques. In order to get comprehensive information we decided to collect behavioural, physiological as well as subjective data as recommended in the literature (3). Most of these measures have been developed in the context of aviation and must be adjusted to shiphandling. Furthermore characteristics of maritime work conditions (4) have to be considered in defining limitations of one-man-manning.

2. MATERIALS AND METHODS

2.1. Subjects

A total of 70 male experienced mates and masters took part in the investigation. They were sent to us by five major shipping agencies during their vacation.

2.2 Instrumentation

The investigation has been carried out at the shiphandling and simulation facility SUSAN in Hamburg. SUSAN is a full-mission simulator with all conning and navigation equipment available on the ship's bridge. Outside view is provided by computer-generated imagery allowing presentation of different waterways, vessels, weather, and time-of-day conditions. Rolling and pitching of the vessel is achieved by a hydraulic motion system.

2.3. Scenario

All subjects had to conn a container vessel from the river Elbe along the German Bight area for two hours. Fig. 1 provides an overview of the vessel's course and all important events that happen during that time. The scenario can be characterized by gradually increasing workload. For practical reasons it is subdivided into six reaches each of them lasting for about 20 minutes. They can be described as follows:

Reach A and B	time for familiarization
Reach C and D	routine jobs
Reach E	risk of collision with other vessels (frigate, VLCC, sailing-boat)
Reach F	rudder failure

We first wanted to make use of the concept of "testing the limits" (5) but had to give up because it was impossible to reach the limits of the mates without creating unrealistically difficult situations.

2.4. Experimental Conditions

Subjects were randomly assigned to seven different experimental groups

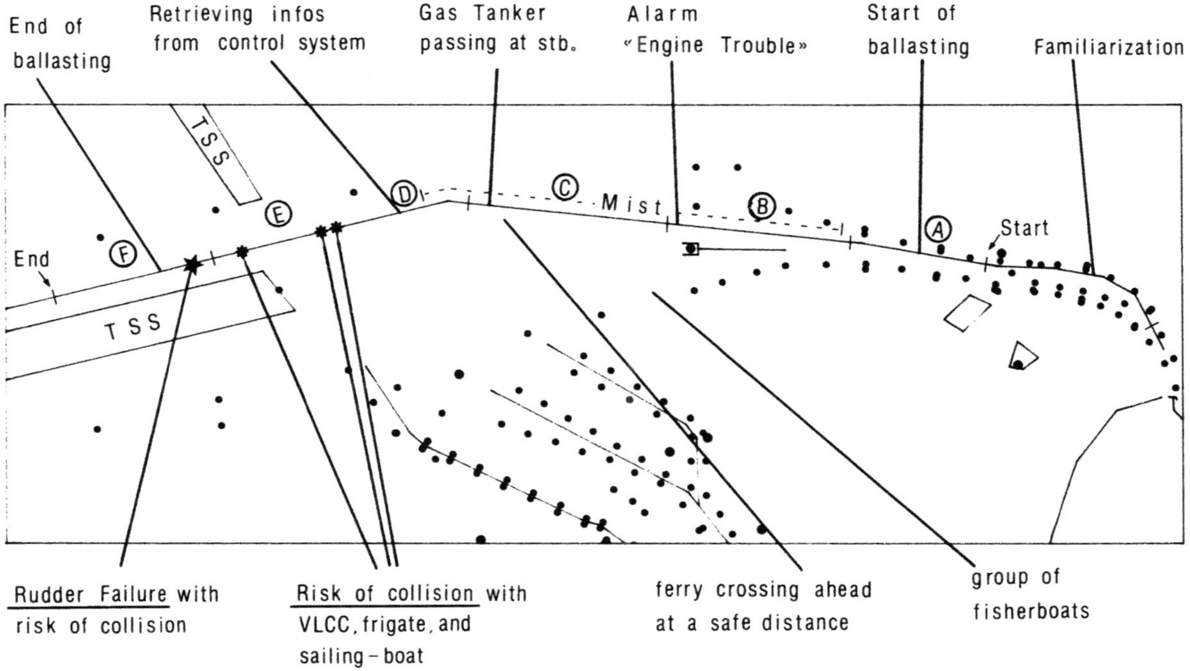

Fig. 1 Experimental Scenario

that differ with respect to bridge equipment, bridge manning, and time of day:

TABLE I Experimental Configurations

	Conventional Bridge		Ship-Surveillance-Center	
	Day	Night	Day	Night
Officer alone	10 (1)	10 (2)	10 (5)	10 (4) 10*(7)
Officer & Lookout	-	10 (3)	-	10 (6)

* these 10 subjects were additonally trained on the ship-surveillance-center
( ) number of configuration

In comparison with a conventional bridge the main new features of the ship-surveillance center are:

- electronic sea-chart
- rasterscan ARPA-radar
- ARCAP (adaptive radar controlled auto pilot)
- control system for internal technical processes (e.g. propulsion, energy, ballast)

The distinction between "day" and "night" refers to the outside view of the simulator. Unfortunately, it was impossible to do real night-time research because of organizational reasons.

The 7 configurations do not significantly differ concerning age and nautical experience:

TABLE II Age and nautical experience
(Means and standard deviations)

Conf.	Age (Years) x	Age (Years) s	Nautical Experience (Years) x	Nautical Experience (Years) s
1	39.7	8.4	15.5	9.5
2	42.7	8.9	15.7	9.4
3	42.0	9.5	17.1	8.0
4	38.9	10.1	15.6	10.0
5	38.3	8.9	13.0	10.2
6	36.1	9.0	10.9	8.8
7	44.5	7.3	21.1	5.8
Total	40.3	8.9	15.6	9.0

2.5. Workload Assessment Techniques

In order to get complementary information we applied different kinds of workload assessment techniques including primary and secondary task, physiological data, and subjective ratings (2,6,7).

2.5.1. Primary Task. As there are no satisfying qualification or procedural standards in navigation it is difficult to decide whenever quality and/or quantity of performance change as a result of workload variations. On the basis of preliminary investigations we selected those few measures that seemed to be meaningful:

- adequacy of communication procedures
- CPA's (closest point of approach) to other vessels
- reactions to instrument failures
- collision avoidance manoeuvres

2.5.2. Secondary Task. To avoid any interference with the mates' primary tasks which require continuous visual observation of the vicinity and of bridge instruments, the secondary task should not involve any visual aspect. We applied mental arithmetics with problems being presented via microphone every 3 minutes and answers given orally. Whenever the mate felt that mental load was too high he could refuse to do the requested calculations. The measures taken were: time needed to answer, number of right answers and of missing answers.

2.5.3. Physiological Data. Inter-beat-intervals (i.e. time between subsequent R-waves in the cardiotachogram) were recorded and heart rate together with two scores of heart rate variability (ST,NM) (8) were calculated. Long-term as well as short-term variations of these parameters were investigated.

2.5.4. Subjective Ratings. Within a post-scenario interview two different techniques of rating workload were used. First, pairwise comparison according to Saaty (9) was applied to find out how much certain events of the run contributed to the overall workload experienced by the officers. Secondly, subjects were asked to draw a workload histogram within a given frame to get an idea of the course of workload.

3. RESULTS

Several of the applied measures turn out to be sensitive indicators of workload reflecting the demands imposed by the scenario. However, few significant differences between the seven experimental configurations can be found. The most valuable results are presented with regard to the six reaches of the experimental run described in chapter 2.3..

3.1 Secondary Task

The number of missing answers increases towards the end of the run with a peak occurring in the rudder failure situation:

Table III Number of missing answers

	Configuration						
Reach	1	2	3	4	5	6	7
A	0	9	1	4	0	1	4
B	7	8	2	1	1	2	5
C	6	9	2	7	1	4	3
D	7	10	4	10	1	5	7
E	7	14	8	9	3	4	4
F	9	12	12	20	11	11	11

The number of missing answers is significantly higher for reach F (rudder failure) compared to the preceding ones in configuration 4 and 5 (Wilcoxon; $p < .05$). The seven configurations only differ within reach A with configuration 2 performing badly (Mann-Whitney; $p < .05$).

3.2 Heart Rate

Mean heart rate values for the six reaches of the run (each lasting for about 20 minutes) are presented in table IV:

Table IV Heart rate (bpm; means and standard deviations)

	Configuration						
Reach	1	2	3	4	5	6	7
A	**94.8**	**105.9**	**99.9**	**86.4**	**85.7**	**88.2**	**94.4**
	(13.3)	(9.0)	(15.1)	(13.6)	(12.6)	(11.2)	(13.7)
B	**95.7**	**108.2**	**99.3**	**88.8**	**86.4**	**90.3**	**94.6**
	(12.9)	(12.8)	(14.6)	(13.4)	(12.8)	(11.7)	(14.3)
C	**97.4**	**108.9**	**98.4**	**90.3**	**86.9**	**89.8**	**96.2**
	(13.5)	(12.1)	(14.5)	(15.8)	(12.6)	(11.6)	(16.3)
D	**97.7**	**108.4**	**98.4**	**90.6**	**86.5**	**92.2**	**96.8**
	(13.6)	(11.3)	(15.2)	(15.8)	(12.6)	(12.3)	(15.5)
E	**97.2**	**108.9**	**99.0**	**91.5**	**88.6**	**93.7**	**98.1**
	(15.2)	(11.3)	(16.4)	(16.4)	(15.1)	(12.6)	(14.9)
F	**96.6**	**109.5**	**99.4**	**92.0**	**88.5**	**92.5**	**96.7**
	(16.8)	(11.4)	(15.9)	(17.8)	(15.6)	(10.4)	(15.0)

Within-group comparisons yield an increase of mean heart rate from reach A to reach D/E for configurations 4 to 7. Concerning between-group comparisons overall heart rate is highest in configuration 2. The lowest heart rate throughout the experimental run can be observed for the configurations 4 to 6 without any significant differences

between these 3 groups (two-way-analysis of variance; $p < .05$).
All configurations show short-term variations of heart rate (ref.: 1 minute) as a result of the rudder failure (two-way-analysis of variance; $p < .05$).
However, instead of looking at group means, it seems to be more interesting to compare individual subjects' heart rate. Fig. 2 and 3 illustrate that even mates belonging to the same configuration do not show corresponding results.

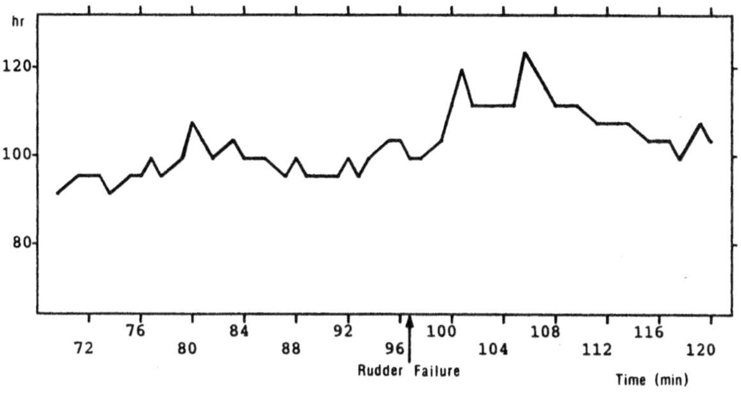

Fig. 2 Heart Rate (bpm) of mate no. 3 - configuration 4

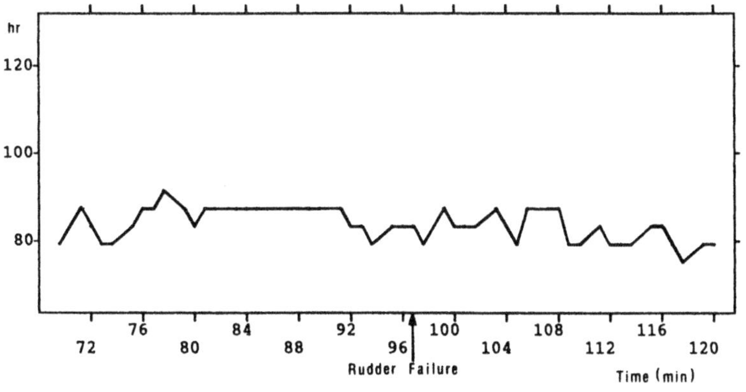

Fig. 3 Heart Rate (bpm) of mate no.8 - configuration 4

Concerning heart rate variability (ST) there are no effects of the different work conditions but in all subjects variability within reach A to C is higher than in reach E to F (two-way-analysis of variance; $p < .05$).

3.3 Subjective ratings

Pairwise comparison according to Saaty also reflects the course of workload built up by the scenario. Workload weightings are given in table V:

Table V Mean workload weightings

	Configuration						
Reach	1	2	3	4	5	6	7
A	.10	.12	.06	.07	.07	.08	.08
B	.09	.10	.09	.07	.07	.08	.07
C	.07	.06	.12	.11	.09	.08	.09
D	.12	.14	.16	.10	.14	.14	.18
E	.26	.19	.11	.23	.26	.25	.21
F	.32	.43	.37	.41	.42	.40	.41

Workload ratings of reaches E and F are significantly higher than for any preceding one (Wilcoxon; $p < .01$). Between-group comparisons do not yield any differences.

4. DISCUSSION

Whereas the primary task does not yield any valuable results, some of the other applied measures can be regarded as being sensitive indicators of workload reflecting the course of the scenario. This is especially true for the number of missing answers in the secondary task, mean heart rate and subjective ratings, all of them showing an increase of load towards the end of the experimental run. Unfortunately, the most difficult situations occur within the last fourty minutes of the scenario because quite a long period of familiarization at the beginning was necessary as ship's bridges are not standardized like e.g. cockpits. Still, the obtained results are not assumed to indicate a confound of workload and amount of time on watch. Mates who are used to keeping watch for at least four hours should not be exhausted after only two hours.

While the appropriateness of the above mentioned measures has been proven they do not differentiate between the experimental configurations in critical situations. Concerning the number of missing answers in the secondary task significant differences are only observed in reach A. Configuration 2 performs badly which might be explained by the fact that in the beginning these mates have to

concentrate on getting familiar with the bridge on their own and at night-time. The observed differences in mean heart rate have to be interpreted very carefully because moving and speaking of the mates could not be avoided nor controlled. This is also true for the expectations of the participants which might influence heart rate. However, within-group variations, especially concerning primary task and heart rate, are quite high. Within this context, subjective impressions of those involved in carrying out the experiments are interesting. They point to the fact that qualification and personality of the mates/masters allow a prognosis of their performance rather than experimental conditions. As this was not expected nor the interest of the study, data on the personality of the participants were not recorded. As mentioned before, there are not even criteria for assessing the qualification of mates. Therefore we can only try to describe these impressions carefully. Best performance was predominantly observed with participants who were quite young, rather introverted and intelligent in the sense of being capable of differentiating between important and unimportant information and tasks. Furthermore the affiliation to the different shipping agencies seemed to have prognostic value. An additional factor was impartiality towards new work conditions.
In summary, the introduction of new equipment and of one-man-manning does not seem to determine safety at sea. Attention should rather be directed towards developing selection criteria for seagoing personnel and towards training of mates concerning operation of modern bridge equipment.

5. ACKNOWLEDGEMENT

I would like to thank J. Froese for discussing the results and conclusions of the paper and J. v. Morgenstern and K. Gevers for providing me with the excellent plot of the scenario.

6. REFERENCES

(1) Froese, J. (1982). Aufgaben und Möglichkeiten für SUSAN als maritime Forschungssimulationsanlage. Schiff & Hafen/Kommando-brücke, 34(9), 265-268.

(2) O'Hara, J.M. and D'Amico, A. (1984). Human Performance and Workload Measurement as Part of a Man-Machine-System in a Maritime Context. CAORF-Report, New York.

(3) Leplat, J. (1978). Factors Determining Workload. Ergonomics, 21(3), 143 ff..

(4) Sablowski, N. (1986). Methodological Problems of Maritime Workload Studies. Paper presented at the CEC-workshop "Workload in Transport Operations" in Bedford.

(5) Sanders, A.F. (1979). Some Remarks on Mental Load. In: Moray, N. (Ed.). Mental Workload - Its Theory and Measurement. Plenum Press, London, 1979.

(6) Meshkati, N. (1985). Mental Workload Assessment: The State of the Art. In: Brown, I.D. et al. (Eds.). Ergonomics International '85, Taylor & Francis Ltd., London, 1985.

(7) Wierwille, W.W. and Williges, R.C. (1978). Survey and Analysis of Operator Workload Assessment Techniques. Systemetrics Inc. Report No. 5-78-101, Blacksburg-Virginia.

(8) Opmeer, C.H.J.M. (1973). The Information Content of Successive RR-Interval Times in the ECG. Preliminary Results Using Factor Analysis and Frequency Analysis. Ergonomics 16, 105-112.

(9) Saaty, Th. L. (1980). The analytic hierarchy process. McGraw-Hill, New York.

FLIGHT DECK AUTOMATION AND PILOT WORKLOAD

ALAN H ROSCOE
Britannia Airways
Luton
Bedfordshire
England LU2 9ND

Abstract

The introduction of 'new technology' automation into the flight decks of modern jet transport aircraft has had a significant impact on pilot workload. This paper presents a number of examples of workload assessment involving automatic flight performed during research and development flight trials and during routine airline operations in Europe. Subjective evaluation by pilots, augmented by recording their heart rates, was used to compare the workload experienced during various degrees of automatic flight with that experienced during non-automatic flight. The examples selected are used to illustrate some of the implications of increasing automation; the possibility of too much automation leading to a decrement in piloting performance is of particular concern.

Introduction

In 1912 the first automatic pilot was flown successfully on a Curtis seaplane at Hammondsport in New York State. This was built by Lawrence Sperry who two years later demonstrated his two-axis autopilot during a historic flight to Paris.

Forty years ago the first fully automatic landing was made in England by an aircraft of the Blind Landing Experimental Unit (BLEU). Twenty years later - in 1968 - a British European Airways Trident carried the 100,000th passenger to be landed automatically when it touched down at London Heathrow on a scheduled flight from Paris. Today most of the latest generation of civil jet transport aircraft are suitably equipped and automatic landings are commonplace.

A quick comparison of the flight decks of these later aircraft - such as the Airbus A310 and the Boeing 767 - with those of a previous generation - like the B737 - will reveal the significant advance in technology that has taken place during the last ten years or so. It is now theoretically possible for aircraft fitted with advanced autopilots and sophisticated flight management systems to fly several thousand

A. Coblentz (ed.), Vigilance and Performance in Automatized Systems/ Vigilance et Performance de l'Homme dans les Systèmes Automatisés, 111–122.

miles from one airport to another without the help of the human pilot. However, in spite of the technicological feasibility, public opinion is not yet ready to accept the pilotless airliner. The role of the pilot, though, has changed markedly over the last two decades, he is now much more of an 'out of the loop' systems manager and monitor than an 'in the loop' active controller.

It is generally accepted that automation on the flight deck has two main advantages - improved performance and reduced pilot workload. There is also an economic 'spin off' (for example, better fuel economy) and so increasing automation tends to be encouraged by the airline industry. But there could be some disadvantages, such as less job satisfaction for pilots, lowered vigilance, under-arousal, and even complacency. In addition, workload may be different with automation but not necessarily lower; for instance, most new aircraft are crewed by two pilots rather than by a crew of three as on many older types.

This paper examines the influence of automation on pilots' workload and on their level of arousal, by referring to examples of practical assessment carried out during the last nineteen years. First, though, it is necessary to define workload. There is no generally accepted definition of the term workload; some ten years ago Ellis and Roscoe (1), using a questionnaire, determined that more than 80% of professional pilots think of workload in terms of effort. This finding led to Ellis and Roscoe proposing a definition of workload based on that given by Cooper and Harper in the introduction to their Handling Qualities Rating Scale (2). "Pilot workload is the integrated mental and physical effort required to satisfy the perceived demands of a specified flight task". This is an interpretation that agrees well with the individual nature of the piloting task, and it is compatible with the use of subjective techniques and with recording heart rate responses.

Method

The most used technique for assessing workload in flight is some form of subjective reporting and currently it is common practice to use a specially designed rating scale (3) (4). However, as subjective opinions are vulnerable to bias and to pre-conceived ideas it is useful to be able to complement subjective data with another measure, such as recording the pilot's heart rate (5) (6). In the examples presented below workload was assessed in this way.

Initially subjective assessment was by in-flight and post-flight comments but from 1977 a rating scale developed at the Royal Aircraft Establishment at Bedford has been used - see figure 1 (7). The 'Bedford' scale, which employs the concept of spare capacity, is based on the 'decision tree' design of the Cooper-Harper Handling Qualities Rating Scale, already familiar to RAE test pilots.

Heart rate has been recorded routinely to augment the subjective assessment of workload. The rationale for using heart rate is based largely on the assumption that it reflects the level of preparedness in the neuromuscular system - neurological arousal - determined by the demands, or more accurately, the perceived demands of the flight task. There is evidence that the failure by pilots to perceive the demands of the flight task correctly has been a causative factor in several accidents; reference in the definition to this aspect of workload reflects its importance.

Heart rate is recorded by means of two electrodes attached to the chest wall. These detect the electrical changes accompanying each heart contraction - the electrocardiograph (ECG). After suitable amplification the ECG signal is recorded onto magnetic tape. It is important to note that the heart rate response is highly individual and thus it is not possible to compare heart rates directly between pilots.

Examples

Workload assessment, using the above technique, was first carried out in 1969 at RAE Bedford during automatic landing (autoland) trials using a Comet four-jet airliner of BLEU. Heart rate levels supported pilot opinions that the workload during autolands was substantially less than during manual approaches and landings.

An unexpected finding was that heart rate responses for autolands in fog (RVRs, of less than 300m) were generally lower than for autolands in clear weather. This lower response highlights the lower workload when there are few visual cues to process. As one pilot remarked "... there is little to do until you hear the screech of the tyres on contact with the runway".

There was some evidence from these early workload assessments at RAE that familiarity with the autoland technique in fog might lead to underarousal and possible complacency. This observation was confirmed in 1976 during the flight trials of manual landings in fog - Economic Category 3 - using a 'fail passive' autopilot system instead of the more usual 'fail operational' autoland system (8). The technique being evaluated consisted of autopilot (with autothrottle) coupled approaches to a decision height of 50ft (HS 748) or 60ft (BAC 1-11), at which time if the approach/runway lights were seen the autopilot would be disconnected and a manual landing carried out. If the lights were not identified a 'missed approach' was flown by the co-pilot.

It was the usual practice to disconnect the autopilot as late as possible for the manual flare. During the trial the guidance system placed the aircraft consistently on the centre line at the touchdown point on the runway and then it was necessary only to close the throttles and gradually increase the attitude to flare the aircraft for a reasonably good landing. In most instances this meant the pilot being in the control loop for less than 15 seconds. Even the assimilation of visual cues and the decision to land had taken only a few seconds in the worst conditions.

During the early flights, in visibilities as low as 150 metres, heart rate increased as decision height was neared and then increased further for the landing (Figure 2). For three of the seven pilots involved in the trials it was observed that after two or three experimental sorties in fog there was an increasing tendency for their heart rates to remain relatively low to a point beyond decision height and to increase only for the landing. This lack of repsonse almost certainly indicated a lowered level of arousal for the decision and possibly an element of complacency. Following discussions with these pilots, who were somewhat concerned at their reduced preparedness, they consciously increased their arousal before decision height by contemplating possible problems - such as an autopilot malfunction or an obstruction on the runway. This increase in arousal was reflected by an appropriate increase in heart rate. It was of interest to note that their workload ratings for the decision and for the landing did not vary with their different levels of arousal before decision height.

Another important aspect of automation was underlined during workload assessment with test pilots flying an experimental VTOL aircraft, the Short SC1 direct lift jet aircraft, in 1971 (5). The introduction of an automatic stability control system was designed to reduce workload during transition between jet-borne and wing-borne flight. Pilot comments indicated a significant reduction in workload but the heart rate responses of two of the five pilots did not show the expected drop with autostablisation. It subsequently became clear that because of doubts about the integrity of the autostabilisor these pilots increased their monitoring of the system, thereby minimising the predicted reduction in workload.

The last examples are from a Britannia Airways study designed to assess levels of workload experienced on the new technology Boeing 767 - with its all digital Flight Management System (FMS) - and compare them with those levels experienced on the older B737 (9). This particular study, which began in 1984, was an extension of one started in 1977 to gain experience of assessing workload on airline pilots flying routine holiday charter flights in Europe in B737 aircraft. Workload ratings by the pilots and by a flight observer seated on the flight deck, together with the pilot's heart rates have been recorded primarily during those phases of flight expected to generate increased levels of workload; namely, the take-off and initial climb, and the approach and landing. Observations and recordings at random during the cruise have allowed comparison of workload levels at relatively undemanding times.

The advantages of an improved autopilot that can be engaged shortly after take-off and of a thrust management system can be seen in the reduced workload for the B767 compared with that for the B737 during a standard instrument departure (SID) - Figure 3.

In addition to the comparison between two aircraft types, workload levels associated with flying the B767 in different modes have also been assessed. Figures 4 and 5 illustrate the different workload levels experienced during hand-flown flight director approach and landings and autolands. Figure 6, as well as showing the benefit of a good autopilot and FMS (top), also illustrates the advantage of a flight director (F/D) fully integrated with the FMS (centre) over a more conventional attitude display (bottom), during standard instrument departures in the B767.

Heart rate responses in the cruise appear to indicate varying levels of arousal according to the amount of mental activity. But neither heart rate nor heart rate variability seem to be sensitive enough to indicate underarousal or reduced vigilance. Personal observation and anecdotal evidence from pilots suggest that an element of underarousal and, possibly, complacency may occur in the cruise - this is particularly likely during long flights at night.

It has been suggested that a lower arousal may even occur during the approach and landing as a result of tiredness towards the end of long duration flights, and that performance may thus be compromised. On the assumption that a lower arousal should be more likely during an automatic approach and landing the same pilot was monitored during autolands following flights of different lengths. Figure 7 shows the heart rates recorded after flights of 2, 5 and 9 hours, and it is clear that the responses are very similar. (The departure heart rates varied because of different operating conditions).

Discussion

Britannia Airways operates six B767 aircraft - with another two due for delivery in the near future - and the workload study has generated considerable interest within the airline. It is common practice for 767 pilots to discuss their operational experiences and their personal views with the author. Many of those comments, together with personal observations on the flight deck and with incident reports, are of special interest in the context of automation and workload.

Pilots have generally welcomed the new technology and they particularly like the Autopilot/Flight Director System (AFDS) and the autothrottles which have a significant influence on workload. Pilots are also enthusiastic about the CRT displays - especially the horizontal map display. Although most pilots convert from the B737 to the B767 without too much trouble the younger pilots seem to have less difficulty adapting to the new techniques than do their older colleagues.

During the first year or so of operation numerous 'glitches' in the FMS, such as failure to capture the localiser and turning onto the wrong heading at waypoints, helped to maintain a high level of awareness. Now, after four years of operating the B767, there is some evidence of decreasing arousal during the cruise and occasionally at other times. Well designed operating procedures, appropriate training, and a general awareness by pilots of the dangers of underarousal appear to be important in minimising the problem. Several pilots have admitted that when in the monitoring role they are more easily distracted by relatively unimportant matters.

Most pilots enjoy hand flying an aeroplane and feel that it helps to maintain their skills, and so it is not surprising that some pilots have expressed concern at the possible reduction in handling skills with increasing use of the AFDS and autothrottles. At present there is no evidence during their six monthly 'base checks' of any decrement in handling performance. However, several 767 First Officers re-converting to the B737 during command training for promotion to Captain have found increased difficulty with instrument scan and speed control.

In addition to demonstrating the influence of automation on workload and arousal the examples presented above have been selected because they also demonstrate a number of different issues that are highly relevant to automatic flight and safety.

In the autoland trials, it was somewhat surprising at first that the degree of workload reduction, as suggested by pilots' comments and heart rate responses, for automated flight compared with manual flight was so great. But on further consideration this reduction probably reflects correctly the reduced demands on the pilot. Nevertheless, it is of some concern that the levels of preparedness during autopilot approaches and autolands in low visibilities might not be appropriate where the failure to respond rapidly to the unexpected could be critical. Increasing familiarity with the autoland or autopilot approach technique, as observed during the Category 3 manual landing trial, is clearly a factor in generating underarousal.

In the Britannia study workload ratings and heart rate responses differentiated between manual landings using the flight director and automatic landings in the B767; even so, heart rates for the latter usually indicated a reasonable level of arousal (Figure 4). Occasionally, though, a pilot's heart rate response for an autoland would be lower than expected (Figure 5), possibly indicating an inappropriately low level of arousal. Such findings seem to be more likely when a pilot has carried out a large number of automatic landings and when he has developed complete confidence in the autoland system.

Increased automation leading, on occasions, to underarousal in the cruise is perhaps understandable, although undesirable; but during an approach to land it is potentially far more dangerous and, therefore, a matter for concern. Surely it must be prudent for a pilot to be prepared for a rapid take-over from the autopilot during every autoland. This preparedness, analogous to a 'warm-up' period, is important when changing from a passive monitor to an active controller and is usually indicated by an increase in heart rate as the runway touchdown area is neared. This response is similar to that seen for a manual approach and landing although not normally of the same magnitude. A good example, is illustrated in figure 4.

Automatic complacency is not new, it has occurred in earlier generation aircraft with less sophisticated systems. For example, the well known accident to a Lockheed 1011 Tristar which, in 1972, gradually descended into the Florida Everglades when the autopilot disengaged whilst the aircraft was orbiting. The flight crew were pre-occupied with a malfunction and failed to monitor the aircraft's flight path. And, in 1979, a DC-10 whilst climbing to cruise altitude stalled and lost 10,000 feet before recovery was effected. The crew had failed to notice that the autopilot was in the 'rate of climb' mode instead of the 'airspeed' mode.

It has already been noted that the introduction of automation may not necessarily reduce workload. The example in which an autostabilisation system fitted to SC1 research aircraft merely transferred workload from controlling to monitoring for two pilots is typical of what can occur. In the early days of the B767 the autopilot system on one of Britannia's aircraft had a tendency to 'trip out' during an autoland approach - sometimes at low level. For some time, even when the problem had been rectified, pilots maintained a healthy level of suspicion - and presumably a higher level of arousal. On two occasions at this time higher heart rates and workload ratings than usual were recorded during the autoland approaches.

The examples cited here, and most of the workload assessment carried out to date, have been concerned primarily with the handling pilot (HP), but it is also important to assess the levels of workload experienced by the non-handling pilot (NHP). In this respect heart rate is of substantially less value and one has to rely more on subjective ratings.

There is some evidence from the Britannia study that the NHP workload during some flight tasks can actually be increased by the level of automation that reduces the workload for his 'handling' colleague. This inbalance is particularly noticeable during emergencies such as an engine failure or hydraulic system failure when automation reduces the HP workload significantly while increasing that of the NHP.

Recently there has been a considerable increase, in Europe and in North America, in the number of anonymous reports from pilots concerning tiredness and fatigue experienced in flight - especially towards the end of long sectors. A large proportion of these reports clearly reflect boredom and monotony in the flight deck rather than genuine fatigue, and automatic flight - even with the previous generation autopilot systems - must be a contributory factor.

With the introduction of extended range two-pilot aircraft - such as the B767-ER - the problem of automation-induced underload and underarousal may become more significant. This situation in the flight deck could well be accentuated by the naturally occurring lower arousal associated with diurnal or circadian changes - the lowest levels occurring between 0300 and 0600 hours local time. Long flights ending around these times may be particularly vulnerable. In addition to well designed flight deck procedures and crew awareness, sensible rostering will become more important in the case of two-pilot extended range operations (EROPS).

Conclusions

There is no doubt that increasing flight deck automation is associated with an increased probability of pilots experiencing underload and underarousal leading to reduced vigilance and even complacency. Beyond a certain level - not yet defined - automatic flight control may well result in a significant and, perhaps, dangerous deterioration in a pilot's performance. The pilot must not be allowed to escape too far from the control loop.

REFERENCES

1. Ellis G A, Roscoe AH — 'The airline pilots view of flight deck workload. A preliminary study using a questionnaire'. Royal Aircraft Establishment Technical Memorandum No FS(B) 465 1982.

2. Cooper G E, Harper R P — 'The use of pilot rating in the evaluation of aircraft handling qualities'. NASA Technical Note 5153 Washington DC 1969

3. Reid G B et al — 'Development of multi-dimensional subjective measures of workload'. In: Proceedings of the International Conference on Cybernetics and Society 1981.

4. Skipper J H, Rieger C A, Wierwillie W W — 'Evaluation of decision-tree rating scales for mental workload estimation'. *Ergonomics* **29** 585-599 1986.

5. Roscoe A H — 'Use of pilot heart rate measurement in flight evaluation'. *Aviat Space Environ Med* **47** 86-90 1976.

6. Roscoe A H — 'The use of subjective ratings augmented by heart rate to assess pilot workload in flight'. Proceedings of CEC Workshop: Workload in Transport Operations Cranfield 1986.

7. Roscoe A H, Ellis G A — 'A subjective rating scale for assessing workload in flight. A decade of practical use'. Royal Aircraft Establishment Technical Memorandum (in preparation).

8. Brown A D, Gurney D J — 'Manual landings in Category 3 conditions'. *The Aeronautical Journal*. Royal Aeronautical Society London August 1979.

9. Roscoe A H, Grieve B S — 'The impact of new technology on pilot workload'. SAE Technical Paper 861773 Warrendale Pa 1986.

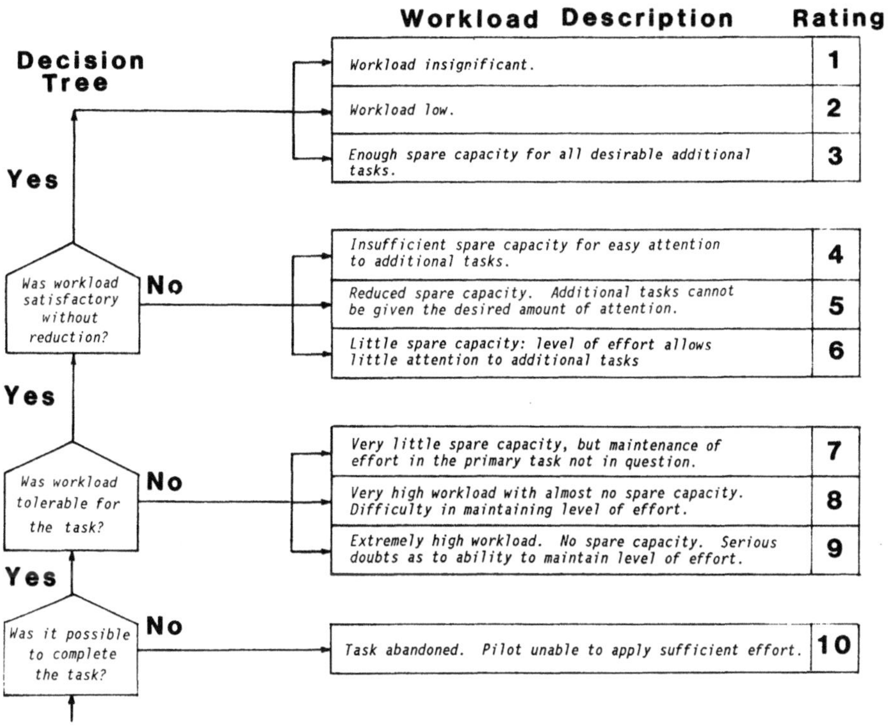

Figure 1. The 'Bedford' Pilot Workload Rating Scale
The pilot starts his decision-making process at the bottom left corner of the decision tree. The workload being assessed is that involved in the execution of the primary task, the effort expended on additional tasks must be included as part of his 'spare capacity'.

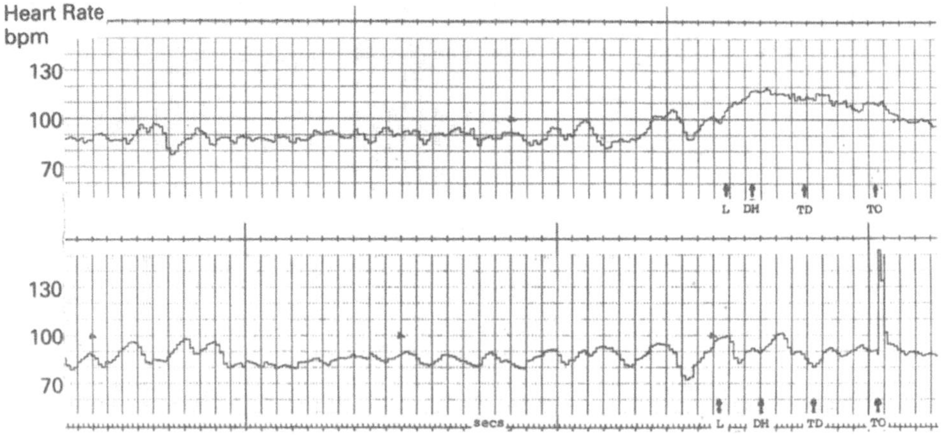

Figure 2. Beat-to-beat heart rates for one pilot recorded during the same experimental sortie in fog (RVR 190m). BAC 1-11. Upper trace - autopilot approach and manual landing. Lower trace - autoland.
L - runway lights seen, DH - decision height, TD - touchdown,
TO - 'rolling' take-off.

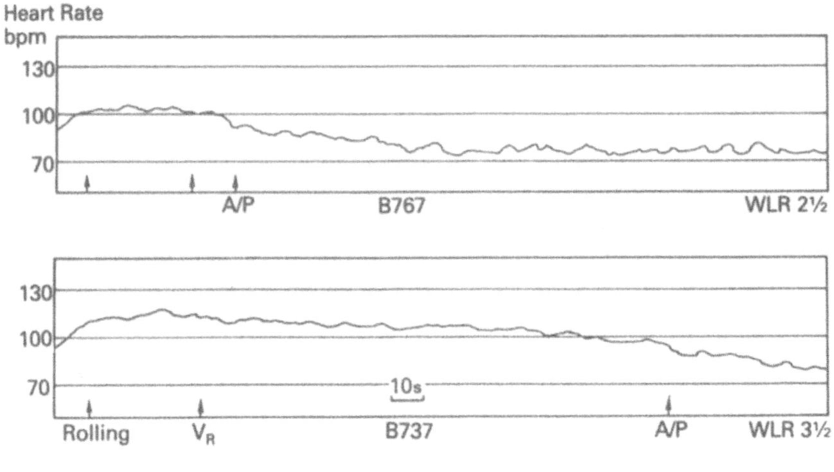

Figure 3. Comparison of heart rate responses (beat-to-beat) and workload ratings (WLR) for standard instrument departures (SIDs) B737 and B767. On the 767 the autopilot is engaged at about 500 ft whereas on the 737 the autopilot is engaged after the flaps are retracted and the aircraft is in trim.

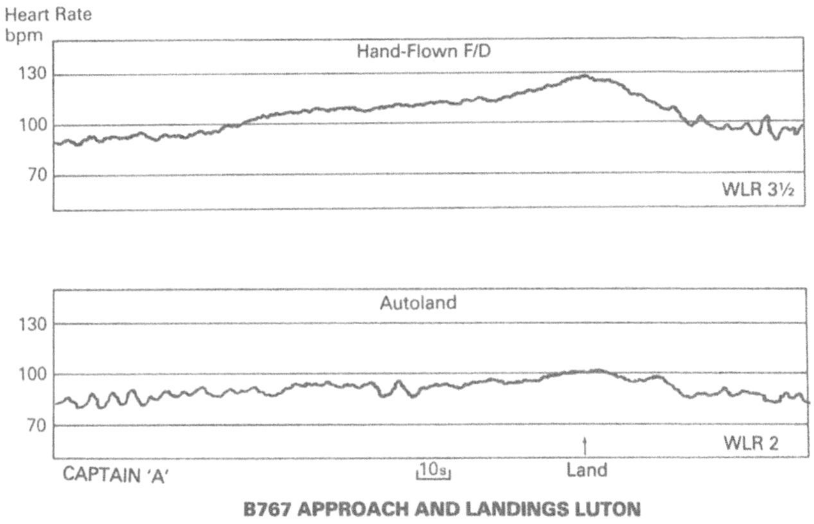

Figure 4

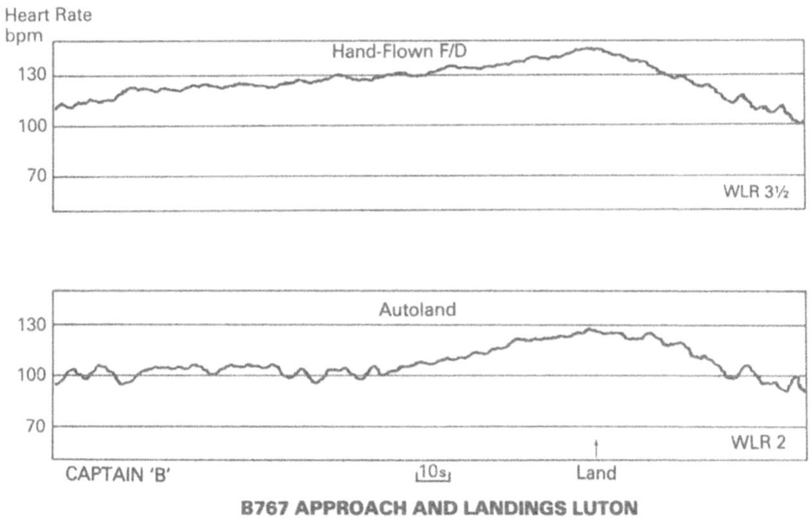

Figure 5

Figures 4 and 5. Heart rate responses and workload ratings (for two pilots) experienced during hand-flown flight director approaches and landings compared with those experienced during autolands. B767.

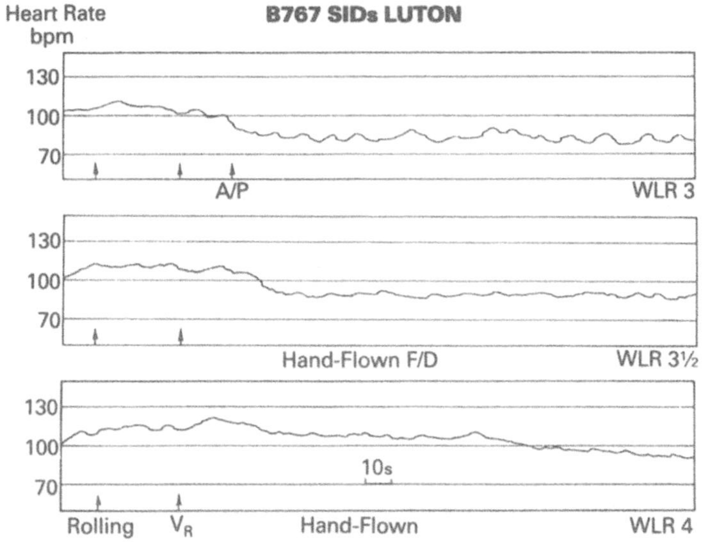

Figure 6. Heart rates and workload ratings for different operating modes during standard instrument departures. B767
Top - Autopilot engaged at 500 ft
Middle - Using F/D integrated with FMS - Hand-flown
Bottom - Using conventional attitude indicator - Hand-flown

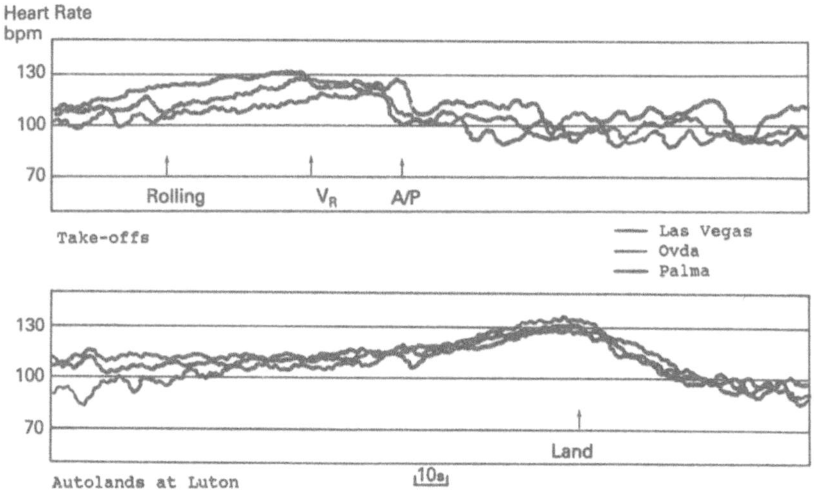

Figure 7. Heart rate responses and workload ratings experienced by the same pilot during autolands at Luton following three flights of different duration. Palma - 2 hrs, Ovda - 5 hrs, Las Vegas - 9 hrs.

APPLICATION OF THE AIRBUS WORKLOAD MODEL TO THE STUDY OF ERRORS AND AUTOMATION

Richard D. Blomberg and Amy L. Schwartz
Dunlap and Associates, Inc.
17 Washington Street
Norwalk, Connecticut 06854
U.S.A.

Jean-Jacques Speyer
Flight Division
Airbus Industrie
Toulouse, France

Jean-Pierre Fouillot
Laboratoire de Physiologie de l'Adaptation
Faculté de Médecine
Cochin-Port Royal
Paris, France

ABSTRACT. The Airbus Workload Model generates second-by-second estimates of pilot subjective mental workload on the seven point subjective rating scale used by Airbus Industrie for aircraft certification. The model uses synchronized measures of aircraft performance, heart rate variability on both pilots, and flight status without intruding on the flying task. As part of the A320 certification campaign, 48 simulated line flights, which involved different levels of automation, were flown. Crew errors were also recorded. This paper demonstrates how the Airbus Workload Model presents a novel tool to study workload and its relationship to both errors and automation. The findings are descriptive in nature because the certification data base was not constructed to support research on these issues. They suggest that there is a relationship between the severity and type of error and workload, and that automation and workload are inversely related.

1. INTRODUCTION

Crew workload is one of the most important human factors issues in aviation. Sustained high workload levels will overtax the crew, thus decreasing the margin of safety and increasing the probability of an accident. Extended periods of unusually low workload may also lead to

A. Coblentz (ed.), Vigilance and Performance in Automatized Systems/ Vigilance et Performance de l'Homme dans les Systèmes Automatisés, 123–137.

problems if the pilots fail to keep abreast of activities in the cockpit. After an extended time at a low state of arousal, a pilot may not be able to perform adequately if forced to respond to an emergency.

One of the major objectives of aircraft automation is to reduce crew workload, but even with highly automated aircraft systems (and assumed low workload), human errors occur. Indeed, it has been argued that high levels of automation tend to encourage gross blunders rather than small errors (Wiener, 1987a). With the anticipated introduction of extremely long range aircraft designed to be flown for extended periods by a crew of two pilots such as the Airbus A340 and the Boeing 747-400, there is an increased need to examine crew workload and its influence on safety. Unfortunately, the complex relationship among subjective mental workload, errors, and automation is poorly understood. Much of the lack of good data on this subject is directly attributable to the absence of a valid, simple, unobtrusive measure of workload which can be used in flight.

Recent efforts related to satisfying the minimum crew complement certification requirements for the Airbus Industrie A320 included the development of a new, continuous measure of pilot workload. This measure has been called the "Airbus Workload Model" because it derives from statistical modeling of actual subjective workload ratings provided by pilots during certification flights. The availability of the Model provides a unique opportunity to examine some of the critical workload issues with data collected during flight.

This paper demonstrates the application of the Airbus Workload Model to two important problem areas: the relationship between workload and errors, and the relationship between workload and automation. The Model allows analysis of these issues in a manner that has not been possible before because it generates second-by-second workload measures without intruding on the flying task. The data used in these analyses were collected during the minimum crew certification flights for the A320. The primary focus of the flights was to demonstrate to the various airworthiness authorities that the aircraft was capable of being flown by a crew of two pilots during typical normal and abnormal flights. Although the extent of automation was varied and some error information was collected, the flights were not structured to control these variables. In fact, the flying problems were intended to simulate regular line operations with their inherent uncertainties rather than to provide a data base for analyzing workload relationships.

There were 48 certification flights in the basic A320 minimum crew campaign. These were structured so that four different airworthiness authority flight crews (pilot or CM1 and copilot or CM2) each flew 12 flight scenarios. Each scenario was designed to examine a specific aspect of the operation of the aircraft. For example, one scenario was conceived to demonstrate, in part, the operation of the aircraft without the autothrust system to manage the engine power settings. All flights simulated actual line operations on the European network of approximately a one hour duration. All flights were between regular passenger airports such as Orly in Paris and

Frankfurt, West Germany. Prior to the campaign, the pilots took part in a program of basic training in the operation of the A320 as configured at that stage of the aircraft development process. This program, although comprehensive, was less extensive than would be given to a typical airline crew. The shortened training time was counterbalanced by the presence of a senior Airbus test pilot in the cockpit at all times.

2. THE AIRBUS WORKLOAD MODEL

One part of the Airbus Industrie approach to minimum crew certification for new aircraft types has been the development of a mathematical model to estimate crew workload ratings. The work on this model, begun in 1982, was a joint effort of Dunlap and Associates, Inc., Airbus Industrie and the Laboratoire de Physiologie at the Cochin Hospital of the University of Paris.

The Airbus Workload Model is a statistical combination of aircraft flight parameter data such as airspeed and roll rate with heart rate variability data on both pilots and flight status measures such as phase of flight. The Model estimates the rating a pilot would have given on the Airbus Rating Scale. This scale is the heart of the Airbus dynamic workload assessment technique.

The dynamic workload assessment technique was first used for the certification of the A300FF. It involves interrogating the pilots for scaled workload assessment ratings throughout the minimum crew certification route-proving flights.

The dynamic assessment approach was continued in the A310 certification campaign using a seven point scale (values 2-8). A value of two on the scale indicates light workload which is very acceptable. A rating of eight is considered unacceptable even instantaneously. The large data set collected during the A310 route-proving flights was then used to develop a mathematical model of the rating process. This model was later validated in separate flights on the A310-300 (Speyer et al., 1987).

The minimum crew certification campaign for the A320 in January 1988 was the first opportunity to use the Model to demonstrate the workload characteristics of a new aircraft. All eight pilots who participated in the campaign were equipped with ambulatory heart rate data recorders using the technique described by Fouillot and his associates (1986). Heart rate data were then used to create the measures of medium and long term heart rate variability required by the model. These were combined with measures derived from data collected by on-board aircraft recorders to calculate second-by-second workload estimates for each pilot.

The estimates from the Model were used to compare workload across scenarios and for various planned and spontaneous maneuvers and operational situations. In addition, because subjective ratings were still requested periodically from the pilots, it was possible to compare the actual quotations with the Model's estimates. This provided an additional statistical validation of the Model in a new aircraft (the A320) with extremely high automation, sidestick controllers and a fly-by-wire control system. The results provided further evidence of the statistical power of the Airbus Workload Model. The 2,183 pairs of actual and estimated workload ratings during flight showed a product moment correlation of 0.534 which is significant at the .001 level. In addition, inspection of the relationship of actual and estimated rating timelines showed an excellent correspondence.

The basic objective of minimum crew certification flights is to demonstrate that the aircraft can be flown by its specified crew complement in a manner acceptable to the airworthiness authorities. The output of the Model provided one set of data which indicated the acceptability of the A320. The more than 48 hours of simulated line flying, however, produced an extensive data base for which continuous (one per second) estimates of workload were available for all participating pilots. This permitted for perhaps the first time ever the detailed examination of workload relationships for an extended period of actual flight operations.

3. ERRORS AND WORKLOAD

One of the major reasons that workload is of interest is because high workload can lead to errors (e.g., Smith, 1979). If the pilot is overloaded, he does not have the spare capacity to attend to anything but the task in which he is engrossed. Conversely, sustained low workload may also lead to errors since the pilot may lose situational awareness and have difficulty getting back "into the loop." The causality between workload and errors probably goes both ways. That is, high workload can lead to errors and certain types of errors can lead to increased workload (Hart and Bortolussi, 1984). Indeed, it is errors that cause aviation accidents, not workload. Workload is a mental construct - an intermediary between the aircraft status and the crew's performance.

Even though errors are so important, little work has been done to try and understand the preconditions under which errors are most likely to occur. As Hawkins (1987) points out, "The concept of pilot error often focuses more on what happened rather than why it happened." Most error analyses involve a post hoc evaluation of an actual accident (e.g., Wiener, 1987b). This is because it is difficult to study errors systematically in-flight since they cannot be reliably and safely induced, and traditional airborne workload measures are not continuously collected. Thus, it has been essentially a chance event if a workload measure was taken at the time of an actual in-flight error. Hart and Bortolucci's work (1984) was done in the laboratory where they presented written descriptions of flight events (including errors) to experienced pilots on a cathode ray tube (CRT), and asked the pilots to rate how the events would impact on workload. In order to investigate more fully the relationship between workload and errors, a continuous measure of workload is needed so that spontaneous crew errors can be related to workload level at the exact point in time when the error occurs.

In the A320 certification flights, crew errors were recorded at the time they occurred or were detected by on-board observers. The observers were expert pilots who were seated on the jump seat in the cockpit during all of the flights. They identified any errors made by the crew and entered them into a data collection system. Errors were categorized by their severity as:

- Minor (M) - Slips or procedural problems which could be fixed promptly;
- Important (I) - More serious deviations from proper procedure which were ultimately corrected or errors not related to safety which went uncorrected by the crew;
- Safety-related (S) - Problems with the potential to degrade safety whether corrected or not.

Although the observer pilot was highly experienced, his training specific to the A320 was quite limited. Therefore, it is likely that some errors were overlooked and others were recorded which were not, in fact, errors on the A320 with its unique and advanced systems.

The 48 flights resulted in 75 errors which occurred at a time for which Model estimates were available. Only one of the 75 errors was deemed of the "Safety-related" type by the observers. Therefore, for the purpose of these analyses this error was combined with those in the "Important" category. Overall, 17 errors were considered "Important" or "Safety-related." The remaining 58 errors were judged to be "Minor."

In order to examine the relationship between errors and workload, timeline plots of each flight were used. These plots show the instantaneous workload estimate for each pilot as a function of elapsed flight time. By locating the time at which each error occurred on these plots, it was possible to assess:

- The relative workload level at the time of the error with respect to the total workload profile for the entire flight. This relative measure, noted as low, medium or high, permitted an assessment of the association of errors with the prevailing level of workload.
- The workload trends or "shapes" prior to and immediately following the time of the error, i.e., increasing, decreasing or steady. This made it possible to assess the existence of lead or lag relationships between errors and the existence and direction of any change of workload.

The original classification of errors (M, I or S) is based on the potential of the error to impact the safety of the flight rather than on its likely influence on or relationship to pilot workload. Therefore, an additional analysis of the errors was performed using a new categorization of errors which was based on the likely awareness of the pilot that an error had occurred. It was assumed that an error of which the pilot was aware might produce a different workload effect from an error which was not noticed by the pilot.

Each error was recoded based on whether it was considered to have occurred with the pilot likely "Aware" or "Unaware" that he had committed it. The pilot was "Unaware" if the error did not result in an alarm or any immediate or long term change in the aircraft performance or flying task which the pilots could detect. These errors included cases such as checklist omissions, slight deviations from target flight levels, and incorrect barometer settings. The fact that the errors were unknown to the pilots at the time they were committed suggests that they might not lead to any immediate increase in workload due to remedial activities or emotional reactions. It does not imply that they were inconsequential to safety. Certainly, omitting checklist items or leveling at the wrong altitude can have potentially disastrous consequences.

The pilot was "Aware" of an error if it resulted in an alarm or the almost sure need for remedial actions at some time relatively soon after the error had been committed. Examples of the second type of error include failure to notice the autopilot was disconnected, no speed holding in climb, or forgetting that both autothrottle systems were lost.

Figures 1 and 2 illustrate the data analysis technique employed. Both plots are for flying pilots in Scenario 2. This Scenario involved a planned loss of both autothrottle systems (ATS), both Flight Control Data Concentrators (FCDC) and both Spoiler and Elevator Computers (SEC). The flights were dispatched with one of each of the devices inoperative, and the remaining unit was "failed" during the flight at the time shown in the Figures.

Figure 1 shows two errors. The first was judged to be Minor (M) in severity. It involved a failure of the flying pilot to hold the aircraft's speed accurately during climb. The pilot was assumed to be "Aware" of this error. Examination of the Figure also shows that the error occurred during a period of relatively low workload. Workload was trending downward during the period immediately before the error was recorded and then showed a slight increase immediately after.

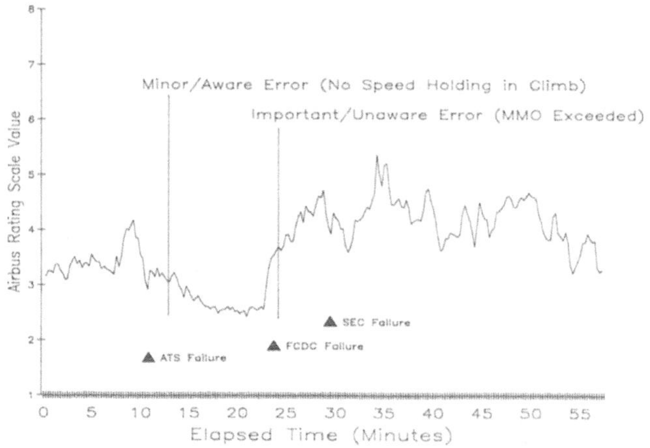

Figure 1. Estimated workload for flying pilot Brest to London - Scenario 2 - January 24, 1988.

The second error in Figure 1 was rated as Important (I) during the campaign. It involved exceeding the maximum operating mach (MMO) for the aircraft. The pilot was judged not to have been aware of this error because there was evidence that he did not know the appropriate MMO value for A320 operations with a double ATS failure. Workload at the time of the error was medium with respect to the level for the total flight. It was increasing both before and after the time of the error.

Figure 2 depicts an error of minor severity early in the flight. This error involved the pilot forgetting that he was operating without assistance from an ATS. In this case, it was assumed that the pilot was aware of the error because corrective action to control the aircraft was required. As can be clearly seen in the Figure, workload peaked at the time of the error. Thus, it was coded as "High" and increasing before the error and decreasing after the error.

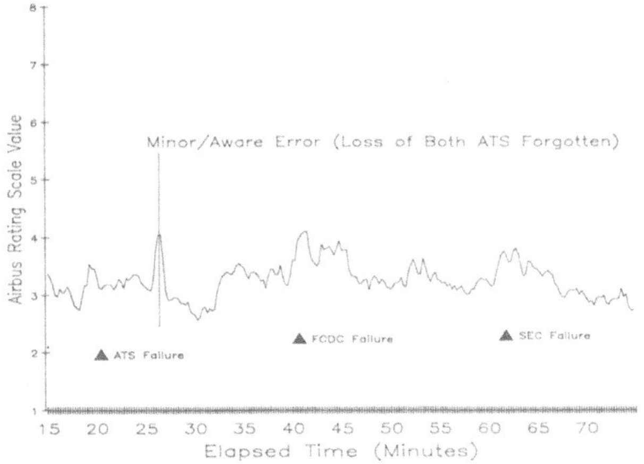

Figure 2. Estimated workload for flying pilot Nice to Frankfurt - Scenario 2 - January 23, 1988.

Table I shows the relationship between the certification campaign error severity measure and the recoded estimation of pilot awareness. There is virtually no difference in the distribution of awareness level as a function of error severity. This suggests that the two measures are independent. It is also interesting to note that for only about 25 percent of either the Minor (M) or Important and Safety (I+S) errors was the pilot considered to have been likely aware of the problem noted by the observer.

Table I. Comparison of error awareness and error severity

ERROR AWARENESS	ERROR SEVERITY: MINOR	IMPORTANT + SAFETY	TOTAL	
LIKELY UNAWARE	43	13	56	# Of Errors
	76.8	23.2	100.0	% Of Likely Unaware
	74.1	76.5	74.7	% Of Severity Category
LIKELY AWARE	15	4	19	# Of Errors
	78.9	21.1	100.0	% Of Likely Aware
	25.9	23.5	25.3	% Of Severity Category
TOTAL	58	17	75	# Of Errors
	77.3	22.7	100.0	% Of All Errors
	100.0	100.0	100.0	% Of Severity Category

Before turning to the detailed analyses of the errors with respect to workload, it is of interest to examine the coded errors generally. Eighty-one percent of all errors occurred during three flight phases: climb, cruise, and descent. This is not surprising since these phases comprise the majority of the time of flight. Overall, the prevailing workload level at the time of an error did not vary greatly. It was slightly more likely to be medium (41% of the errors) than either low or high (29% each).

Errors associated with sustained low workload levels were not seen in these data. However, the nature and length of the certification flights and the numerous flying problems included in the scenarios did not provide an opportunity for extended periods with only low task demands presented to the pilots.

The error severity was examined with respect to the workload level at the time of the error. Table II shows the distribution by workload level for the two types of error. It can be seen from this Table that the M errors were approximately evenly distributed among low, medium and high relative workload times of the flights. The I+S errors, on the other hand, were most likely to occur during periods of medium and high prevailing workload and relatively unlikely to be observed during periods of low workload. The difference observed, however, was not statistically significant.

Table II. Error severity versus workload level

ERROR SEVERITY	WORKLOAD LEVEL: LOW	MEDIUM	HIGH	TOTAL	
MINOR	19	22	17	58	# Of Errors
	32.8	37.9	29.3	100.0	% Of Minor
	86.4	71.0	77.3	77.3	% Of Workload Level
IMPORTANT + SAFETY	3	9	5	17	# Of Errors
	17.6	52.9	29.4	100.0	% Of Important+Safety
	13.6	29.0	22.7	22.7	% Of Workload Level
TOTAL	22	31	22	75	# Of Errors
	29.3	41.3	29.3	100.0	% Of All Errors
	100.0	100.0	100.0	100.0	% Of Workload Level

Table III shows the pilot awareness measure as a function of the prevailing workload level. The pattern is quite uniform with approximately one third of each awareness category occurring at each of the three workload levels. There is, however, a slight tendency for more errors during medium workload conditions. These results suggest that the prevailing workload levels during the A320 certification flights were not systematically related to the pilots' awareness of the errors they committed.

Table III. Error awareness versus workload level

ERROR AWARENESS	WORKLOAD LEVEL: LOW	MEDIUM	HIGH	TOTAL	
LIKELY UNAWARE	16	24	16	56	# Of Errors
	28.6	42.9	28.6	100.0	% Of Likely Unaware
	72.7	77.4	72.7	74.7	% Of Workload Level
LIKELY AWARE	6	7	6	19	# Of Errors
	31.6	36.8	31.6	100.0	% Of Likely Aware
	27.3	22.6	27.3	25.3	% Of Workload Level
TOTAL	22	31	22	75	# Of Errors
	29.3	41.3	29.3	100.0	% Of All Errors
	100.0	100.0	100.0	100.0	% Of Workload Level

The severity or awareness of a pilot's error might also be associated with the trend in workload. Table IV shows the observed workload trend immediately prior to each error as a function of the error severity. Examination of the Table shows a slight, but not significant, tendency for the I+S errors to be associated with increasing workload and for the M errors to be associated with decreasing workload. However, the sample size for the I+S errors is too small to support any valid conclusions based on these subtle differences.

Table IV. Error severity versus workload trend before error

ERROR SEVERITY	TREND BEFORE ERROR: STEADY	INCREASE	DECREASE	TOTAL	
MINOR	15	15	28	58	# Of Errors
	25.9	25.9	48.3	100.0	% Of Minor
	78.9	68.2	82.4	77.3	% Of Workload Trend
IMPORTANT + SAFETY	4	7	6	17	# Of Errors
	23.5	41.2	35.3	100.0	% Of Important+Safety
	21.1	31.8	17.6	22.7	% Of Workload Trend
TOTAL	19	22	34	75	# Of Errors
	25.3	29.3	45.3	100.0	% Of All Errors
	100.0	100.0	100.0	100.0	% Of Workload Trend

The association between awareness of an error and workload trend was also examined. Table V shows these data. It can be seen in the Table that almost half of the errors for which the pilot was judged aware occurred during a period of increasing workload. Conversely, half of the "Unaware" errors took place during times when workload was decreasing. The observed pattern approached statistical significance by the Chi Square test.

Table V. Error awareness versus workload trend before error

ERROR AWARENESS	TREND BEFORE ERROR: STEADY	INCREASE	DECREASE	TOTAL	
LIKELY UNAWARE	15	13	28	56	# Of Errors
	26.8	23.2	50.0	100.0	% Of Likely Unaware
	78.9	59.1	82.4	74.7	% Of Workload Trend
LIKELY AWARE	4	9	6	19	# Of Errors
	21.1	47.4	31.6	100.0	% Of Likely Aware
	21.1	40.9	17.6	25.3	% Of Workload Trend
TOTAL	19	22	34	75	# Of Errors
	25.3	29.3	45.3	100.0	% Of All Errors
	100.0	100.0	100.0	100.0	% Of Workload Trend

The trend in workload after an error occurred was also examined. In general, the errors were followed approximately equally by periods of steady, increasing or decreasing workload. The trend after the error was also not differentially related to the pilot's awareness of the error. Thus, there is no indication in these data that an error noticed by a pilot impacts immediately subsequent workload any differently from an error which goes unnoticed.

In order to obtain a complete description of the workload conditions surrounding an error, it would be best to examine all combinations of the trends before and after the error and the prevailing workload level as a function of error severity and awareness. Unfortunately, a total of 75 errors is not sufficient to support this type of analysis. However, various subsets of the data were examined to see if any results were suggestive of areas which might be profitably explored in future studies.

Important and Safety errors showed an interesting pattern when workload was increasing or decreasing rather than steady just prior to the error. With increasing workload preceding the error, fully 71 percent of the errors exhibited *decreasing* workload immediately after the error. This observation suggests that I+S errors may be associated with peaks in workload. When workload was decreasing before the error, it was more likely to change to steady (50%) or increasing (33%) immediately after the error than to continue decreasing (17%). This is consistent with the possible generation of workload by the higher severity errors.

It was also thought provoking to discover that *all* of the Important and Safety errors which occurred at a "High" workload level were associated with increasing workload just before the time of the error. Thus, there is the suggestion that when workload is high and increasing and an error occurs, that error will be severe. This is further supported by the observation that only 24 percent of the Minor severity errors occurred under the same conditions of workload level and trend. Unfortunately, the small sample sizes (there were only five I+S errors at High workload) preclude reaching any firm conclusions.

The foregoing analyses have described some of the possible associations between workload and errors. There is an indication that severe errors may occur during periods of high and increasing workload. There is also the suggestion that they take place near a peak in workload. The workload decrease after the more severe errors may be the result of the pilot "giving up" and shedding tasks. It may also simply be an artifact of the certification flying situation in which there is constant monitoring and possible assistance available to the crew from other specialists in the cockpit.

4. AUTOMATION AND WORKLOAD

The previous section showed how the Airbus Workload Model can be a useful tool for investigating spontaneous errors. This application of the Model is possible because moment-by-moment workload measures are provided. This type of microanalysis is not possible with most other workload measurement techniques.

Workload data can also be used for a more molar analysis to investigate the relationship between workload and automation. In the A320 certification flights, the scenarios were arranged to vary the flying problem presented to the pilots. In the process of creating scenarios, the automation level was not systematically varied. However, three of the 12 scenarios had relatively constant and different levels of automation for the majority of their in-flight periods.

Scenario 3 involved flying the A320 with most of its high automation navigation and flight controls disabled. This involved using standby navigation instruments with no assistance from the Flight Management and Guidance System (FMGS) of computers. The result was a relatively low automation level compared with the inherent design of the aircraft. Conversely, Scenario 12 involved no planned failures and the full automation level of which the A320 is capable. Since the aircraft design is state-of-the-art, the automation level for Scenario 12 can be considered high.

The flying conditions for Scenario 6 were somewhere between the two extremes and, thus, it can be considered a moderate automation situation. Although all navigation and flight controls were available, the flight warning system was rendered inoperative. Thus, the pilots had to engage in systems monitoring without the aid of the CRT displays in the Electronic Centralized Aircraft Monitoring System (ECAM) and without gongs and displayed red warnings and amber cautions. The autopilot was also lost for a small portion of the flight.

In order to assess the impact of automation, the average workload for each crew by flight phase was calculated. This was accomplished by numerically averaging the second-by-second Model estimates for each of the two pilots and aggregating by phase. Since each crew flew each scenario once, all eight pilots in the campaign are represented in the data for each scenario. The total number of Model estimates entering the calculations is quite large as each flight spanned approximately one hour, and an estimate was generated every second. A typical cruise phase, which was the time when automation level was most consistent, lasted for about 1,000 seconds.

Figure 3 shows the average calculated workload by flight phase (takeoff through descent) for each of the three scenarios under consideration. The automation level by scenario was only uniformly low (Scenario 3), medium (Scenario 6) or high (Scenario 12) for the climb, cruise and descent phases. Approach and landing were not necessarily consistent with the designated levels. For example, the manual navigation tasks which made Scenario 3 a low automation situation are largely irrelevant to the latter stages of approach and virtually all of landing. The takeoff phase was at a high level of automation for all three scenarios. It is shown in Figure 3 simply as a reference.

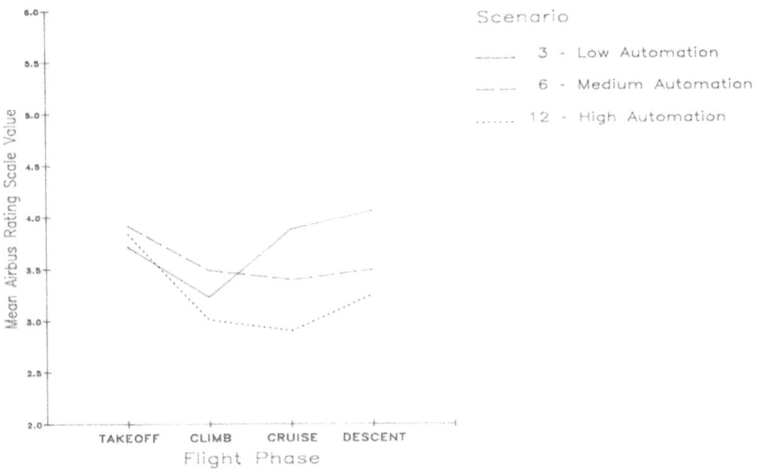

Figure 3. Estimated workload for 3 scenarios with varying automation levels.

It can be seen in Figure 3 that the three scenarios array as expected. Scenario 3 with the lowest automation level shows the highest average workload. Scenario 12 has the lowest workload and Scenario 6 lies in between. The estimates for the takeoff phase, however, are all quite close together as predicted by the equivalent flying problems presented to the crews.

As with the results related to errors, these findings are only suggestive. However, they do point to a potentially strong inverse relationship between workload and automation level. This is precisely the situation which theory would predict.

5. DISCUSSION

The Airbus Workload Model was developed as a tool for aircraft certification and development. This paper has shown that this tool has even broader applications and can be used to gain a better understanding of the relationship between workload and other variables such as errors and automation. The results discussed here demonstrate the feasibility of using the Model for these purposes.

One of the great strengths of the Model as a measure of workload is its ability to collect data continuously throughout the course of an actual flight. This permits research to be undertaken on actual flight events rather than artificial situations in the air or simulator. For example, the errors studied from the A320 certification flights were spontaneous, "real" errors. The problems associated with their analysis are associated with the detection and classification of the errors themselves and not with the workload metric. No other documented workload measurement technique presents a researcher with a similar opportunity to examine workload microscopically and unobtrusively throughout the entire duration of a real flight.

The analyses performed for this paper suggest several studies of workload using the Model which would add significantly to the understanding of pilot workload, its origins and consequences. Among these are:

- A study of sustained periods of low workload such as those which might be associated with extended range commercial flights in a high automation aircraft. Nothing contained herein even begins to shed light on this potentially important problem.
- A more detailed analysis of the relationship of errors to workload in normal flight operations. This might best be accomplished through the cooperation of an airline company. In addition to monitoring workload with the Model, this study would require the development and application of a reliable, unobtrusive and cost-effective system to detect and categorize naturally occurring errors. With the use of the Model, a study of this type might finally begin to shed light on the question of the causal relationship between workload and errors.
- A controlled examination of the relationship between automation and workload and among automation, workload and errors. Experimental control is needed for this study to vary systematically the level of automation as a function of other relevant parameters such as task type, e.g., navigation, systems monitoring.
- A study of the effect of pilot human factors on workload and errors. This might examine, among other things, the influence of fatigue, emotional problems, age and experience on the overall level of workload and its pattern at the time of errors.

Much of the research on pilot workload has been relegated to the laboratory and the simulator because of the operational problems associated with most workload measurement techniques. The development, validation and acceptance of the Airbus Workload Model for use in certification has also provided the research community with a potentially valuable measurement tool. This soon to be patented approach (French patent application number 87.07048 of 20 May 1987, European patent application number 88401200.6 of 21 June 1988 and U.S. patent pending) opens new avenues in the study of pilot workload which can greatly add to our knowledge and understanding of the flying task.

6. REFERENCES

Fouillot, J.P., Tekaia, F, Speyer, J.J., Regnard, J. and Blomberg, R.D. (1986, October 27-28). Monitoring ambulatoire de la frequence cardiaque du pilote d'avion de transport. In *Proceedings of Workshop on Effects of Automation on Operator Performance*. Commission of the European Communities, Medical and Public Health Research Programme. Laboratoire d'Anthropologie Appliquée, Université René Descartes, Paris.

Hart, S.G. & Bortolussi, M.R. (1984). Pilot errors as a source of workload. *Human Factors*, 26, 545-556.

Hawkins, F.H. (1987). *Human factors in flight*. Brookfield, Vermont: Gower Technical Press.

Smith, H.P.R. (1979). *A simulator study of the interaction of pilot workload with errors, vigilance, and decisions*. NASA Technical Memorandum 78482.

Speyer, J.J., Fort, A., Fouillot, J.P. and Blomberg, R.D. (1987, June). Assessing pilot workload for minimum crew certification. *The Practical Assessment of Pilot Workload*. AGARDograph No. 282. North Atlantic Treaty Organization Advisory Group for Aerospace Research and Development, London.

Wiener, E.L. (1987a, December). Management of human error by design. *Human Avoidance Techniques Conference Proceedings*. Washington, D.C.

Wiener, E.L. (1987b). Fallible humans and vulnerable systems: lessons learned from aviation. In J.A. Wise and A. Debons (Eds.), *Information Systems: Failure Analysis*. Berlin: Springer-Verlag.

TOWARDS DESIGN-INDUCED ERROR TOLERANCE

J.J. SPEYER
Flight Division
Airbus Industrie
31707 BLAGNAC
FRANCE

SUMMARY. The purpose of this paper is to review the span between initial design aims and subsequent flight evaluation and measurement.

Except for the recommendations of Fitts, Wanner, Wiener and Curry, few fundamental design guidelines appear to be available in the scientific field of human factors, illustrating a posteriori that this activity is an art in infancy. With new technology emphasis was gradually put on human factors by all those concerned, manufacturers, airworthiness authorities, pilots and airlines. Measuring the impact of new technology on the operational interface could precisely help setting up this human factors capability. Which in turn should eventually influence on design guidelines and specifications.

A practical review is presented of the operational objectives and technological modules that marked the outgrowth of the Airbus family of commercial aircraft.

The emerging role of the pilot is becoming more that of a systems monitor than that of a controls handler. Several views suggest that the pilot be brought back into a more active role to avoid automation -or design- induced errors.

What has been achieved with the early 1980's A310/A300-600 cockpit however is, in our view, still in the vicinity of the Wiener and Curry philosophy. In the latter 1980's A320/A330/A340 design aims have been intended to cover even more error -tolerance or- protection against incidents of the Wanner scheme.

Reviewing aircrew comments on design aims and achievements from flight evaluation became a practice in the early 1980's with crewmember opinions. The success of the questionnaire technique prompted Airbus Industrie to use it again for its Fly-by-wire proof-of-concept experiment on the A300 testbed.

More systematic efforts on man-machine interface analyses were put in the wake of the crew complement question which triggered the development of several evaluation methods.

A. Coblentz (ed.), Vigilance and Performance in Automatized Systems/ Vigilance et Performance de l'Homme dans les Systèmes Automatisés, 139–165.

One of these, the Performance Criteria Methodology, developed with Dunlap & Associates, statistically investigates the impact of new technology features such as the EFIS, the FMS and Fly-by-wire/Sidestick. Performing analyses of variance on basic flight parameter measures allowed to demonstrate marked smoothness and stability improvements flight efficiency, reduced task and workload.

The classic "pilot error" is giving way for a more caution consideration of the underlying factors in pilot behavior. Models of human operator behavior would be needed for incorporation of error tolerance or protection into the design of the automation device itself. Aware of the need for a better knowledge of the man-machine relationship Airbus Industrie in cooperation with Dunlap & Associates and the Cochin Faculty of Paris developed and validated a predictive crew workload rating model based on heart rate variability, aircraft performance parameters and situational information (scenario, flight phase conditions, etc...).

Cleared for experimental the Airbus workload model was used in January 1988 for the A320 minimum crew certification to generate second-by-second estimates of subjective workload. Crew errors were also recorded and classified according to their severity and awareness as reported by Blomberg and al in another paper. Further work will however be necessary to model crew vigilance, to investigate how automation, workload, level of arousal, performance and errors all relate to each other and to establish procedures to regulate these in new technology aircraft.

1. System-induced errors : human factors

Current aircraft control and monitoring systems have reached unparalleled levels of reliability and redundancy. The success of the integration and evolution process achieved through systems' interfacing and advances in technology suggested the human operator in the safety loop, a goal long sought in suggest the eventual inclusion of the profession (1).

Emphasising the necessity to consider human factor issues in a broader systemic context was an early paper by Wiener in 1977, titled "Controlled Flight into Terrain : System-Induced Errors" (2). The contradiction suggested in this title aims to indicate the difficulty of integrating two levels of system management : aircraft control (inner loop) and aircraft monitoring (outer loop).

In the aviation transport area alone, some 75% of accidents are catalogued as pilot error, the remainder being evenly split between purely technical and environmental causes. Wiener makes the point that a majority of these accidents are design-induced, referring to human engineering concepts (2), air traffic control (3) and automation (4).

Stressing the failure of the human factors profession to play a more significant role in the design process, he addresses the following interface aspects :

- vigilance problems and crew overload/underload
- inadequate workplace design and extra-cockpit communication
- faulty pilot-controller communication and crew coordination
- unadapted, excessive or wrong automation
- unheard or cancelled warning devices and visual illusion
- confusing terminology and charts.

Politics aside, this author, also a pilot and management scientist, took a critical view of the systems design of air traffic control and, considering the long-debated issue of collision avoidance systems, confined himself throughout a series of papers to the topic of cockpit automation, its promises and problems (4).

The central issue was taken to be that it is highly questionnable whether system safety was going to be enhanced by automating tasks and functions normally performed by the pilot.

To illustrate their qualms, Wiener and Curry examined a number of aviation accidents and incidents and the role of automation in the causal chain towards breakdown :

- failure of automatic equipment
- automation-induced error compounded by crew error
- crew error in equipment set-up
- crew response to a false alarm
- failure to pay attention to an automatic alarm
- failure to monitor
- ensuing loss of proficiency.

This is the controlling versus monitoring issue again, the distinction between automation of control and automation of monitoring being represented by two orthogonal dimensions as illustrated in Figure 1.

Should the pilot monitor the machine or should the machine monitor the pilot ?

A recurrent theme, the implications with regard to design philosophy for control and monitoring system being (4) :

- that the designer must ask to what extent the human should be included in the inner and outer control loops and to what extent automation should assist him in multi-attribute decisions,

- that the designer must consider the single channel behaviour of the human and the integration of the numerous devices, alerting and warning systems and conditions to be monitored.

  Said Don Engen, former FAA administrator, at the SAE in 1985 (5) "Human tolerance to poor design can be high until routine operation commences, then the tolerance level will decline in a random fashion, usually when least expected, and always when not wanted ... The answer for the best system lies with highly competent test pilots and design engineers who provide systems free from error".

  The classic misnomer "pilot error" is giving way to a more cautious consideration of the underlying factors in pilot behaviour (6) (7).

2. Design guidelines : rules and philosophies

Few fundamental guidelines are available in the scientific field of human factors. They all appear to come "a posteriori" illustrating that this activity is an art in infancy.

Except the general guidelines of Wiener, Curry and Wanner, we are still short of standards sought by the industry (8) (9) (10) since it is still evolving at a rapid pace.

Wiener and Curry conclude their survey of cockpit automation by proposing 15 recommendations for designing and using automated systems (Table 1). In need of a philosophy regarding automation, Wiener has persistently advocated the respect of at least three (4) (11) :

. to allow the crew more freedom to fly the plane and use automation in the manner it wishes, but surrounded by a multidimensional warning and alerting system that informs the crew if they are approaching some limit ; as long as within the limits, the crew would have freedom to conduct the flight according to its style ; this concept of an "electronic cocoon" is called flight **management by exception,**

. to use forecasting models and trend warnings to predict a penetration of the cocoon rather than waiting for it to happen and alarms to be activated when the system reaches a critical point ; in order for the crew to have maximum control of the situation, values for the forecasting parameters could be selected by the crew ; this concept is called **exceptions by forecasting,**

. to allow the crew to inform the machine of its strategic goal or intention which would allow the computer to check crew inputs and system outputs to determine if they are logically consistent with the overall goal ; if not, an exception message would be issued ; this concept is called **goal sharing**.

Developing his argumentation with regard to human factors and flight safety, Wanner (12) considers three classes of incident that may be playing a causal role in aircraft incidents or accidents :

. **pilotability incidents** whereby the crew erroneously allowed the functioning point to stray outside the authorized envelope with difficult or impossible reversion,

. **perturbation-sensitivity** incidents whereby external perturbations (windshear) or internal perturbations (failures or fire) let the displaced functioning point go beyond limits, whether limits after perturbations are new or unchanged,

. **manoeuvrability incidents** whereby the execution of a procedure (to modify trajectory or avoid obstacle) and the displacement of the functioning point as a result of incidents of the two preceding types brings the resulting functioning point close to or past its limits.

The goal of safety rules is precisely to lower the probabilities of occurrence of these three types of incident.

Wanner identifies five basic human attributes that allow him to derive a set of proposed design rules mainly covering pilotability incidents :

. single channel behaviour or functioning by the operator implying sequential work procedures,

. need for information input and drive for forecasting and extrapolation of events and situations,

. possibility of performance compensation by increasing workload to cope with higher task difficulty,

. use of mental representation model whether still learning or experienced,

. inability to estimate risks reliably in any given situation.

Wanner proposes eight pilotability rules to alleviate incidents of that type (Table 2). Perturbation-sensivity and manoeuvrability rules are, in his view, easier to establish since they do not implicate the human operator as such but rather system redundancy and integrity and operational knowledge of mission and environment.

As will be seen in the following paragraphs, these "a posteriori" rules and philosophies are finally not too far from the operational orientation that has inspired the functional layout and organisation of our new cockpits and associated systems. Design being an inextricable exercise of compromise, art and science, any "à la lettre" correspondance should however not be expected "a priori".

3. New technology : pilots and facts

In fairness to all, Wiener admits in a later paper (13) that the negative side of automation should not be overstated. The number of automation- or design-induced incidents brought about by new technology aircraft has been very limited and their consequences very small. In 1982, Gannet (14) stated that "the final report card on the results of the most recent application of digital technology and human performance principles to air transports will not be available for several years. It is firmly believed that, in the long run, the benefits far outweigh the costs".

In 1986, Caesar (15) stated that, with 8,000 big jets built so far, the average loss percentage since 1959 was 3.8%. Among these, was not a single of the so-called electronic cockpits like the A300FF, A300-600,, A310, 757, 767 or MD-80, then numbering 800 since being gradually introduced from late 1979.

Most automation works with extremely high reliability and pilots are very satisfied with new control and monitoring interfaces (16, 17, 18, 19) when compared to former equipment. The "good old days" fallacy is even repeatedly discredited with recent incidents and accidents having occurred on earlier technology aircraft (20, 21, 22, 23).

Most crews interviewed in field studies expressed high praise for the new "glass cockpits" which provide "a more dynamic source of information and greater awareness of the aircraft with respect to the operating environment, making conventional electromechanical cockpits obsolete" (19).

- According to Wienner and others, incompatibility with the air traffic control system limits the usefulness of modern flight management computers. The ATC system, designed for the slow climbs and fast descents of older aircraft, forces new ones to routinely use speedbrakes for descents and prevents effective use of the flight management computers for vertical navigation.

- Although this tends to disappear with experience, some pilots have the impression that higher monitoring and programming demands associated with the flight management and guidance systems may force them towards more head in the cockpit time and less availability for extra-cockpit scanning in terminal areas. Some others may, in turn, be concerned about losing their classic monitoring scan and manual flying skill from over-reliance on the automated cockpit and resist this "complacency tendency" by self-imposed flying programs and training exercises to maintain proficiency.

- All 17 pilots and observers that participated to the A310 crew complement certification (18) volunteered to answer an anonymous questionnaire on the quality of the man-machine interface. As a group, these pilots appeared to cover an experience-level corresponding to the operation of almost all commercial jet aircraft introduced to that period.In answering the questionnaire, pilots were asked to compare the A310 with the aircraft they considered most representative of their experience. For each question a scale from 1 to 6 was presented, the 1 showing low appreciation, requesting improvement, the 6 showing high appreciation, recognizing high quality. This questionnaire covered the whole spectrum of aircraft design issues with regard to :

  . functional grouping of instruments/controls

  . general presentation of flight systems information

  . crew comfort

  . overhead panel layout

  . PFD information presentation, speed scale, mode annunciators, etc.

  . ND information presentation involving Arc mode, Map mode, Plan mode, Radar...

  . ECAM information presentation involving warning display, system display, etc.

  . FMS control and display unit, manipulation, pagination, etc.

  . AFS flight control unit, autothrottle operation, thrust rating panel, autopilot/flight director procedures, etc.

  . checklists and procedures

. outside appearance/mechanical ergonomy
. tasksharing and worksharing
. workload and error-inducing potential
. intelligibility of hardware/procedures/mental ergonomy.

More than 75% of all questions received very high marks in the appreciation scale-range of 4 or rather 5, pilots refraining from making extreme position judgements. Where some improvement was desirable for the future they felt free to rate in this direction.

Examples in this area concern seating, software, procedures, use of space on the overhead panel, FPV-information, attention-getting capability of some alarms on the PFD and on the ECAM, interpretation of error messages on the FMS, homogeneity of abnormal/emergency checklist with ECAM WD.

This provided a tremendous and often too-scarce opportunity for us to do some homework having learnt throughout the years that more contact was going to be necessary to listen to and understand the pilot community. A step in the right direction was prompted by the crew complement issue.

4. Cockpit management : errors and resources

From the observation of errors in our A310 crew complement certification campaign, it was concluded that their mere existence - most having been minor slips, blunders, mishaps and problems in resource management or insufficient familiarity with the aircraft - made possible to establish learning loops (18). Some of the recommended improvements concerned operational procedures and the training syllabus with particular emphasis on system knowledge and insight. Specific operations engineering bulletins were also issued for ECAM, EFIS, FMS and engine monitoring. Special care was devoted to understandable tasksharing in the cockpit smoke or fire procedure.

Analysing newly trained crew's errors revealed that most of these related to procedures/checklists and navigation, with autoflight and tasksharing taking a minor share. No direct relationship could be established between test scenario difficulty and the number or importance of errors. The succession of failures and problems embedded in the simulator and flight scenarios never brought crews to such workload levels that they could but commit errors, one of the underlying verifications in carrying out these tests. Which underlines also the tolerance of the system with regard to various blunders, mistakes, misperceptions and omissions.

Not unrelated to the fact that this area was considered too speculative until recently, a majority of airline accidents in recent years appears to be attributed to human factors communication problems. Most of these seem to be caused by some aspects of inadequate cockpit resource management due to a problem of information transfer or communication - be it man-man or man-machine, system-induced or not (24).

Caesar (15) devides these human failures into the following categories.

. Active failure : 35%
Non-adherence to rules and SOPs. Lack of discipline or vigilance inadequate flight management - short cuts,

. Passive failures : 30%
Crew misunderstanding and communication problems
Distraction - coordination breakdown - fatigue

. Judgement errors and decision-making : 30%
  Underestimation of actual operational conditions
  Bad landing - wrong alternative

. Crew incapacitation : 5%
  Subtle or obvious, requiring take-over.

In the aftermath of a landmark study on the interactions of workload with errors, vigilance and decisions, performed by the late Pat Ruffell Smith (25, 26), NASA concluded that these problems can be classified into the following categories :

. social and communication skills

. leadership and management skills

. planning, problem solving and decision-making skills

. human and machine characteristics.

Directing an automatic system rather than doing the job manually while coordinating with the other crewmember needs skills different from those required in conventional cockpits. And the changing nature of piloting, the increasing complexity of the environment (ATC congestion, regulations and procedures), the vast number of computer-based devices replacing manual control and mental calculation, jointly demanded the development of a new attitude towards automation.

Any feeling of "being along for the ride", ideas of underload, lack of vigilance, boredom and complacency, any sense of "being out of the loop" needs to be properly addressed by means of cockpit management techniques that help to foster the attitude that automation is just another resource to manage (27). While at NASA, John Lauber (28) emphasized that the major impact of the inclusion of microprocessor and display technology in new generation transport aircraft would be the quality and quantity of real-time information available to the crew and the way this information would be manipulated and used to conduct a flight. As a result of previous accidents with conventional technology aircraft due to shortcomings in crew communication and coordination, the NTSB recommanded that the FAA urge operators to indoctrinate flight crews in the principles of flight deck resource management.

It is, however, not the manufacturer's responsibility to ensure that the airlines maintain adequate cockpit resource management standards. Rather, the manufacturer has to offer high quality flight decks and associated procedures that make good communications possible.
In other words the manufacturer is committed to deliver a product that should be, so to speak, immune to information transfer problems (24).

5. Error tolerant systems : monitoring performance, monitoring the operator

Pointing beyond the sterile cockpit, Wiener insists (13) that the long-standing recommendation that humans serve as monitors of automatic devices must be reconsidered. In his view, shared by others, the pilot should be brought back into a more active role in the control loop (29), aided by decision support systems, especially in view of very long range operations. At one airline, crews on these flights may be required to perform routine in-flight tasks and calculations to sustain alertness, with flight management system backup (30). The vast amount of literature on vigilance has indeed shown that the human, after all, is not a very effective monitor, is less likely to detect system fault or wrong set-up, and is more likely to commit large blunders. For this reason, the famous controller versus monitor doctrine - placing the pilot rather as a monitior than as a controller - is being re-examined.

Potentially weak vigilance of the pilot and vulnerable aircraft systems suggested a growing need for the machine to monitor itself and the operator, diagnosing errors performed by the human. The need to develop automatic systems that monitor human operator performance rather than the reverse has also been suggested in the aerospace medical world (31).

Not that familar, Wiener takes a cautious but nevertheless interested view with regard to potential promises in information technologies, decision support systems, expert systems or artificial intelligence. Pelegrin, on the other hand (32), advocates increased automation, using these technologies to include automatic reconfiguration after failure, ongoing maintenance and systems status backdating and updating, pilot assistance in abnormal situations not necessarily resulting from failures and crew error tracking and interrogation systems.

The error-protection capabilities of map displays and fly-by-wire flight control discussed in the following paragraph have somewhat paradoxically been accompanied by a renewed interest in error-tolerant systems as safety nets. Trying to automate error away may merely displace the problem of pilot error, since the human will still prove to be the ultimate cross-check.

Models of human operator behaviour are needed for incorporation of such error-tolerant capability into the automation device itself. Aware of the need for better knowledge of the man-machine relationship, Airbus Industrie has, in cooperation with Dunlop & Associates and Cochin Faculty, developed and validated a predictive crew workload rating mode (33) based on heart rate variability, aircraft performance parameters and situational information (scenario, flight phase conditions etc.) as shown in Figure 2. Further work will, however, be necessary to model crew vigilance, to investigate how automation, workload, vigilance, performance and errors all relate to each other.

Seifert and Brauser have proposed two strategies for future flight deck design (34) equating the pilot's capabilities and needs :

. provision of sufficient feedback to allow the pilot to detect and correct unintentional performance errors before they affect system performance,
. introduction of fail-safe and fail-operational ergonomic design by means of human error detection and correction functions into the man-machine interface intelligence.

Extensive analysis on human errors including preflight and inflight data, physiological and psychological parameters, opinion ratings, environmental factors, mission requirements, organizational and technical factors, proficiency factors, pilot errors and their consequences (35) has revealed that human errors, systematic or random, can be interpreted as a four-dimensional structure : vigilance errors, perception errors, information processing errors and sensory errors. In particular, it has been found that determinants and background variables of human factor incidents and accidents provide a model of basic man-machine system interaction patterns which is useful and predictive of human error occurrence.

The new goaal for cockpit design proposed by Seifert and Brauser may be posed : to provide mutual monitoring of man and technical system.

6. Airbus technology : evolution in design

The pilot's role evolution is inevitably linked to that of the aircraft and its cockpit triggered, as it has been, by economic and technological imperatives.

Whether in technical areas or operational functions, continuity as well as progress elements have been pervasive throughout the A300/A310/A320 family (Table 3).

The forward-facing crew cockpit of the A310 and A300-600 stems from a cascade evolutionary process. Its design proceeds from A300 experience which itself reflected experience with the Caravelle, Trident, BAC I-II and Concorde. It is also the offspring of our customers, traditionally kept involved from the start of the design process through so-called task forces (36, 37). The FFCC concept itself is a direct result of multiple discipline research programmes that included a human factors orientation such as the NASA Terminal Configuration Vehicle Program, the French PERSEPOLIS project and the BAe Advanced Cockpit Design.

Early consideration has been given to making the original A300 a two-pilot crew aircraft. The technology then available did not permit such an approach with the necessary degree of confidence.

The overall design goal being to ensure adherence to the need-to-know and fail-safe principles which the 1970's A300 cockpit did not fully satisfy :

. side panel not fulfilling fail-safe criteria
. number of dials and displays giving unnecessary information during a large part of the flight
. necessity to scan in different places.

The innovations of this first twin wide-body proved a major credit to Airbus :

. automatic throttle from brakes-off to touchdown
. automatic angle of attack protection and speed reference system against windshear incidents
. automatic two and one-engine go-around
. thrust computer linked to autothrottle allowing derated engine operation/increased engine life.

The necessary evolution became possible in the 1980s with the help of technology leaps providing concentration, standardisation and flexibility (38). Design goals of the FFCC were formulated as follows :

. better crew integration and cross-monitoring.
. better man-machine interface
. lower risk of flight/system management error
. automation of routine actions
. simplification of tactic actions in normal, abnormal and emergency procedures
. augmented availability for strategic flight actions
. clear presentation of predigested information for easier and quicker diagnosis and correction
. reduction of crew workload.

These goals were, as we know, effectively reached by :

. pushbuttons combining control, visual feedback and visual malfunction alert

. systems controls all on an overhead-mounted panel improving visibility, accessibility, use of space for all,

. combining previously separated systems and subsystem panels into synoptic presentations (source to distribution), according to frequency/urgency of use (normal, abnormal, emergency)

. associating all the above with a rigorous dark cockpit philosophy to simplify procedures and understanding of system operations

. introducing basic, transparent system automation to avoid repetitive actions in normal operation or after a single failure (electrical, APU, fuel, pressurization, avionics cooling, air bleed etc.)

. introducing digital flight management (AFS) with preselected functions and reliability enabling less crew intervention, clear and unambigous operation, improved flight envelope protection (FAC, FLC) and automatic landing potential to CAT III.

First to embark on this new lead was the digital A300FF, first wide-body certificated for operation by two pilots embodying all the new goodies except display units for flight and systems monitoring offered on the A310 and A300-600 :

. replacing electromechanical ADIs for short-term flight path control with electronic PFDs allowing a more efficient scan and direct access to predigested information (improved speed scale with critical ranges, margins, trend and speed error, flight path information with path angle and vector, mode annunciation and autoland indications), illustrated in Figure 3.

. replacing electromechanical HSIs for medium-term navigation with electronic NDs allowing a more efficient scan of weather, navigation progress versus constraints, beacons and waypoints monitored by the FMS with several display options available (Rose, Map, Arc and Plan).

. introducing Flight Management Systems with the purpose of :

  - full area navigation and guidance
  - full vertical navigation and guidance with profile optimisation for maximum economy
  - instant prediction and flight planning facility,

. introducing Electronic Centralized Aircraft Monitoring (ECAM) eliminating the need for frequent system scanning in normal operation with :

  - temporarily used systems/functions monitoring
  - routine system monitoring
  - system parameter trend monitoring,

and simplifying information processing in abnormal/emergency operation through :

  - improved failure analysis
  - orientation of corrective actions
  - minimal need of paper checklists
  - improved understanding of post-failure configuration.

The Flight Warning Computer whose logics have made it possible to provide ECAM operational features is a real breakthrough achieving in addition :

- the reduction of discrete sounds to a bare minimum, basic attention-getters alerting the crew of warning display messages
- the adaptation of the warning to flight phase with inhibitions where safety mandates (Figure 4).

We are convinced that, with selective and tailored information according to the situation, the major contribution is towards protecting the crew against mishaps, blunders or more serious errors. What has been achieved with the A310/A300-600's cockpit is also in the vicinity of Wiener's philosophy on automation (flight management by exception, forecasting and trend possibility, goal sharing of strategic intent) and Wanner's philosophy of human attributes and pilotability incidents.

In the step that was to be taken in the mid-1980s, design aims have essentially been extended to cover more error-tolerance or protection against pilotability, pertubation-sensitivity and manoeuvrability incidents by safeguarding functioning points and envelopes. In this respect a major emphasis was put on windshear (39) as evidenced by Table 4. Continuing developments in digital technology and data transmission have also led to greater possibilities of systems integration. Former breakthroughs with FFCC technology have been refined in several areas but essentially kept as mainstays for the future.

The most prominent innovation on the A320 being to introduce fly-by-wire (FBW) as primary means of flight control in conjunction with sidesticks, (40, 41). A natural decision since it is available on Concorde with an analogue electrical control system in all axes, mechanically backed up, and to a more limited extent on the A310 with digital signalling of the roll spoilers and aileron with no back-up.

For the A320 to become fly-by-wire with minimal mechanical back-up is furthermore based on considerable partner companies' experience with military aircraft. But Airbus Industrie's decision on this hinged upon considerable flight testing work with its R & D testbed. From September 1983 onwards this aircraft went through an evaluation program involving 48 pilots from airworthiness authorities and airlines. It was equipped with a sidestick controller, from which output signals were fed through a black box into the existing AFCS to give primary FBW-control in roll and pitch. The unanimous acceptance of this configuration was quantitatively confirmed by a small test program involving manually flown experimental circuits that posed a variety of operational problems. All performance measurements of smoothness and stability favoured the FBW-system compared to a conventional system by at least 20%, pilot control inputs being reduced by about 50% (Figure 5) to allow better performance and more availability for general cockpit management (42).

Originally derived from NASA studies on self-correcting flight control concepts, a C* law was chosen as the fundamental mode of control in pitch, with roll control based on roll rate. C* provides short-term direct flight path control by modulating the load factor using a blend of pitch rate and load factor at low speed and pitch speed without autotrim below 200 ft normal flare.

Its characteristics feature is to maintain the aircraft at 1g as long as the stick is neutral which permits a stable trajectory through turbulence without any need for the pilot to interfere. Likewise, this concept automatically eliminates pitch oscillations when performing angle of attack changes, and eliminates the need for pitch trim with thrust or configuration changes.

Safety features and procedures offered by the A320's FBW/sidestick combination with regard to Wanner's incident scheme are many (40,41) :

- . envelope protection - overspeed with marked stability beyond high speed warning
- . stall protection so that
  - : below 1.13 Vs the control law changes to an angle of attack control law,
  - : below 1.1 Vs the classic alpha floor (A300, A310) commands full thrust
  - : with maximum angle of attack at full stick at 1.06 Vs resulting in the spectular flight path demonstrations shown by the A300 testbed
- . windshear protection by applying full stick with no fear of stalling (Table 4) resulting in a safe escape, and
- . manoeuver protection limiting the manoeuvre demand systems to 2.5 g to protect against breaking up at high h in the case of collision avoidance
- . attitude protection with automatic recovery in pitch and roll if disturbances take the aircraft beyond its limits
- . engine failure compensation after take-off the aircraft rolling to a small (7 degree) bank and taking a mild (1 degree/sec) turn towards the failed engine, hands-off
- . electronic sidestick coupling backed by warning lights allowing a training captain to modify a student's flight path without any risk of resistance and helping take-over situations to be handled diligently in case of neighbour pilot's incapacitation or derangement.

Another novelty on the A320 concerning protection of functioning envelope is the Full Authority Digital Engine Control (FADEC) which is deemed to succeed electronic engine control introduced on the A310/A300-600 models.

Offering full engine control and protection under both stabilized and transient conditions of operation, the FADEC provides a wide array of operational functions, some of which were still handled by pilots until recently :

- . gas generation control
- . engine limit and overshoot protection
- . power management and derating
- . engine starting (automatic)
- . thrust reverser control
- . fuel return valve control
- . engine condition monitoring.

Similar to sidesticks complementing the FBW-controls, FADEC is associated with a pair of fixed-gate throttle levers. Thrust control through FADEC manages to provide the best power suited to each flight phase either in manual, according to throttle position, or in automatic, associated with throttle detent and in response to ATS/FMGS demands. Engagement of ATS modes (SPD, MACH, THRUST, RETARD) is automatic according to AP/FD mode engagement, the FADEC receiving AFS commands via an ARINC transmission bus.

On the A320, the AP/FD and the ATS are fully merged with the FMS into the FMGS, an integration enabled since flight control laws are taken over by ELAC/SEC computers leaving the elaboration of horizontal and vertical guidance to a fully integrated FMGS. Short-term flight control inputs to the FMGS are made through the FCU on the glareshield, longer-term inputs are introduced via the MCDU to select upper navigation and performance modes and advisory functions.

The systems which the FMGS dialogues with (Figure 6), FAC, FADEC, EIS, ELAC/SEC, ADIRS, FCDC etc. - are themselves interfaces with other calculators EIU, EFIS, ECAM etc., branching out towards almost all other aircraft systems. A good knowledge of these systems implies firts of all the understanding of functions and interactions.

Another newcomer to Airbus cockpits, the Radio Management Panel will faultlessly manage the frequency selection for all VH/HF and transponder and radio communications as well as ILS/VOR and DME radio navigation aids in case of FMGS failure.

Apart from the innovations just cited, the experienced observer will note :

. the uncluttered layout (from 35 instruments on the A310 to 12 on the A320)
. the use of six interchangeable display units for EFIS/ECAM, systems whose architecture notes a further trend towards integration (EIS)
. the provision of altitude/vertical speed on the primary flight display, more refined Nav, Rose and Map modes on the navigation display, primary engine pictorials with checklist items on the ECAM upper display, the ECAM lower display presenting refined system or status pages
. the further reduction of pushbutton and overhead system panels through :
  - a radical integration of avionics (from 74 calculators on the A310 to 32 on the A320) and a logical rearrangement of the center row dedicated to engine-related systems and functions (fire controls, hydraulics, fuel, electrical and air conditioning)
  - deletion of several functions switches or levers (no manual temperature and flow control, no fuel manifold valve control etc.),

. the further reduction of system management tasks after single failures presently of three types :
  - no action, disconnection performed automatically (examples : air conditioning temperature overheat, pressurization control)
  - confirmation action to cover the case of automatic disconnection failure (examples : ELAC/ SEC flight control computers, electrical generators)
  - mandatory actions not covered by automatic disconnection (examples : fuel and hydraulic pumps),
. the subsequent simplification of most abnormal procedures mainly consisting of confirmation checks with some confirmation or bakc-up actions but less than on former models.

7. Conclusion

Some years ago, it was anticipated that the 1990's era cockpit (10, 24) can be imagined to be an all-digital, all-electronic flight deck with high-resolution multifunction displays and man-machine interactive systems that will monitor flight operations.

It was anticipated that crews would only be there for aggregate information treatment and strategic decision-making, concentrating solely on mission supervision and management, leaving interactive electronic systems with routine data processing, memory work, control tasks, procedural decision - making and a quasi-human aptitude to monitor, detect and plan.

The high-tech command post of the future with table tops, joysticks and pushbutton controls at fingertips has virtually arrived, yet the A320 pilot will fly as the horseman drove his coach in the not-too-distant past.

Performance and initiative being what he is best at, it is up to man to decide whether he or the machine are to tolerate errors. The best anticipation strategy being to avoid, detect or correct by design. The practical orientation taken by Airbus Industrie in this matter conveys the promise of even smarter intelligence to be in store for the future.

NOMENCLATURE

AP/FD	:	Autopilot/Flight Director
AFS	:	Automatic Flight System
AFCS	:	Automatic Flight Control System
ATS	:	Automatic Throttle System
ECAM	:	Electronic Centralized Aircraft Monitor
EIU	:	Engine Interface Unit
EFIS	:	Electronic Flight Instruments
ELAC	:	Elevator and Aileron Computer
EIS	:	Electronic Instruments
FAC	:	Flight Augmentation Computer
FADEC	:	Full-Authority Digital Engine Control
FBW	:	Fly-By-Wire
FCC	:	Flight Control Computer
FCU	:	Flight Control Unit
FFCC	:	Forward-Facing Crew Concept
FLC	:	Flight Limitation Computer
FMGS	:	Flight Management and Guidance System
FMS	:	Flight Management System
FPV	:	Flight Path Vector
FWC	:	Flight Warning Computer
INS	:	Inertial Navigation System
IRS	:	Inertial Reference System
MCDU	:	Multipurpose Control and Display Unit
ND	:	Navigation Display
PFD	:	Primary Flight Display
RMP	:	Radio Management Panel
SEC	:	Spoiler and Elevator Computer
SPD	:	Speed mode of AFS-ATS

References

(1) Blue Book, EUROPILOTE and ALPA, 1980 p.10

(2) E.L. WIENER, University of Miami,
Controlled Flight into Terrain Accidents :
System-Induced Errors, Human Factors, 1977, 19(2) 171-181

(3) E.L. WIENER
Midair Collisions : The Accidents, The Systems, and the Real Politik,
Human Factors, 1980, 22(5)
521-533

(4) E.L. WIENER, R.E. CURRY
Flight-deck Automation : Promises and Problems,
Human Factors, 1980, 23(10), 995-1011

(5) D. ENGEN, FAA
Aircraft Accidents are Caused. A brief Human Factors Analogy related to recent accidents.
SAE paper 831412 Oct.1983
Second Aerospace Behavioral Engineering Technology Conference

(6) J.H. STOKLOSA, NASA Ames
Accident Investigation of Human Performance Factors,
Second Symposium on Aviation Psychology,
Columbus, Ohio State University, Proceedings 1983, 429-436

(7) R.L. SCHLEEDE, NTSB
Identifying Design-Induced Human Factors,
Human Factors in Managing Aviation Safety 37th Annual
International Air Safety Seminar, Oct.84,
Zurich, Switzerland

(8) R.B. O'DONNELL, USAF
Contributions of Psychophysiological Techniques to Aircraft
Design and other Operational Problems,
AGARD - AG-244, July 1979

(9) M.R. HOAGLAND, ALPA
Winging it in the 1980s :
Why Guidelines are needed in Cockpit Automation,
SAE paper 841634,
October 1982

(10) J.J. SPEYER, A. FORT, Airbus Industrie
Human Factors Approach in Certification Flight Test,
SAE paper 821340, October 1982

(11) E.L. WIENER
Cockpit Automation : in Need of a Philosophy,
SAE paper 851956, October 1985

(12) J.C. WANNER
Facteur Humain et Sécurité
Académie Nationale de l'Air et de l'Espace
Colloque International sur la Sécurité Aérienne, November 1984

(13) E.L. WIENER
Beyond the Sterile Cockpit,
Human Factors, 1985, 27(1), 75-90

(14) J.R. GANNETT
The Pilot and the Flight Management System,
Symposium on Behavioral Objectives in Aviation Automated System,
SAE, October 1982

(15) Capt H. CAESAR, Lufthansa
Long-Term Statistics and its Impact on Safety Management
and Operational Procedures.
39th Annual International Air Safety Seminar, October 1986, Vancouver

(16) E.L. WIENER
Human Factors of Cockpit Automation : A field study of flight
crew transition,
NASA Contractor Report No.177333, 1985, Moffett Field, CA.

(17) R.E. CURRY, NASA
The Introduction of New Cockpit Technology : A Human Factors Study,
NASA Tech. Memo No.86659, 1985, Moffett Field, CA.

(18) J.J. SPEYER, A. FORT
A310 Two-man Crew Workload Analysis and Questionnaires,
Airbus Industrie, Flight Division, AI/V-F 383/83

(19) B.D. NORDWALL
Electronic Cockpits are Making Conventional Cockpits Obsolete,
Aviation Week and Space Technology, November 3, 1986

(20) F. MACHADO
The destruction of KAL 007 : How did it happen ?
International Journal of Air Safety, 1984(2), 5-14

(21) G. BRUGGINK
Reflections on the Potomac,
International Journal of Aviation Safety, 1983(1), 5-12

(22) Report on accident of MH 684 at Kuala Lumpur,
Report on accident of SAS 901 at JFK, New York,
International Journal of Aviation Safety 1985(3,3), 218-222

(23) Aircraft Accident Report,
China Airlines Boeing 747-SP, N4522V, 300Nm Northwest of San Francisco,
CA, 19 February 1985 (NTSB-AAR-86/03)

(24) J.J. SPEYER, A. FORT
Communications : Major Human Factors in Cockpit Design,
SAE paper 831530, October 1983

(25) H.P. RUFFELL SMITH, NASA
A simulator study of the Interactions of Pilot Workload with Errors,
Vigilance and Decisions,
NASA Technical Memorandum 78482, January 1979

(26) J.K. LAUBER, NASA
Resource Management on the Flight Deck,
Proceedings of a NASA/Industry Workshop held at San Francisco, CA
Report NASA CP-2120, March 1980

(27) J.J. SPEYER, A. FORT
The Inside Track in Resource Management,
Second Aerospace Behavioral Engineering Technology Conference,
SAE paper 831530 bis, October 1983

(28) J.K. LAUBER
Aviation Technology and the Flight Crew,
International Aeronautical Symposium, Tokyo, Japan, August 1982

(29) E.L. WIENER
Fallible Humans and Vulnerable Systems : Lessons learned from Aviation,
NASO, Workshop on Failure Analysis of Information Systems, Bad Windsheim,
Germany, August 1986

(30) Aviation Week and Space Technology, December 1st, 1986
Airline Observer

(31) R. MOLLARD, G. IGNAZI, Université René Descartes
Automatisation et Performance : Dessins et Perspectives,
Proceedings of CEC - workshop on Effects of Automation on Operator
Performance, Laboratoire d'Anthropologie Appliquée, Université
René Decartes, October 1986

(32) M. PELEGRIN, CERT-ONERA
Difficulties and Risks of the Semi-Automatic Control (namely piloting modern aircraft),
Proceedings of CEC-Workshop on Effects of Automation on Operator Performance,
Laboratoire d'Anthropologie Appliquée,
Université René Descartes, October 1986

(33) J.J. SPEYER, A. FORT, Airbus Industrie, R.D. BLOMBERG, Dunlap and Associates,
Dr J.P. FOUILLOT, Cochin Faculty,
Assessing Workload for Minimum Crew Certification
in AGARD AG-282, The Practical Assessment of Pilot Workload,
Dr A.ROSCOE June 1987

(34) R. SEIFERT & K. BRAUSER, MBB
New Flight Deck Design in the Light of the Operational Capabilities,
AGARD, Conference Proceedings No.347, May 1983

(35) K. GERBERT & R. KEMMLER, German Air Force Institute of Aerospace Medecine,
The Causes of Causes : determinants and background variables of human
factor incidents and accidents,
ERGONOMICS, 1986 Vol 25, No.11, 1439-1453.

(36) Capt B. SCHMITT, Swissair
Using Operational Experience in designing effective flight decks,
Flight Safety Foundation,
37th Annual International Air Safety Seminar, Zurich, Switzerland,
October 1984

(37) Capt H. CAESAR, Lufthansa
Design Philosophies of New-Technology Aircraft and Consequences for its Users,
Flight Safety Foundation,
38th Annual International Air Safety Seminar, November 1985,
Boston USA

(38) P. BAUD & D. IVANOFF, Airbus Industrie
Cockpit Display of A310,
International Aeronautical Symposium, Tokyo, Japan, August 1982

(39) A. FORT
Windshear,
The Airbus Industrie Approach,
39th Annual International Air Safety Seminar,
October 1986, Vancouver, Canada

(40) B. ZIEGLER, Airbus Industrie
M. DURANDEAU, Aerospatiale
Flight Control System on Modern Civil Aircraft,
14th ICAS Congress, Toulouse, France, September 9th-14th, 1984

(41) S.G. CORPS, Airbus Industrie
Airbus A320 Side Stick and Fly-By-Wire - An update.
SAE paper 861801, October 1986

(42) R.D. BLOMBERG, Dunlap & Associates
J.J. SPEYER, Airbus Industrie
Performance Analysis of Fly-By-Wire
Airbus Industrie Paper, 1984

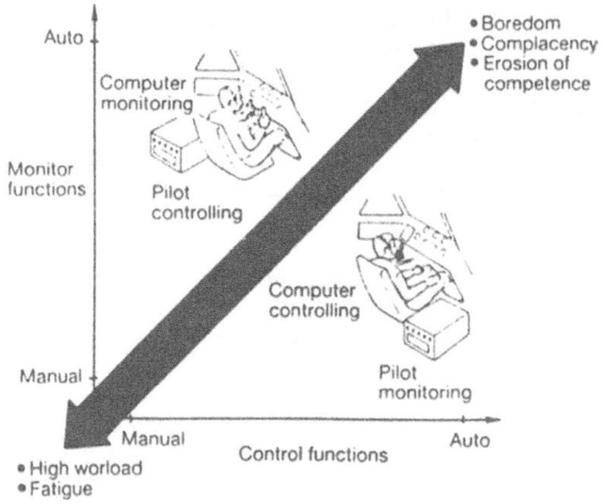

Fig 1
Monnitoring Versus Controlling

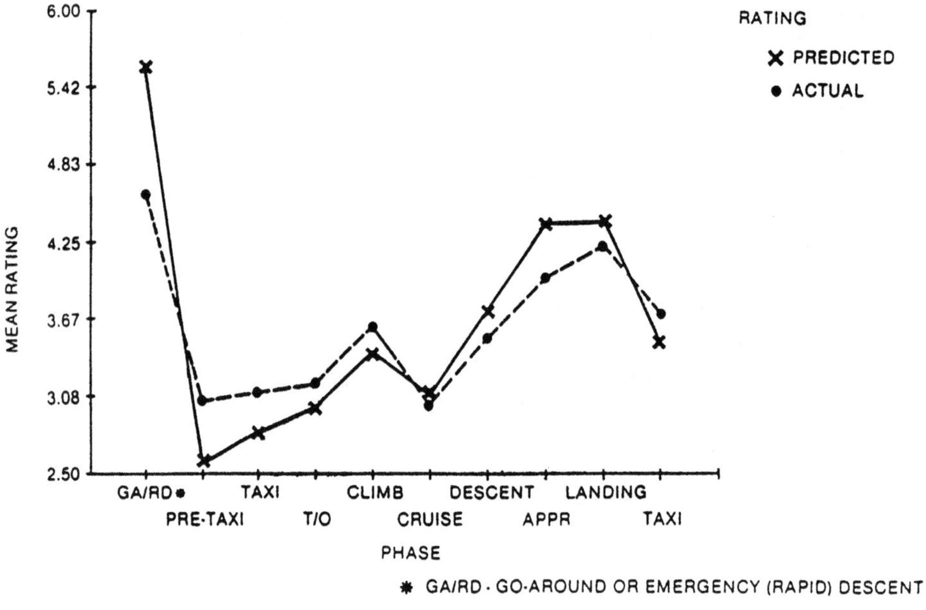

Fig 2
Actual VS. predicted by phase
(Validation flights)

PFD SPD scale schematic

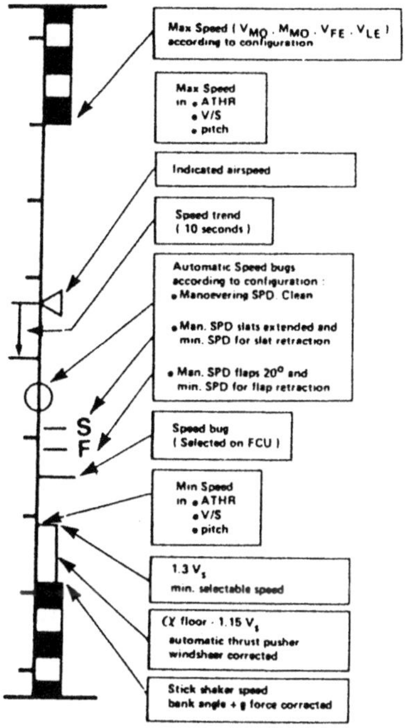

Fig 3

A300 - 600 / A310 Electronic
Flight Instruments (EFIS)

- ensures proper warnings only when necessary.
- avoids warning without interest for a flight phase ; an inhibition logic filters the presentation of warnings according to the flight phase.
- allows the crew to cancel warning presentation when the corrective actions have been done (CLEAR functions).

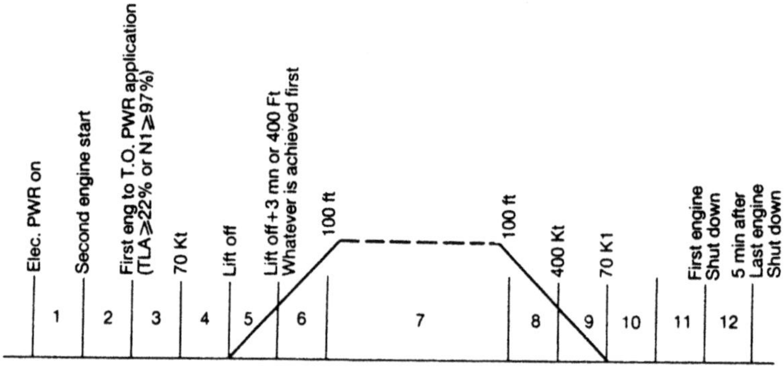

Fig 4

Electronic Centralized Aircraft Monitor (ECAM) warning inhibition

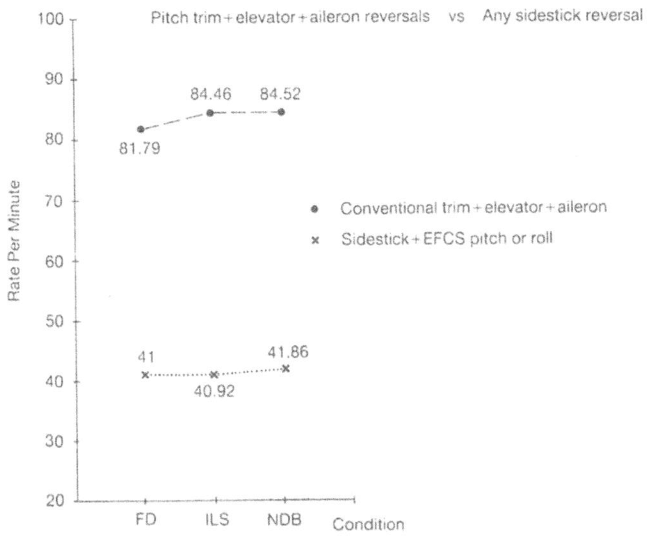

Airfield pattern

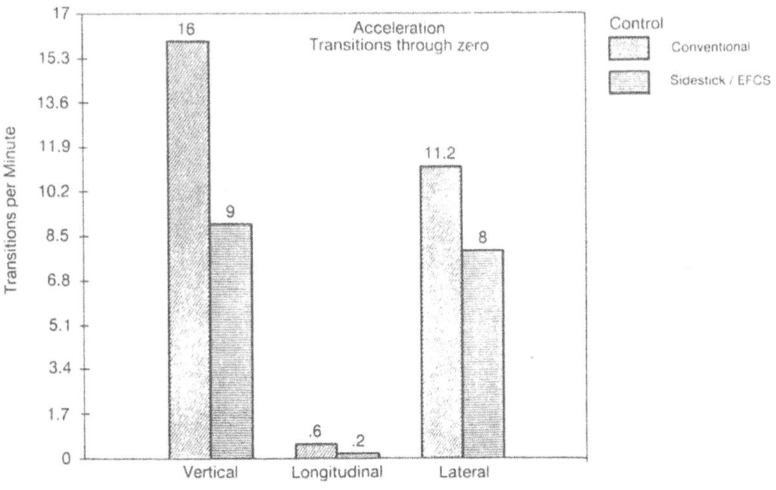

Acceleration type

Fig 5
A300 flight experiment - Airfield pattern and acceleration type

4 MODES
SPD
MACH
THRUST
RETARD

Engagment of ATS mode is automatic according to AP / FD mode engagement*

AP / FD MODE	ATS MODE
V / S-FPA ALT(ACQ / HOLD)	SPD / Mach
Expedite	Thrust / Retard
Descent / Climb	Retard Thrust / SPD / Mach
SPD / Mach	Thrust / Retard
Approach final descent glide flare	SPD SPD Retard
TO / GA	ATS Armed

(FADEC control)

*If no AP / FD is engaged ATS can be used in SPD / Mach mode
*In case of excessive angle of attrack configuration ATS Alpha floor mode is engaged : The GA limit thrust is acquired and held.

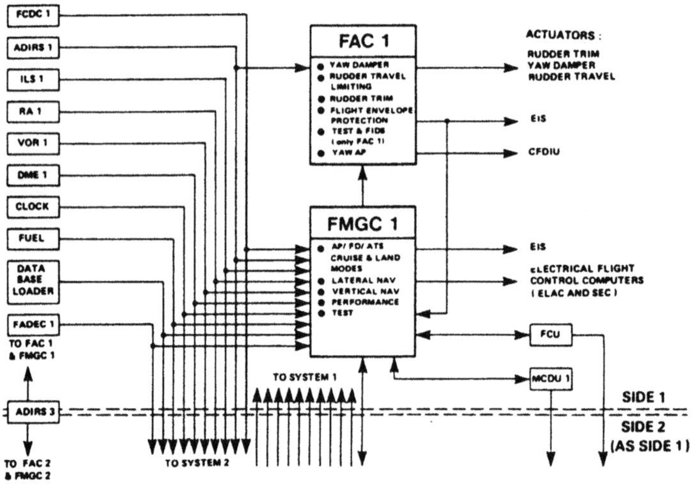

Fig 6

On A320
- Automatic Flight System (AFS) and Automatic Throttle System (ATS) modes.
- Automatic Flight System (AFS) Block diagram.

Control Tasks

1) System understandability to facilitate operation and malfunction diagnosis.
2) Automatic system to perform functions in the way the operator wants them to be done.
3) Task demands to remain at reasonable levels to ensure available time for monitoring.
4) Training so as to use automation as an additional resource.
5) Operations to allow more or less automation according to operator style, desires and needs.
6) Overall system insensitivity to different options and styles of operation.
7) Independent set-up and error-checking equipment/procedures to be available when appropriate.
8) Extensive training to ensure proper operation, set-up, anomaly detection, diagnosis and treatment.

Monitoring Tasks

9) Operator training, motivation and evaluation of monitoring activity.
10) Operator involvement should be managed to resist distraction by providing meaningful primary tasks.
11) Alarm rates to be kept within acceptable limits.
12) Alarms to indicate conditions responsible for their display.
13) Provision of ways and means to diagnose the automatic system and warning system operation and alarm validities when automatic response time is not critical.
14) Degree of emergency to be indicated by level of urgency of alarm.
15) Understandability of alerts and combinations of alerts through training techniques and hardware.

Table 1
Wiener and Curry recommendations on automation

1) Useful information is necessarily pertaining to the functioning point and its position relative to corresponding limits.
2) Information difficult to interpret or implying the execution of complex mental processes must be avoided.
3) All useful information needs to be provided, excessive or useless information must be avoided, redundant information must only be used to perform reasonableness checks.
4) Information allowing forecasting/extrapolation of functioning point movements needs to be provided, particularly with regard to longer-term process control and strategic anticipation.
5) Vigilance retrieving and meaningful state information must be provided even with stable/routine control and management conditions.
6) Visible, audible and understandable alarms need to be provided so as not to remain unattended by the operator ; their number should be limited to those strictly necessary using logical hierarchisation, caution and warning of limit exceedances and proper selection of threshold limits.
7) The number of actions should be limited but close-looped for feedback since open-loop tasks are innately if not exceedingly difficult to monitor ; actions should always prepare tactics and strategies depending on the understanding of system state and evolution.
8) The use and usefulness of levers must be addressed carefully.

Table 2
Wanner's pilotability rules

AIRBUS TECHNOLOGY : EVOLUTION IN DESIGN

	A 300 FF	A 310	A 320
* PUSHBUTTON TECHNOLOGY AND ADVANCED SYNOPTICS OVERHEAD PANEL LAYOUT	X	X	X
* ADVANCED SYSTEMS AUTOMATION AND DARK COCKPIT PHILOSOPHY	X	X	X
* DIGITAL AUTOMATIC FLIGHT GUIDANCE SYSTEM (AFS)	X	X	X
* ELECTRONIC CENTRALIZED AIRCRAFT MONITOR (ECAM)		X	X
* ELECTRONIC FLIGHT INSTRUMENTS (EFIS)		X	X
* FLIGHT MANAGEMENT SYSTEM (FMS)		X	X
* FLY-BY-WIRE FLIGHT CONTROLS AND SIDE-STICKS (FBW)			X
* FULL-AUTHORITY DIGITAL ENGINE CONTROL (FADEC)			X

COCKPIT EVOLUTION

	A 300	A 310 AND A 300-600	A 320
GENERAL ARCHITECTURE	CONVENTIONAL	FFCC/OPTIMISED SYNOPTICS	DESIGN/INTEGRATED UNITS
FLIGHT AND NAVIGATION	• STYLISED ELECTRO-MECHANICAL	• CATHODE RAY TUBES EFIS (PFD AND ND) • FMS/IRS ASSOCIATION	• IMPROVED DISPLAY UNITS (ALT. VV....)
SYSTEM INFO PRESENTATION	• SYSTEM PANELS AND ASSOCIATED SYSTEMS	• CATHODE RAY TUBES ECAM (WD AND SD) • PUSHBUTTONS	• IMPROVED DISPLAY UNITS (ENG. IND....)
FLIGHT MANAGEMENT CONTROL	• OPTIMISED AP/FD • COUPLED AP/INS	• PILOT INTERFACE {TCC, FCC} • INTEGRATED MODES {AP/FD, ATS} • COUPLED AP/FMS	• FBW-SIDESTICK • FMGS INTEGRATING AFS, ATS & FMS • SIMPLIFIED THRUST CONTROL : FADEC
SYSTEMS ARCHITECTURE	• INCREASED REDUNDANCY • EASE OF OPERATION	• DIGITAL TECHNOLOGY • ADAPTED AUTOMATION	• SYSTEMS AND FUNCTIONS INTEGRATION • INCREASED AUTOMATION

Table 3

Airbus evolution in design and in cockpit

Airbus Industrie efforts against windshear since 1972

	A300	A310 & A300-600	A320
Angle of attack protection	×	×	×
Speed reference system	×	×	×
Beam deviation on ILS	×	×	×
Stick shaker speed and speed Trend information		×	×
Wind speed and direction info		×	×
Flight path vector		×	×
Visual warning		×	×
Stall protection	Stick shaker	Stick shaker	Full protection against stall
High Angle of Attack Control			×

AIRBUS EVOLUTION : WINDSHEAR PROTECTION

	DETECTION ALERT	GUIDANCE	PROCEDURE
A 300	RAW DATA BEAM DEVIATION AUTOTHROTTLE ACTIVATION	SPEED REFERENCE SYSTEM (PITCH BAR)	MAXIMUM THRUST STAY ABOVE STICKS-HACKER SPEED MONITOR GROUND SPEED
A 310	SAME PLUS : SPEED TREND SPEED MARGIN FLIGHT PATH VCTR WARNING IN PFD	SAME	SAME PLUS : VISUAL SPEED CUES FLIGHT PATH VECTOR MONITORING
A 320	SAME	SAME NOTE : PITCH LAW WILL TRY TO MAINTAIN FLIGHT PATH	FULL STICK AFT.

Table 4
Airbus windshear protection

TARGET AIMING PROPERTY OF RIFLE FIRING

Kazuhiko YAMASAKI and Kiyota KOREISHI
T.R.D.I. Japan Defense Agency
2-2-1 Nakameguro Meguro-Ku Tokyo-To
153 JAPAN

ABSTRACT. Effects of firing posture, body cooling, wind and work load on aiming of rifle were studied using live and simulative firing. The scattering values of aiming were affected by postures. It became larger in the order of prone with bipod, prone without bipod, kneeling and standing. In the cooling condition the shell temperature decreased about 15 degrees centigrade, the effect was not significant. When the wind velocity was 8.8 m/sec in a wind tunnel, the scattering values became larger significantly. Effects of work load equivalent to about 65 percent of maximum oxygen intake were smaller than the wind effects.

1. INTRODUCTION

A soldier has to perform a vigilance task in the field usually,furthermore he should fire a rifle and hit a target as occasion demands. In the field firing a soldier would suffer many disturbance such as shivering, wind, rain, etc.

The purpose of the present study is to examine the effects of several disturbance on rifle firing. So far as the authors know there are few reports similar to the present study except the studies on the firing posture (e.g. Dees et al., 1971; Giordano, 1979).

The TAF(target aiming function) test is one of the performance test in order to evaluate the function of concentration maintenance. It has been utilized for the evaluation of factory workers' fatigue or vocational aptitude (Takakuwa, 1968; Kumashiro and Saito, 1978). In the TAF test a subject aim a small still light, and exert all possible efforts to minimize the motion of a line of sight.

In a rifle firing, on the other hand, the timing of trigger is more important than minimizing the motion of a line of sight. However, assuming that to maintain a rifle still was difficult under some disturbance, the scattering of bullets would become larger.

A. Coblentz (ed.), Vigilance and Performance in Automatized Systems/ Vigilance et Performance de l'Homme dans les Systèmes Automatisés, 167–174.

2.METHOD

2.1. Classification of Experiment

Present study divided into two experiments (table 1). In the experiment I, subjects performed live firing in the field and simulative firing using a mock-up rifle in a hutment. In the experiment II, simulative firing was tried in the wind or just after the work.

Table 1. Classification of experiment

	Code	Firing	Factor	Posture
EXP.I	1S	live	field	standing
	1K			kneeling
	1P			prone without bipod
	1Pb			prone with bipod
	2S	simulation	control	standing
	2K			kneeling
	2P			prone without bipod
	3S		cooling	standing
	3K			kneeling
	3P			prone without bipod
EXP.II	1S	simulation	no wind	standing
	1K			kneeling
	2S		slow wind, frontal	standing
	2K			kneeling
	3S		dorsal	standing
	3K			kneeling
	4S		fast wind, frontal	standing
	4K			kneeling
	5S		dorsal	standing
	5K			kneeling
	6S		slow walking	standing
	6K			kneeling
	7S		fast walking	standing
	7K			kneeling
	8S		running	standing
	8K			kneeling

2.2. Subjects

All subjects were healthy adult males (table 2). One subject in the experiment I was a civilian and others were soldiers. One civilian and six soldiers in the experiment II were labolaroty personnel. All subjects had rare chance of live firing.

Table 2. Physical characteristics of subjects.

		Age(yrs)	Ht(cm)	Wt(kg)
Exp. I	mean	30.3	168.7	69.7
n=6	s.d.	6.2	4.9	12.6
Exp. II	mean	34.2	165.5	62.3
n=6	s.d.	1.5	3.4	5.4

2.3. Live and Simulative Firing

2.3.1.Firing. All subjects used one real or one mock-up rifle prohibited the foresight and backsight adjustment. Five times of firing were tried at 3-4 second intervals in each condition. In the control phase (2S,2K,2P), three times of test sets were achieved, accordingly fifteen times of firing data were obtained. The target distance was 100 m in the live firing.

2.3.2. Simulation appalatus. A mock-up rifle had the same shape and weight as a rifle for live firing. A small CCD camera (Toshiba, TC-1100) and a small light were attached to the tip of muzzle. The target located close to the muzzle had a flexible motion in the direction of barrel .

When the trigger was pulled a small light glimmered. The moving of muzzle and glimmering were recorded by VTR (Sony, EVA1). The coordinates of positions of muzzle were obtained by slow playback of VTR (Sony, GCS50). The accuracy of measurment was 0.1 mm .

2.3.3. Firing posture. Three postures (standing, kneeling and prone) were adopted in the experiment I, and two (standing and kneeling) in the experiment II. In the prone live firing, with or without bipod conditions were adopted. Before the measurement of the simulative firing, the subjects practiced having each firing posture and aiming to their satisfaction.

2.4. Conditions of Environment and Work Load

2.4.1.Experiment I. In the live firing the air temperature was about -10 degrees centigrade and the wind velocity did not exceed 3 m/sec. In the control phase, the room temperature was about 15 degrees centigrade.

In the cooling condition, the subjects stood still under the open sky putting on summer combat wear, helmet, earmuffs, army gloves and boots for 60 minutes. The weather conditions were same as in the live firing. Before the cooling and after, the skin temperatures of the forehead and the back of right hand were measured using a handy CA-type thermometer (Anritsu, AM-7001).

2.4.2.Experiment II. Room temperature was 24-26 degrees centigrade and relative humidity was about 70 percent. The wind velocity was 4.7 and 8.8 m/sec in the slow and fast wind condition, respectively. The frontal or dorsal side of body faced to windward.

The work load increased gradually using a treadmill. Each work duration of slow walking, fast walking and running was 7, 5 and 4 minutes, respectively. The subjects aimed just after each work's end, and commenced next work again after about 60 seconds. The oxygen intake was measured by Douglas bag method using a gas analyzer (Fukuda,BM-79), and heart rate was obtained from a record of electrocardiogram.

2.5. Treadmill-Wind Tunnel System

In the experiment II a treadmill-wind tunnel system was used. Figure 1 shows the outline of the system executed by Shimazu Inc.

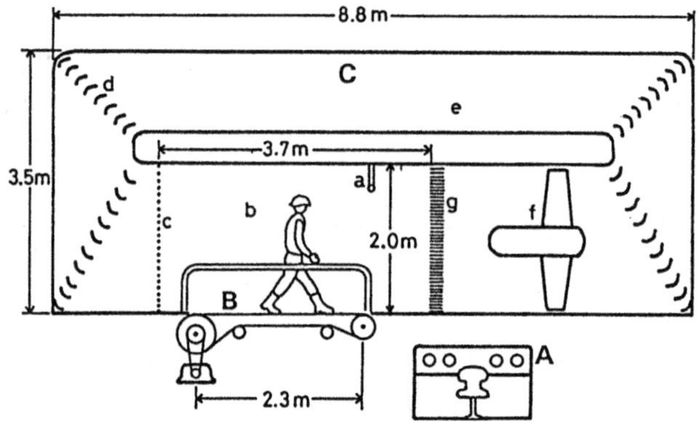

Figure 1. Treadmill-Wind tunnel system. A: control pannel, B: treadmill C: wind tunnel, a: sensor of wind velocity, b: working area, c: protective net, d: deflecting plate, e: return duct, f:fan, g:rectifying grid.

2.6. Data Analysis

Because of forbidding to adjust the sight, the scattering values (standard deviation) was compared. The paired two sided t-test was applied for the statistical analysis.

3. RESULTS

3.1 Experiment I

Figure 2 and table 3 shows the main results of the experiment I. The scattering area became larger as the center of gravity of body became higher, namely, the minimun scattering value was found in the prone posture.

Table 4 shows the changes of the mean skin temperature. There was seen a difference of 15 degrees centigrade between before and after the cooling, however the scattering area did not change significantly.

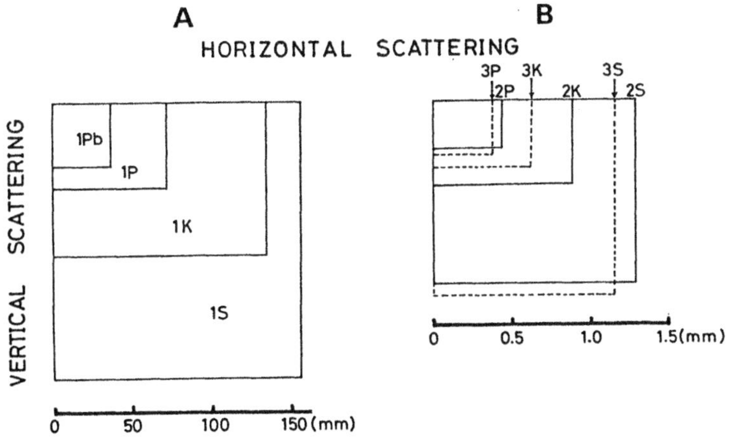

Figure 2. Scattering areas in live firing (A) and in simulative firing (B). One side of rectangle is equal to s.d. in each case.

Table 3. Comparison of the scattering areas in the exp.I.

p<0.05	p<0.01	p<0.001
1Pb<1S 1P<1S		
2K<2S 3P<3K 3P<3S		2P<2K 2P<2S

Paired two sided t-test was applied. The scattering area is defined as a value multiplied vertical s.d. by horisontal s.d.

TABLE 4. Skin temperature (centigrade) in the cooling test.

	before		after		difference	
(n=6)	forehead	hand	forehead	hand	forehead	hand
mean	33.9	30.3	19.0	15.7	-14.9	-14.6
s.d.	1.0	2.6	1.3	1.9	1.4	3.5

3.2 Experiment II

3.2.1 Phisiological load. Table 5 shows the phisiological indices of load during rest and work. Oxygen intake and heart rate during running was about eightfold and twofold larger than during rest, respectively.

TABLE 5. Physiological indices of load in the exp. II.

		Rest	Slow Walking	Fast Walking	Running
Velocity	mean	-	5.56	7.46	9.36
(km/hr)	s.d.	-	0.25	0.06	0.33
O_2 Intake	mean	288	1047	1814	2212
(ml/min)	s.d.	87	180	129	181
Heart Rate	mean	80	107	147	165
(beats/min)	s.d.	9	9	16	14

3.2.2. Scattering area. Figure 3 shows the scattering areas in the experiment II. The main results of data analysis are shown in table 6. Apparently the scattering area was smaller in the kneeling posture than in the standing one in the same way as the experiment I.

A noticeable fact is the effects of wind was enormous remarkably in the standing and kneeling postures. Furthermore the horizontal scattering was larger than the vertical one (tabel 7). The consistent difference between the directions of wind was not seen.

The effects of work load were not significant, however the scattering area had a tendency to be larger as the work load became harder.

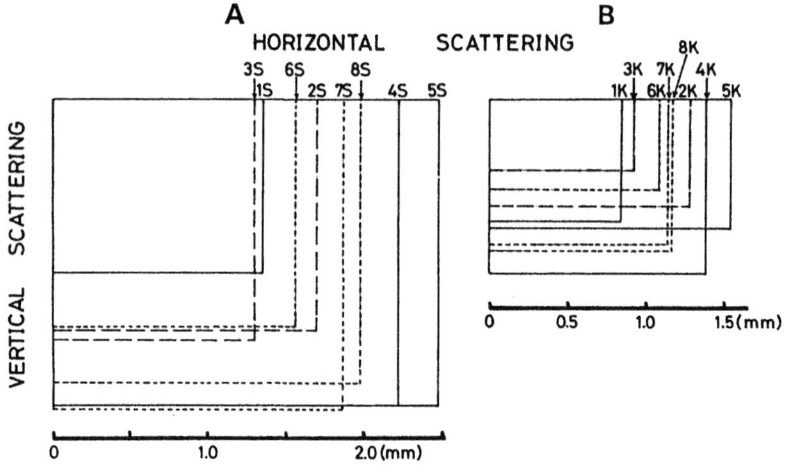

Figure 3. Scattering areas in the standing posture (A) and the kneeling posture(B) in the experiment II. One side of rectangle is equal to s.d. in each case.

Table 6. Comparison of the scattering areas in the exp. II.

$p<0.05$	$p<0.01$	$p<0.001$
1S<4S 2S<5S 3S<4S 3S<5S	1S<5S	
3K<4K 3K<2K		
2K<2S 4K<4S 7K<7S	5K<5S	all K<all S

Paired two sided t-test was applied. The scattering area is a value multiplied vertical s.d. by horisontal s.d.

Table 7. Comparison of vertical and horizontal scattering.

$p<0.05$	$p<0.01$	$p<0.001$
2K 4K 5S 4S+5S	3K	2K+3K 4K+5K

Paired two sided t-test was applicated. In all cases of this table the horizontal scattering was larger than the vertical scattering significantly.

4. DISCUSSION

Present study shows the effects of postures and wind on the rifle aming property can not be disregarded. The effects of cooling or work load were not significant, however these effects would be clear supposing that the subjects received the harder degree of them.

Wind drag is given by the equations:

$$\text{Drag(kg)}=0.5\cdot\rho\cdot Cd\cdot A\cdot V^2$$

where ρ is air density, Cd is drag coefficient, A is projected area, and V is relative wind speed. In the experiment using a wind tunnel, the blockage ratio (projected area / cross sectional area) can not be disregarded (El Sherbiny, 1972; Brownlie et al., 1987).

Mentioning above, the drag was calculated at about 1.1 and 3.8 kg in the case of slow wind (4.7m/sec) and fast wind (8.8m/sec), respectively. Furthermore the wind velocity in the field which would yield the same effect was faster than in a wind tunnel. It could be estimated that the velocity of 8.8 m/sec in the present wind tunnel system was equivalent to about 11 m/sec in the field.

Hunt et al.(1976) measured the side wind effects on walking in a wind tunnel, and found that for the steady uniform wind, effect on performance became noticeable above 6 m/sec, and walking became difficult above 13 m/sec. The facts above mentioned are not so different from the results of present study.

In the experiment II, the mean value of subjects' maximum oxygen intake (VO2 max) was about 3400 ml/min following the estimation nomogram reported by Sato et al.(1976). Consequently, the load of slow walking, fast walking and running are equivalent to about 30, 50 and 65 percent of the maximum ability, respectively.

It is interesting that the wind drag of 3.8 kg was more effective than the work load corresponding 65 percent of maximum ability. In the actual field condition the wind velocity fluctuates usually, the effects of wind would be larger.

It can be concluded from the present study that the aiming ability more depends on the firing postures and the uncontrollable effect by oneself such as wind than the controllable effects such as shivering and breathing.

5. ACKNOWLEDGEMENTS

The authors wish to thank Colonel Fukuda and Lieutenant Colonel Tsumura for their considerable assistance.

6. REFERENCES

Brownlie, L., Mekjavic, I., Gartshore, I. ,Mutch, B. and Banister, E., 1987, 'The influence of apparel on aero-dynamic drag in running', Ann. Physiol. Anthrop., 6, 133-143.

Dees, J. W., Magner, G. J. and McClusley, M. R., 1971, 'An experimental review of basic combat rifle marksmanship: marksman,phase1',TR-71-4 (AD-722-394)

El Sherbiny, S., 1972, 'Effect of wall confinement on the aerodynamics of bluff bodies', Msc.Thesis, University of British Columbia.

Giordano, D. J., 1979, 'Aiming point displacement from firing a rifle from the open-bolt position', TM-2-79 (AD-A068-215)

Hunt, J. C. R., Poulton, E. C. and Mumford, J. C., 1976, 'The effects of wind on people; new criteria based on wind tunnel experiments', Building and Environment 11,15-28.

Kumashiro, M. and Saito, K., 1978, 'Studies on the learning effects of target-aiming performance (in japanese)', Japanese.J.Ind.Health, 20, 212-217.

Sato, M., Yamasaki, K. and Tsuda, T., 1976, ' Estimation of maximal oxygen intake from submaximal works at air temperatures of 20 ℃, 30 ℃, and 40 ℃(in Japanese)', J.Anthrop.Soc. Nippon, 84,259-265.

Takakuwa,E.,1968, 'Evaluation of mental stress and the function of concentration maintenance(TAF)(in Japanese)',Japanese J.Ind.Health, 10,387-390.

ETUDE DE LA VARIABILITE CARDIAQUE DE PILOTES D'AIRBUS.

J.P.Fouillot*, F.Tekaia*, R.D.Blomberg**, M.Benaoudia*, J.J.Speyer***
* Laboratoire de physiologie, Faculté de médecine Cochin, 24, rue du Faubourg Saint Jacques, 75014 Paris, France
** Dunlap and Associates, Norwalk, Connecticut 06854, USA
*** Flight Division, Airbus Industrie, Toulouse, France

RESUME

Au cours de la campagne de vols de certification en équipage minimal de l'AIRBUS A310 puis A320, nous avons étudié la variabilité du rythme cardiaque des pilotes en fonction des phases de vol simultanément observées, ainsi que de l'évaluation subjective de la charge de travail des pilotes.
Dans une première étape, trois types de mesure de la variabilité du rythme cardiaque en fonction du temps ont été effectuées, les tendances à long terme du rythme cardiaque,la variabilité cardiaque à long terme, la variabilité cardiaque à court terme correspondant à la variabilité battement à battement.
Nous avons mis en évidence une correspondance entre ces variables et l'évaluation subjective de la charge de travail. Les évaluations élevées correspondent à une fréquence cardiaque plus élevée et une variabilité cardiaque basse; les évaluations basses correspondent à une fréquence cardiaque plus basse et une variabilité cardiaque élevée.
Dans une deuxième étape le spectre de puissance des fréquences cardiaques recueillies au cours de la campagne A320. Ces résultats pourraient objectiver une plus grande sensitivité de la variabilité cardiaque, à la charge de travail du pilote.

SUMMARY

HEART RATE VARIABIITY OF AIRBUS AIRCREWS

During the minimal crew certification flight campaign of Airbus A310 and A320, we studied heart rate variability with reference to flight phases simultaneously observed and to pilot workload subjective ratings.
In a first step, three types of heart rate variability time domain measures have been processed, long trends in cardiac periods, long term variability corresponding to simus arrhythmia, very short term expressed by beat to beat variability.
A correspondance has been found between these heart rate variability (HRV) data and pilots ratings simultaneously recorded. High pilot workload ratings correspond to higher HR and lower HRV and low ratings

A. Coblentz (ed.), Vigilance and Performance in Automatized Systems/ Vigilance et Performance de l'Homme dans les Systèmes Automatisés, 175–189.

correspond to lower HR and higher HRV.
In a second step, we have processed power spectrum of heart rates obtained during A320 certification. These results could show a better sensitivity of heart rate variability to pilot workload.

1. INTRODUCTION

La charge de travail imposée aux pilotes des avions commerciaux est une source de préoccupation majeure en raison du lien généralement admis entre le niveau de la charge de travail et l'augmentation de la probabilité d'erreur du pilote ainsi que la diminution de l'aptitude à faire face aux situations extrêmes. La démonstration d'un niveau suffisamment lors de la charge de travail requise fait ainsi partie de la procédure de certification d'un avion.
Au cours des vols de certification en équipage minimal des AIRBUS A300 FF, A310, A320, la charge de travail a été évaluée selon une méthode subjective. Il est cependant difficile de solliciter en toute circonstance une évaluation de la part du pilote. A partir des données enregistrées au cours des vols de la campagne A 310, un modèle statistique de prévision de la charge de travail a été élaboré (SPEYER). Le modèle dont l'application à l'étude des conditions de survenue d'erreurs est présentée plus loin (BLOMBERG), comprend des paramètres de vol de l'avion, des mesures de la variabilité cardiaque des deux pilotes et des mesures définissant l'état de l'avion. De nombreux paramètres physiologiques sont susceptibles d'être modifiés par la charge de travail, cependant l'application à des conditions de vol réel réduit le choix à quelques variables parmi lesquelles la fréquence cardiaque et la variabilité de la période cardiaque occupent une place de choix. Il nous fallait concilier la recherche de variables susceptibles d'être sensibles aux différentes situations de charge de travail avec une méthodologie compatible avec des conditions simulant des vols commerciaux sur lignes européennes.

2. METHODE

La méthode utilisée devait être robuste, non intrusive, le matériel miniaturisé et autonome de façon à autoriser une liberté totale de mouvement particulièrement aux escales. L'équipement devait pouvoir être posé en dix à quinze minutes car deux équipages se succédaient, effectuant chacun 3 à 4 vols par jour. Chaque campagne de vols comprenait 48 à 50 vols. Le monitoring de l'ECG remplissait ces conditions. Deux méthodes ont été utilisées successivement :

2-1 Au cours de la campagne de certification AIRBUS A 310, l'ECG a été enregistré sur un enregistreur magnétique miniaturisé MEDILOG 4-24 d'un poids de 440g. Sur une deuxième voie est enregistrée un signal de fréquence 60 Hz destiné à servir d'horloge externe lors de la lecture effectuée à une vitesse 60 fois plus rapide. La mesure des périodes cardiaques est effectuée avec une précision de deux millisecondes au moyen d'un miniordinateur MITRA 225.

2-2 Au cours de la campagne de certification AIRBUS A320, la mesure de la période cardiaque a été effectuée au moyen d'une centrale d'acquisition portable VITALOG PMS8. Nous avons modifié le programme d'origine afin de mémoriser les périodes cardiaques de chaque pilote pendant une durée de deux à quatre heures, celle-ci dépendant du rythme cardiaque. L'application d'un procédé de compression de données ne concerne alors que les artefacts éventuels. Cette méthode à l'avantage de permettre un traitement rapide, puisque les fichiers de périodes cardiaques sauvegardés à chaque escale à bord de l'AIRBUS A320 pouvaient transférés le soir même sur micro-ordinateur HP VECTRA afin d'être utilisés au calcul d'indices de variabilité cardiaque.

2-3. L'observation des taches de pilotage et des conditions de vol.
Une observation chronométrée a été réalisée au moyen d'une grille d'observation permettant de définir la séquence de vol considérée au moyen d'un code à 4 chiffres.
Le premier chiffre définit la condition de vol, normal, panne ou incident, check-list relative à la panne ou l'incident, check-list de routine ou communication. Les deux chiffres suivants définissent le type de panne, d'incident, de check-list, ou de communication. Le quatrième chiffre définit la situation de vol.
L'observateur était situé en cabine et pouvait suivre le déroulement du vol au moyen de deux écrans de contrôle transmettant les images de deux caméras grand angle prenant le tableau de bord et le panneau supérieur, de la phonie, des principaux instruments de bord. Le découpage temporel des séquences était effectué au moyen de l'horloge numérique incrustée sur le téléviseur de contrôle. Le fichier "Evénements" est ainsi constitué du découpage de l'activité de l'équipage en une suite de segments temporels ou séquences dont chacune, identifiée par un code, correspond à une étape déterminée de son activité. L'observation est synchronisée avec l'enregistrement en envoyant un signal de marquage d'événement à l'instant t0 du début de l'observation chronométrée.

2-4. Fréquence cardiaque et variabilité de la période cardiaque.
De nombreuses études ont retrouvé une augmentation de la fréquence cardiaque du pilote pour certaines phases de vol telles que décollage ou atterrissage, au cours de manoeuvres délicates, mais en dehors de ces trois situations ou le stress est notable, les études faites sur des vols réels ou simulés, montrent des résultats divers et parfois opposés.
En fait l'analyse de la fréquence cardiaque moyenne ne constitue qu'une des informations apportées par la période cardiaque. Celle-ci varie d'un battement cardiaque à l'autre. Une oscillation importante de la période cardiaque, sans modification du déroulement séquentiel de l'ECG, est qualifiée d'arythmie sinusale. La relation entre cette arythmie sinusale et la ventilation a été décrite pour la première fois par LUDWIG en 1847, par HERING en 1871. L'inspiration entraine une diminution de la période cardiaque, l'expiration une augmentation. La ventilation n'est cependant pas le seul phénomène à déterminer une arythmie sinusale .

La variation de la pression artérielle, la régulation de la température cutanée sont également à l'origine de variations de la période cardiaque que les méthodes d'analyse fréquentielle du rythme cardiaque ont mis en évidence (SAYERS, MULDER).
A partir des mesures de périodes cardiaques, nous avons calculé des indices de variabilité qui se répartissent en trois groupes :
- les variations à long terme de la période cardiaque, qui surviennent dans un délai de 1 minute ou plus ;
- les variations à court terme, battement à battement ;
- les variations intermédiaires dites également à long terme, et correspondant à l'arythmie sinusale.
Nous avons fait un choix parmi les indices proposés par MULDER et coll. et avons ajouté un groupe d'indices permettant d'éliminer la composante lente. Nous avons utilisé des moyennes exponentielles afin de lisser les données brutes mesurées. Les pilotes ayant tendance à baser leur jugement subjectif sur leurs impressions intégrées pendant une certaine période de temps, nous avons calculé les moyennes exponentielles à l'instant de l'évaluation subjective de la charge de travail, ceci permettant d'attribuer un poids élevé aux périodes cardiaques précédant immédiatement l'événement et un poids de moins en moins élevé au fur et à mesure que l'on remonte le temps.
La formule générale pour une moyenne exponentielle avec une constante de temps T secondes est :

$$AO = \frac{A1 + DT\,(SO - A1)}{T}$$

ou : AO est la valeur de la moyenne exponentielle calculée
A1 est la valeur précédente de la moyenne exponentielle
DT est l'intervalle de solution, c'est-à-dire, la période entre les points de données successifs (périodes cardiaques)
T est la constante de temps ou le facteur de lissage de la moyenne en secondes
SO est la valeur du paramètre pour le temps 0.
Il n'existait aucune base de sélection de la valeur unique et adéquate de T. Nous avons calculé toutes les moyennes exponentielles avec des constantes de temps (T) de 8, 16, 32, 64 et 128 secondes et avons choisi parmi ces résultats les valeurs qui apportaient une information différente.

2-4-1. Variabilité cardiaque à long terme
* Moyenne exponentielle brute des périodes cardiaques (E)
* Moyenne exponentielle centrée réduite des périodes cardiaques (EC)
La moyenne et l'écart-type des périodes cardiaques ont été calculés pour chaque vol et chaque pilote pendant l'ensemble du vol. La moyenne exponentielle a été calculée sur les différences entre les périodes cardiaques considérées et la moyenne divisée par l'écart-type.
* Moyenne exponentielle des différences entre les périodes cardiaques et une période moyenne de référence (EP)
Cette moyenne a été calculés pour chaque vol et chaque pilote pendant les deux minutes de vol de montée, deux minutes après le décollage.

Cette phase de vol apparaissait en effet comme la séquence où la période cardiaque était la plus stable, compte tenu du fait qu'il ne survenait aucun incident avant la phase de croisière.

2-4-2. Variabilité cardiaque à court terme.
* Racine carrée des différences entre deux RR consécutifs, (SD); (Root mean square of successive differences : RMSSD) .

$$RMSSD = \sqrt{\frac{(RRi - RRi+1)2}{N - 1}}$$

avec RRi = iième intervalle de l'enregistrement
N = nombre d'intervalles de l'enregistrement
Cet indice a été calculé sur un nombre entier de périodes pendant les 8, 16, 32, 64, 128 secondes précédant l'instant d'évaluation.

2-4-3. Arythmie sinusale.
* Mesure de décalages entre périodes cardiaques (LM)
Les différences absolues cumulées ou valeurs de décalages constituent parfois d'excellentes mesures à court terme. Les différences absolues cumulées des 16 et 32 périodes cardiaques précédant l'instant d'évaluation, ont été calculées.
* Racine carrée de la somme des différences entre chaque période et la moyenne des seize périodes cardiaques
précédentes (SM). (Root mean square of differences from the mean, RMSM).

$$RMSM = \sqrt{\frac{(RRi - RR)2}{N}}$$

avec RRi = iième intervalle de l'enregistrement
RR = moyenne des intervalles RR
N = nombre d'intervalles de l'enregistrement
Cet indice a été calculé sur un nombre entier de périodes pendant les 8, 16, 32, 64, 128 secondes précédant l'instant d'évaluation.
* Coefficient et résidu de la régression (ER)
La composante à long terme de la variabilité cardiaque peut être obtenue en calculant la droite de régression linéaire des périodes cardiaques, sur les 32 ou 64 battements précédant l'évaluation subjective de la charge de travail. La déviation standard résiduelle ER est calculée à partir des différences entre les périodes de 16 ou 32 RR précédant l'évaluation et la droite de régression des RR .

2-4-4. L'analyse des données.

L'étude de la correspondance entre les indices de variabilité cardiaque et l'évaluation subjective de la charge de travail a été effectuée au moyen de la méthode d'analyse factorielle des correspondances (A.F.C.) développée par BENZECRI , et appliquée par F. TEKAIA à ces données. Cette méthode a pour but de synthétiser l'information contenue dans un tableau de données et de visualiser les relations entre les éléments de deux ensembles (ensemble des lignes et ensemble des colonnes) en en donnant des représentations graphiques planes. L'analyse factorielle des correspondances s'applique à un tableau de données de nombres positifs, dans lequel les individus I sont représentés en ligne et les variables J en colonnes. Il est permis de représenter l'ensemble des variables et des individus dans un système de base orthogonale définie par des axes factoriels. Dans ce système d'axes, deux variables sont déclarées proches l'une de l'autre, si leur "comportement" est semblable par rapport à l'ensemble des individus. De même deux individus sont également proches l'un de l'autre si leurs comportements sont en rapport étroit avec l'ensemble des variables. Cette notion de proximité ou de voisinage entre les variables (ou les individus) caractérise leurs rapports respectifs. Dans le système d'axes déduits à partir des p premiers facteurs données par l'analyse factorielle des correspondance d'un tableau ayant les dimensions NI sur NJ, une variable Vj est définie par la position des coordonnées Ga(j)

$$Vj = \{ Ga(j);\ a = 1, p\}$$

De même un individu Im est défini par ses coordonnées Fa(m)

$$I(m) = \{ Fa(m);\ a = 1, p\}$$

Distance entre les variables: La distance orthogonale entre les deux variables Vj et Vk est définie comme suit:

$$d\ (Vj\ ,\ Vk) ** 2 = \Sigma \{ (Ga(j) - Ga(k)) ** 2;\ a = 1,\ p\}$$

Distance entre les individus: La distance orthogonale entre les individus Im et In est donnée par la formule suivante:

$$d\ (Im\ ,\ In) ** 2 = \Sigma \{ (\ Fa(m) - Fa(n)) ** 2;\ a = 1,\ p\}$$

La distance ainsi définie permet de quantifier la notion de proximité dans l'espace vectoriel. Pour définir les plus proches voisins d'une variable donnée (ou individu), on peut procéder conformément à deux critères différents, l'un, en fixant la distance maximale, l'autre en fixant le nombre maximal de voisins. Nous avons utilisé ici la variable dans le but de chercher les voisins les proches. (On peut suivre le même raisonnement et obtenir des résultats similaires avec des individus).

a - Voisins situés dans une hyper-sphère de rayon R:

Si l'on considère l'espace factoriel, une grande sphère centrée à une variable Vj et d'un rayon R; la distance maximale choisie par l'utilisateur pour trouver les voisins de Vj et le rayon R. Les variables les plus proches de Vj seront localisées dans cette sphère. La position des points les plus proches de Vj sera définie comme suit:

$$(Vj) = \{ Vk;\ d\ (Vj\ ,\ Vk) <= R \}$$

b - Les N plus proches voisins:

Un autre critère permet de fixer le nombre maximum N des voisins à partir de la variable Vj, il est donc nécessaire de calculer la distance entre Vj et le reliquat des variables. On retient donc seulement les N variables ayant les plus petites distances de Vj.

Dans les deux cas, la recherche des plus proches variables demande de calculer la distance du vecteur {d (Vj,Vk);k=1,NJ} , entre la variable Vj et les (Nj - 1) autres variables.

Matrice des distances entre les variables:

Généralement, la matrice des distances D (Nj , Ni) entre les variables paires est déterminée en utilisant les éléments djk de D (Nj , Ni) et évaluée comme suit:

djk = d (Vj , Vk) **2 = Σ { (Ga(j) - Ga(k)) **2; a =1, p}

Chaque colonne de cette matrice représente le vecteur distance d'une variable donnée à partir des (Nj-1) autres variables.
Notons que cette matrice est symétrique, c'est à dire djk=djk.
Recherche des plus proches voisins:
Dans le but de trouver les plus proches voisins d'une variable donnée Vj on procède comme suit:
a - compter la matrice D (NJ , NJ) entre les NJ variables de la table des données dans l'espace factorielconsidéré
b - classer dans un ordre ascendant la colonne j ème de la matrice D (NJ , NJ) correspondant aux distances de Vj aux (NJ-1) autres variables.
c - Dans le cas où l'on cherche les voisins de Vj localisés dans la distance R, et choisir à partir des colonnes rangées j ème. les Vk variables telle que la distance de Vj demeure inférieure ou égale à R. Il sont donc localisés dans l'hypersphère de centre Vj et de rayon R.
d - Pour chercher les N plus proches voisins de Vj, sélectionner les N premières variables correspondant aux N premières distances de la colonne j th.
Relation quantitative à partir des voisins les plus proches :
Cas de régressions multiples:
La description des plus proches voisins montre un rapport qualitatif entre les variables. Cette technique peut être particulièrement appliquée pour expliquer la relation entre les variables quantitatives et qualitatives. Elle peut être aussi appliquée en régressions multiples:
Pour réduire le nombre des variables indépendantes.
Pour tester la validité du modèle d'extrapolation.
a - Quand les variables indépendantes sont nombreuses, le cas de régression multiple exige de réduire ce nombre dans le but de garder la stabilité des résultats. Pour achever cette réduction il faut replacer ces variables indépendantes selon les facteurs orthogonaux Gx ainsi que le décrit CAZES. Un deuxième moyen est de considérer uniquement les variables indépendantes localisées le plus près possible des variables dépendantes.
b - Pour extrapoler un modèle fourni par une régression multiple, il est nécessaire de s'assurer de la signification de cette extrapolation des nouveaux sujets, en respectant le modèle d'établissement, dans ce but il est nécessaire que les nouveaux sujets soient proches de la position des sujets déjà utilisés pour établir le modèle initial.

3. RESULTATS

Nous avons appliqué cette méthode à l'étude de la relation entre l'estimation subjective de la charge de travail et la variabilité de la période cardiaque. Les estimations de la charge de travail et les mesures de la fréquence cardiaque des deux pilotes ont été enregistrés pendant la campagne de certification Airbus A310.

L'estimation de la charge de travail par le pilote a varié entre 1 et 5, la valeur 6 ayant été regroupée avec la valeur 5. La variabilité de la période cardiaque au moment de chaque évaluation de la charge de travail a été caractérisée au moyen de 8 indices choisis pour leur signification particulière parmi l'ensemble des indices décrits précédemment. Six d'entre eux ont été calculés à partir des périodes cardiaques enregistrées 32 secondes avant l'évaluation. Nous avons utilisé 3032 temps d'évaluation de la charge de travail et divisé chaque indice de variabilité en sept classes, allant des faibles valeurs aux valeurs élevées. Les limites des classes et leurs effectifs sont présentés dans la table 1. Ainsi la première classe de la moyenne exponentielle brute E321 est définie par l'ensemble des valeurs inférieures ou égales à 527. Cette classe contient 437 séquences. La seconde classe nommée E322, est définie par l'intervalle (527<E322<562); elle contient 414 séquences etc.. . La même notation est utilisée pour caractériser les classes des autres indices. L'évaluation de la charge de travail du pilote, RAP1 correspond à la valeur 1, RAP2, la valeur 2, RAP5, la valeur 5. A la suite de la codification de la charge de travail et des divers indices de variabilité, 64 modalités sont donc déduites. La table de données initiale a des dimensions de (3032,64), et la figure 1 montre le plan factoriel (F1,F2) où les 64 modalités sont représentées. Cette figure montre que les modalités varient selon la pente spécifique, allant des plus petites aux valeurs les plus élevées d'une modalité.
Par exemple, les modalités de l'indice SM3 suivent une courbe parabolique de SM31 à SM37, tandis que la charge de travail RAP* estimée par le pilote varie en sens inverse (RAP1* à RAP5*).
La proximité des différentes modalités peut être étudiée, sachant que deux modalités sont d'autant plus proches qu'elles sont intercorrélées. La matrice des distances entre les variables, dans un espace vectoriel formé avec les sept premiers facteurs, permet de chercher les plus proches voisins des évaluations de chaque charge de travail. On remarque que les évaluations de charge de travail élevées du pilote ont comme plus proches voisins, des classes d'indice de variabilité faible correspondants à des fréquences cardiaques élevées. Les évaluations de charge de travail du pilote les plus faibles ont comme plus proches voisins des classes d'indices élevés correspondants aux fréquences cardiaques basses. La table 2 montre le vecteur distance de l'évaluation 5 (RAP5); les classes des indices les plus proches sont DS33, C322, LM32, E322, ER32, E323, et D326, D327, qui correspondent à une fréquence cardiaque élevée et une variabilité cardiaque plus faible. On peut noter que l'évaluation 2 (RAP2) est la plus éloignée de la 5. La table 3 montre le vecteur distance de l'évaluation 2 (RAP2). Les indices des classes les plus proches correspondent à une fréquence cardiaque basse et sont SM34, E327, DS35, DS34, ER35. La méthode des plus proches voisins permet donc d'établir le rapport existant entre les variables discrètes et continues. Dans cette étude, cette méthode a été utilisée pour étudier la corrélation entre l'évaluation subjective de la charge de travail et la variabilité de la période cardiaque. Nos résultats montrent que les évaluations élevées de la charge de travail

d'un pilote sont proches des classes de fréquence cardiaque élevée et des classes de variabilité faible, alors que les évaluations de charge de travail basse sont proches des classes de fréquence cardiaque basse et des classes de variabilité élevée.
Cette méthode permet également d'analyser les rapports entre les individus (séquence de vol) et les variables (variabilités de la période cardiaque). La mise en éléments supplémentaires des barycentres de l'ensemble des points correspondant à un même pilote selon qu'il assure ou non la fonction de pilotage, ou à un scénario de panne permet de représenter les tendances de la variabilité cardiaque face à ces conditions expérimentales.
Nous avons projeté Fig 3 les barycentres de l'ensemble des séquences de chaque capitaine, selon qu'il assurait la fonction de pilotage (F), ou non (NF). Pour l'ensemble des pilotes sauf un l'ensemble du vol se traduit par déplacement du barycentre des séquences NF vers des classes de fréquences cardiaques plus élevées et une variabilité de la période cardiaque plus basse.
La projection des barycentres de chaque scénario (Fig.4) montre que le scénario 6 réalisant une descente d'urgence, correspond à des classes de fréquence cardiaque plus élevée et de variabilité cardiaque plus basse. Les scénarii 1.A à 1.C étaient des scénarii sans panne au cours de vol. Le scénario 1.C apparaît cependant la plus proche du scénario 6. En fait au cours de ce scénario les pilotes devaient contrôler la commande des gaz, ce qui peut expliquer la tendance à présenter une fréquence cardiaque plus élevée, une variabilité cardiaque plus faible.

4. DISCUSSION

La sensibilité de la variabilité cardiaque à la charge de travail a été constatée par de nombreux auteurs. Ainsi JEX et ALLEN ont constaté chez certains sujets, une diminution de la variabilité cardiaque lors d'une tache de tracking. ETTEMA et ZIELNUIS ont observé également que deux indices de variabilité cardiaque diminuent lors de l'exécution de taches. Les constatations de KALSBEEK, BOYCE, MULDER vont dans le même sens, MULDER notant la sensibilité du spectre de puissance de la fréquence cardiaque à la charge de travail de l'opérateur, en particulier la bande corrélée aux variations de pression artérielle à court terme. Cependant tous les auteurs n'ont pas la même unanimité. MOBBS ne constate aucune corrélation entre la variabilité cardiaque et la difficulté de la tache. SAYERS conclut dans le même sens, en soulignant cependant l'intérêt de l'étude du spectre de la fréquence cardiaque. HICKS et WIERWILLE ne constatent aucune relation avec la difficulté de la conduite automobile.
Une première analyse de cette brève revue de la littérature donnerait à penser à une certaine incohérence de résultats, si l'on ne soulignait pas les différents biais susceptibles d'atténuer la signification des variations de la période cardiaque par rapport à la difficulté de la tache de pilotage, ou de rendre peu comparables les différentes études.
Le choix des modes de représentation de la variabilité du rythme cardiaque (fréquence cardiaque moyenne, indices de variabilité cardiaque, analyse spectrale de la fréquence cardiaque), le type de

tâche étudiée peuvent expliquer en grande partie ces divergences.
La sensibilité à la charge de travail des variables que nous avons choisi apparaît à travers la constatation d'une correspondance entre ces indices de variabilité cardiaque et l'évaluation subjective de la charge de travail. Le choix de la moyenne exponentielle et surtout la combinaison de trois modes de variation de la période cardiaque pourrait expliquer ce résultat, alors que chaque indice considéré isolément ne montre aucune corrélation avec l'estimation subjective de la charge de travail. L'étude du spectre de puissance de la fréquence cardiaque permet de discriminer les composantes fréquentielles déterminant l'arythmie sinusale, et présente une plus grande sensibilité à l'effort mental d'après les résultats préliminaires obtenus à l'issue de la campagne A320.
Cette approche multifactorielle de la variabilité de la période cardiaque objective donc une plus grande sensibilité de de cette variable à la charge de travail, qui contribue à justifier l'introduction d'indices de variabilité cardiaque dans un modèle de prévision de la difficulté de la charge de travail.

5. BIBLIOGRAPHIE

BENZECRI, J-P. et BENZECRI, F.
Pratique de l'analyse des données. Tomes I et II. Dunod ed., Paris, (1980).

BLOMBERG, R.D, SCHWARTZ, A.L., SPEYER, J-J., FOUILLOT, J-P.,
Application of the Airbus workload model to the study of errors and automation, présenté aux journées d'études prospectives OTAN "Vigilance et Performance dans les systèmes automatisés, PARIS, (1988).

BOYCE, P.R.
Sinus arrhytmia as a measure of mental load. Ergonomics, 17, PP 177-183, (1974).

ETTEMA, J.H., ZIELHUIS, R.L.
Physiological parameters of mental load. Ergonomics, 14, PP 137-144, (1971).

FOUILLOT, J-P., DROZDOWSKI, T., TEKAIA, F., REGNARD, J., IZOU, M-A., FOURNERON, T., LEBLANC, A., RIEU, M.
Methodology of heart rate ambulatory monitoring recordings analysis, in relation to activity: applications to sports' training and workload studies.
ISAM-GENT-1981, proceedings of the Fourth International Symposium on Ambulatory Monitoring and the Second Gent Workshop on Blood Pressure, Academic Press Edit., pp. 377-383, (1982)

HERING, E.
Uber eine reflectorische bezichung zwischen lunge und herz. S.B. Akad. Wiss. Wien., 64, PP 333-353, (1871).

HICKS, T.G. and WIERWILLE, W.W.
Comparison of five mental workload assessment procedures in a moving base driving simulator. Human factors, 21, 129-143, (1979).

KALSBEEK, J.W.H.
Sinus arrhythmia and the dual task method in measuring mental load. In W.T. Singleton, J.G. Fox, and D. Whitfield (Eds.). Measurement of man at work. London: Taylor and Francis, pp. 101-113, (a),(1973).
KALSBEEK, J.W.H.
Do you believe in sinus arrhythmia?, Ergonomics, 16, 99-104.(b), (1973).
MOBBS, R.F., DAVID, G.C., and THOMAS, J.M.
An evaluation of the use of heart rate irregularity as a measure of mental workload in the steel industry. London, England: British steel corporation, BISRA, OR/HF/25/71, (1971).
MULDER, G. and MULDER-HAJONIDES VAN DER MEULEN, W.R.E.H.
Mental load and the measurement of heart rate variability. Ergonomics, 16, 69-83, (1973).
MULDER, L.J.M.
Model based measures of cardiovascular variability in tj•he time and the frequency domain. In: Orlebeke J.F., Mulder G., Van Doornen L.J.P. edit., Psychophysiology of cardiovascular control. New-York, Plenum Press, 333-352, (1985).
SAYERS, B.M.
Analysis of heart rate variability. Ergonomics, 16, 17-32, (1973).
SPEYER, J.J., FORT, A., FOUILLOT, J-P., BLOMBERG, R.D.
Assessing pilot workload for minimum crew certification. The practical assessment of pilot workload. AGARDograph No.282, NATO Advisory Group for Aerospace research and development, London, (1987).
TEKAIA, F., FOUILLOT, J-P., DROZDOWSKI, T., REGNARD, J., SPEYER, J-J., RIEU, M.
Incidences des contraintes psychiques et intellectuelles sur la fréquence cardiaque du pilote. Cah. Anal. de Données, No.6, pp.175-182, (1981).
WIERWILLE, W.W.
Physiological measures of aircrew mental workload. Human factors, 21(5), 575-593, (1979).

Cette étude a pu être réalisée avec l'aide d'Airbus Industrie et de la D.R.E.T. (contrat N° 81/34/594).

<u>Tableau 1</u> : Limites de classes des variables analysées

```
    cl       527     562     589     617     657     705
E32 --------I-------I-------I-------I-------I-------I-------
    nb  437     414     437     422     453     445     424
    cl        -1     -.59   -.28     .06     .42     .91
C32 --------I-------I-------I-------I-------I-------I-------
    nb  425     420     411     462     441     442     431
    cl       -45     -16      2      20      38      65
D32 --------I-------I-------I-------I-------I-------I-------
    nb  441     433     446     423     413     447     429
    cl       18.7     24     29.5     36      42      51
SM3 --------I-------I-------I-------I-------I-------I-------
    nb  438     422     441     457     403     425     446
    cl        9.4    12.7    15.7    19      23.5    36
SD3 --------I-------I-------I-------I-------I-------I-------
    nb  445     436     449     404     420     442     436
    cl        2       3       4       5
RAP --------I-------I-------I-------I-------
    nb  367    1813     736     116
    cl        5      10.5    16.2    23.5     34      52
DS3 --------I-------I-------I-------I-------I-------I-------
    nb  437     437     408     438     437     448     425
    cl       195     263     328     410     515     679
LM3 --------I-------I-------I-------I-------I-------I-------
    nb  424     447     445     423     425     425     443
    cl        35      52      67      84     107     145
ER3 --------I-------I-------I-------I-------I-------I-------
    nb  421     426     423     450     429     444     439
```

Tableau 2 : Vecteur distance de la variable RAP5

Nom :		RAO5	RAO4	RAP4	DS33	C322	LM32	E322	D326	ER32
Distance :		193	236	271	604	625	627	632	647	653
Nom :	E323	DS31	D327	DS32	ER33	SM32	SD32	C321	E326	DS34
Distance :	665	670	687	708	708	732	750	758	774	791
Nom :	DS35	RAP3	RAO3	E325	E324	D325	C323	ER34	E321	D324
Distance :	793	794	808	812	819	859	872	874	880	895
Nom :	C324	DS36	LM33	SM33	ER35	SM36	D323	ER31	SM34	SM35
Distance :	905	952	985	992	1031	1085	1092	1099	1119	1168
Nom :	ER36	C325	SD33	SM31	SD31	SD36	C326	SD34	D322	DS37
Distance :	1179	1240	1261	1287	1322	1366	1366	1374	1389	1405
Nom :	SD35	LM35	ER37	SD37	LM36	LM31	LM34	E327	SM37	LM37
Distance :	1413	1429	1439	1457	1478	1486	1503	1508	1634	1663
Nom :	C327	D321	RAP2	RAO2						
Distance :	1905	2031	2269	2474						

Tableau 3 : Vecteur distance de la variable RAP2

Nom :		RAO2	RAO3	RAP3	SM34	E327	DS35	DS34	ER35	C325
Distance :		17	410	427	437	437	468	508	511	514
Nom :	DS36	E324	ER36	ER34	LM34	SM35	D323	E325	D325	DS32
Distance :	527	532	554	564	587	628	629	647	659	661
Nom :	SM33	C324	C326	E326	D324	D322	E323	DS31	E322	SD33
Distance :	680	682	682	686	699	709	712	718	721	721
Nom :	SM36	ER33	DS33	C323	SD34	SD36	LM35	SD35	ER32	LM33
Distance :	723	732	734	742	747	751	787	848	849	938
Nom :	D326	LM36	ER37	C322	DS37	E321	D321	C327	SM37	RAP4
Distance :	943	965	1006	1026	1041	1062	1092	1122	1122	1219
Nom :	SM32	RAO4	SD32	SD37	LM32	LM37	ER31	D327	C321	LM31
Distance :	1262	1273	1326	1328	1347	1370	1577	1654	1731	1890
Nom :	SD31	SM31	RAP5	RA05						
Distance :	1937	1948	2269	2690						

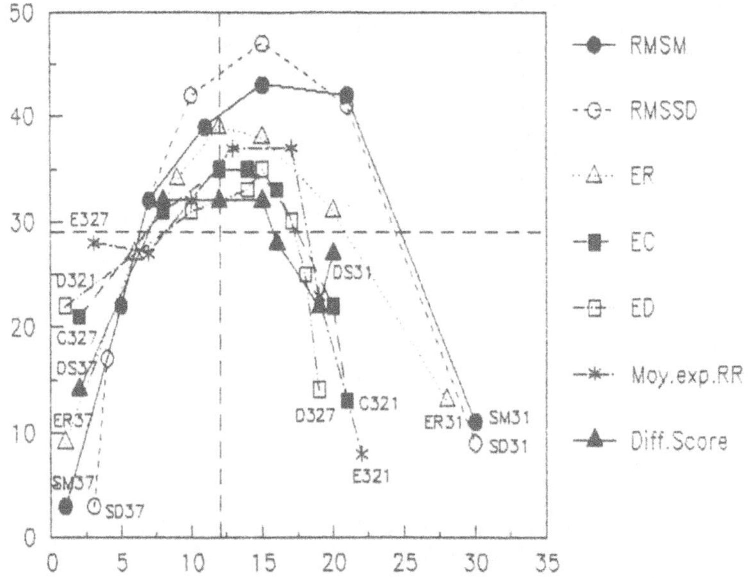

Fig.1 - AFC variabilité cardiaque/charge de travail.
Projection des indices de variabilité cardiaque

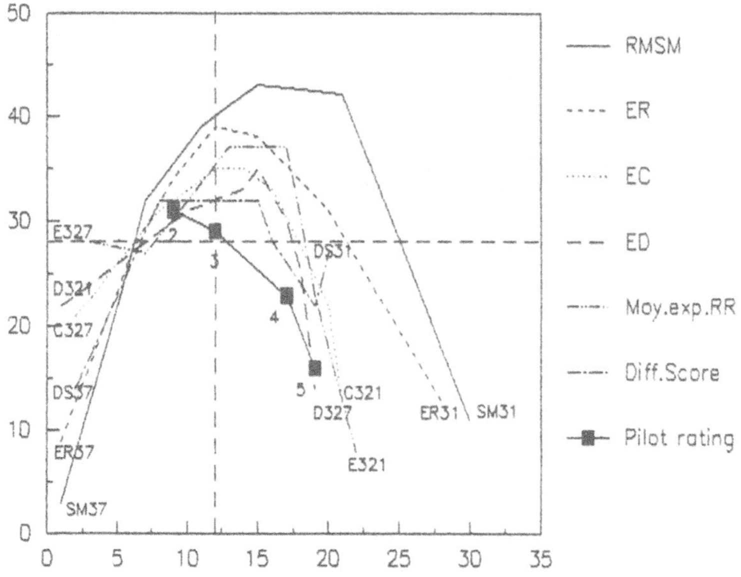

Fig.2 - AFC variabilité cardiaque/charge de travail
Projection des barycentres des évaluations de charge de travail

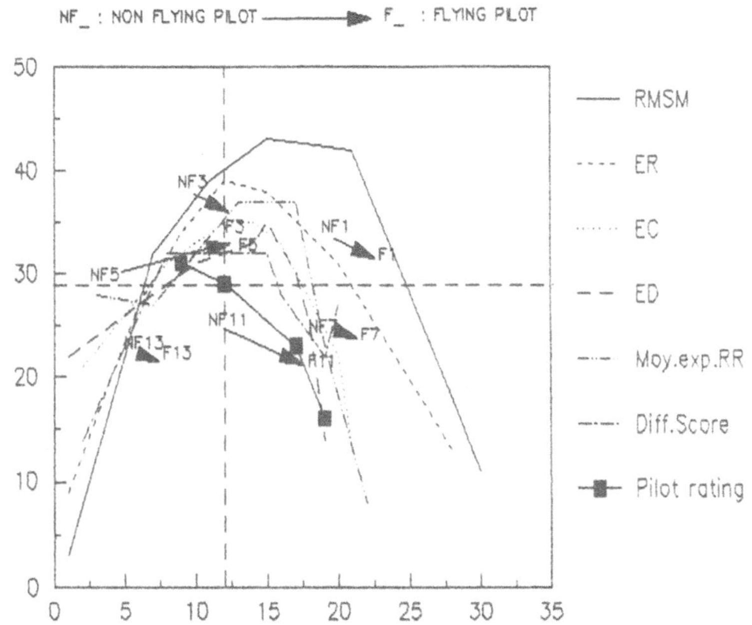

Fig.3 - AFC : Projection des barycentres des séquences des commandants de bord exerçant (F) ou n'exerçant pas (NF) la fonction de pilotage.

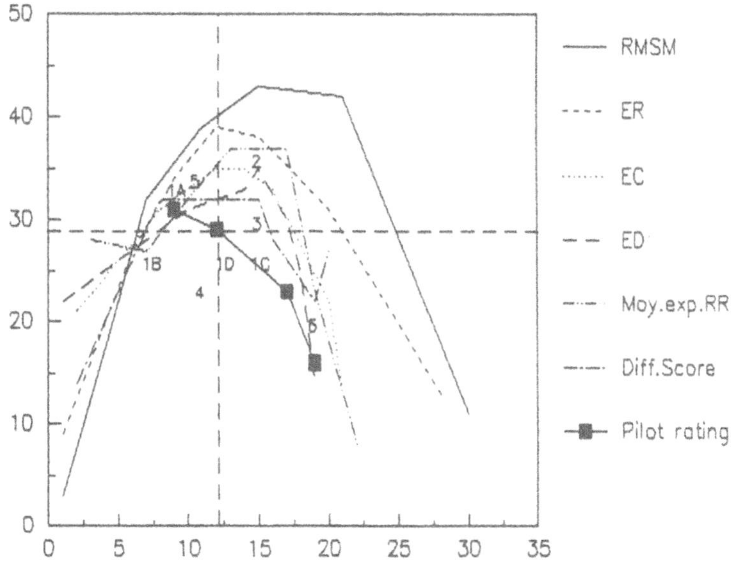

Fig.4 - AFC : Projection des barycentres des scenarii.

WORKLOAD MANAGEMENT FOR WORKERS ON IRREGULAR SHIFTS

H. Ugajin, T. Ikeda, and S. Yoneyama
Railway Technical Research Institute
2-8-38 Hikari-cho, Kokubunji-shi
Tokyo 185
Japan

ABSTRACT. This study constructs a model evaluating the amount of workers' fatigue at an arbitrary point in time using a train driver's job as an example to investigate its validity in order to manage the workload for workers on irregular shifts. The parameters for constructing this model were assumed rather subjectively, but the results show its usefulness for its objective of evaluating the amount of fatigue. A number of problems remain before this model can actually be applied. The use of this for duty rosters can, however, help to avoid an excessive workload.

1. INTRODUCTION

A train driver's job can be given as a typical example of irregular shift work. This job has the follwoing characteristics according to its dependence on the train schedule and the need to balance the workload among drivers.

1) Shifts differ according to the commencement time at work.
2) The number of working hours differ according to the shift.
3) Shift intervals vary between the shifts.
4) Since a shift is made up of various patterns of work and breaks, workload cannot be in a simple proportion to working time.

Since train drivers are directly responsible for the safe operation of trains, they must maintain their best mental and physical condition while on the job. Backup systems for drivers' errors are still, with the exception of the Shinkansen, insufficient, and low arousal levels can lead directly to such dangers as human error, signal violations, and excessive speed. Moreover , irregular shifts bring about noticeable levels of fatigue in drivers leading to a bad situation in which the driver's performance declines during the operation of the train.

The Japan Railways Group has set a standard which must be depended on to assure efficiency and avoid excessive workloads in the formation of train drivers' work cycle (about two weeks per cycle). For example, the standard for running period is 40 hours a week, 6.67 hours a day, the maximum distance for operation being 220 km for one driver and 330

A. Coblentz (ed.), Vigilance and Performance in Automatized Systems/ Vigilance et Performance de l'Homme dans les Systèmes Automatisés, 191–199.

km for two drivers. There are, in addition, standards for the length of break time at the destination point and for the interval between shifts.

The above standard was formed from a collection of fragmentary knowledge including many experiential judgements. The time at which the train driver's fatigue peaks on the job and the judgement as to whether this falls within allowable limits are not included. There had been some debate in the past as to whether adjusting the product of the running time and running distance to a lower fixed value would be valid as an index for managing workload, but this has never been put into effect.

The following describes research carried out aimed at contributing a more scientific basis for the construction of schedules which do not lead to excessive fatigue for train drivers (Ikeda et al. 1986). This was done by measuring the amount of fatigue at an arbitrary point in time based on data concerned with train drivers' daily lives.

2. MODEL

2.1. Modelling

This model was constructed in order to evaluate the Cumulative Workload Value (CW) at the point that an individual completes his daily activity elements (work, sleep, etc.). We assumed that, in the case of the average worker, the amount of fatigue at this given point is determined by the accumulated amount of previous workload (positive or negative). CW is a ratio scale which varies according how the upper limit is set with the lower limit fixed at zero.

This model attempts to evaluate the CW at the end of number k activity when individual activities (work, rest, etc.) after sufficient sleep, which makes CW 0, can be described on a time scale.

The basic concept for this model is illustrated by the following two tables.

First, an extract of the Workload Grading Table (Ohshima 1979) is shown in Table I. The Table shows the workloads for all other work periods with 100 as the value for the workload for a continuous eight-hour work period beginning at 9:00 a.m. The foundation for this is a mass of data measuring changes in a worker's vitality level in accordance with circadian rhythms. Each point in the scale reflects the effect of the work time period and the cumulative effect of hours worked. The weak point, however, is that the hardness of the work itself is assumed to be fixed.

On the other hand, Table II illustrates an extract of the Sleep Effect Grading Table (Ohshima 1979, Ikeda 1988). Like the Workload Grading Table, the Table shows the effect of sleep on the recovery from fatigue for all other time periods with the effect of eight hours of sleep beginning at 10:00 p.m. set at 100. Sleep functions as a negative workload. A rating of 100 for workload is offset by a rating of 100 for sleeping effect. This also has the weak point of assuming that changes in depth of sleep according to sleeping conditions are fixed.

TABLE I
Workload Grading Table (Ohshima 1979, extract)

Work Duration (h)	Commencement Time at Work (h) 0	3	6	9	12	15	18	21
2	32	35	27	24	24	24	26	29
4	67	64	51	48	48	49	54	60
6	100	90	76	73	73	77	85	94
8	130	117	103	100	102	109	120	132
10	158	145	131	129	134	144	159	165
12	188	175	161	162	170	183	197	196
14	216	203	191	195	206	222	228	224
16	242	230	221	228	243	253	254	250
18	266	257	251	261	275	278	278	274
20	292	286	283	296	302	302	302	298
22	320	317	318	325	326	326	326	323
24	350	350	350	350	350	350	350	350

This shows the workloads for various work shifts with 100 as the value for the workload for a continuous eight-hour work period beginning at 9:00 a.m.

TABLE II
Sleep Effect Grading Table (Ohshima 1979, Ikeda 1988, extract)

Sleep Duration (h)	Sleep-Onset Time (h) 0	3	6	9	12	15	18	21
2	26	26	23	18	18	18	20	23
4	54	52	44	38	38	40	44	51
6	78	72	61	55	56	60	68	76
8	98	87	76	70	73	80	91	99
10	111	99	88	84	89	100	111	117
12	123	111	101	99	108	120	130	132
14	135	123	115	116	128	140	147	144
16	147	137	131	136	148	158	160	156
18	160	152	150	156	167	173	172	168
20	174	169	170	176	184	185	184	180
22	190	189	190	194	197	197	196	194
24	209	209	209	209	209	209	209	209

This shows the effect of sleep on the recovery from fatigue for various time periods with the effect of eight hours of sleep beginning at 10:00 p.m. set at 100.

The model, then, can be expressed as follows.

$$W_k = \sum_{i=1}^{k} A_i E_i T_i \tag{1}$$

Wk: the CW at the end of k-th activity Bk
Ai: the specific cumulative fatigue (recovery) effect for Bi
Ei: the cumulative fatigue (recovery) effect for a work period including Bi
Ti: the time duration of Bi

Wk does not take a negative value. In the process of addition, Wk is set at 0 when Wk < 0 and the following caluculations are invalid until a positive value appears.

The value Ai is determined on the basis of the relative relationships among individual activities' strength of cumulative fatigue (recovery). If it is an activity which brings about cumulative fatigue, the result is Ai > 0, and if it is an activity which results in recovery from fatigue, the result is Ai < 0. In addition, when neither of these can be determined Ai = 0 is used. In that case either the Workload Grading Table (+) or the Sleep Effect Grading Table (-) is applied corresponding to the value of Ai for Ei. When considering the cumulative effect of a work period or a rest period, it is more rational to apply each table to these work periods or rest periods as a series rather than divide these times into individual activities. Daily activities can be classified into four types according to their individual characteristics as 1) on the job (work), 2) on the job (rest), 3) off the job (from getting up to beginning work and time after from finishing work until going to bed), and 4) off the job (time sleeping at home). In this case, Ei is the final cumulative fatigue (recovery) effect for the time period including Bi averaged into units of time. In other words, Ei is constant as long as it belongs to the same time period.

2.2. Assuming the Parameters

In order to assume the parameter Ai and determine part of Ti, the following data was collected from the answers to a questionnaire distributed among foremen of train drivers (N=116).

1) A subjective rating value for the relative amount of fatigue for basic types of work. The value of the intensity for "main line driving" was set at 1.00 conditioned on a continuous one-hour period from 10:00 a.m.
2) A subjective rating value for the relative amount of recovery for a typical rest periods. This was set at 1.00 for a continuous period of four hours from midnight for time spent "sleeping at home."
3) As for time needed to get ready, there are a) standard time needed from getting up to leaving home for work and b) standard time needed from the time of arriving home until

TABLE III
Indices of cumulative fatigue effect or fatigue recovery effect estimated by train drivers by activity (Ai)

	Activities	Ai
ON-THE-JOB:	main line driving	1.00 #
	training	1.00 *
	taking car into or out of the shed	0.94 *
	shunting	0.94
	watching the state of power car	0.94 *
	preparation for the next work	0.56 *
	preparation for turning back	0.56
	moving on foot	0.56 *
	travel to take up duty	0.46
	waiting for train	0.37
	rest at destination	-0.58
	sleep at destination	-0.73
OFF-THE-JOB:	commuting by train	0.38
	getting ready	0.00 #
	rest or sleep at home	-1.00 #

#: Fixed. *: Assumed value applied from similar activity.

bedtime. At the same time this is the data for assuming the amount of sleeping that is possible when drivers are off the job. Time to get ready is a broad concept which singles out activities including mealtime and family life needed to maintain a healthy life on the job.

The geometric mean for each activity is found from the results of the above 1) and 2) and used for the assumed value of Ai (Table III). As for activities not covered by the data collected, the assumed value is applied from similar activities in the table.

The geometric means for time needed to get ready are a) 83 min. and b) 150 min. Based on a statistical report covering 58 workplaces commuting time was set at 33 min. Moreover, time durations for on-the-job activities were assigned according to individual work schedules under evaluation.

2.3. Adequacy of Model

The CW model was applied to a particular work example and the results of the CW calculation for the time that each activity was completed are shown in Figure 1 as the change in CW for that axis in time. For the purpose of simplification commuting time was set at 60 min. for this example.

Since there is no generalizability for fixing the CW at 0 for the time to get up and out of the necessity for the separate examinations of the cumulative fatigue effect and fatigue recovery effect, it was decided to use the follwoing two indices derived from each work.

1) Peak Workload Value (PW): The maximum CW while on the job subtracted from the CW at the beginning of work.
2) Aggregate Workload (AW): The CW at the beginning of work is set at 0 and the negative values (rest effect) in the process of calculation are ignored to find this CW at the completion of work.

In order to examine the adequacy of the model, it was decided to establish the relationship between PW and the subjectively rated value (SW) of the size of the workload, and the one between AW and SW. The SW was found for 155 work examples from the relatively subjective value (steps were taken to offset individual differences in evaluations and differences in work schedules themselves) gotten from the train drivers (N=116 as previously mentioned) for the size of the workload for each job in a single work schedule. The resulting correlation coefficient for PW and SW was a low r = 0.656, but for AW and SW it was a most satisfactory r = 0.849. This was the highest value gotten (Table IV) for evaluating workload including experiential or common indices such as working hours.

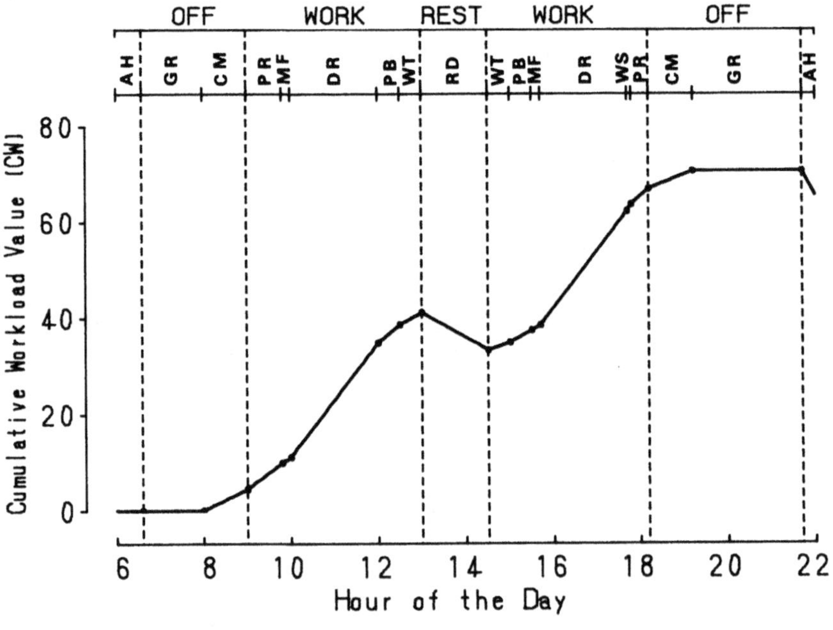

Figure 1. Changes in Cumulative Workload Value (CW) throughout a work shift. Descriptions of activities are abbreviated as follows.
AH: rest or sleep at home, GR: getting ready, CM: commuting (by train), PR: preparation for the next work, MF: moving on foot, DR: main line driving, PB: preparation for turning back, WT: waiting for train, RD: rest at destination, and WS: watching the state of power car.

TABLE IV
Correlation coefficients for subjectively estimated workload score and various workload indices

	Workload Indices	r
Cumulative Workload: (proposed)	Aggregate Workload (AW)	0.849
	Peak Workload Value (PW)	0.656
Typical ones:	Working Hours	0.800
	Running Hours	0.774
	Compulsory Hours *	0.694
	Mileage	0.364

* Compulsory Hours means the sum of working hours and off-duty hours in split shift.

3. DISCUSSION

Conclusions which can be made about the CW model at this stage are that the evaluation of each workload using only the effect of accumulated fatigue correspond well with the perceptions of those on the job for the relative relationships between shifts which is one point indicating the validity of this model. By further refining this model, it will be possible to quantitatively grasp the size of train drivers' workload so that an overly heavy load on any particular shift can be avoided when the work schedule is drawn up, and a standardized individual workload can be achieved precisely through a simple process. The two evaluation tables which are fundamental to this model lack a scientific basis, but the results gotten from the practical application of this model are highly significant. Soon a program for evaluating workload using this model will be put to use on trial basis, an experience which should provide further useful data.

As for the effects of activities assumed to foster the recovery from fatigue and changes in the CW accompanying the work process, there is no evidence to support the adequacy of the model. In order to verify this, the collection of data on fatigue for even smaller units of time and the actual measurement of the vitality level across a single shift, along with numerous other procedures, need to be undertaken. Thus, if the adequacy of the model can be supported for the CW peak which arises during a shift and the fatigue recovery effect of rest breaks, it can be used as a method to hold down the CW peak below a fixed upper limit making it possible to find a rational method for inserting rest periods in particular shifts.

An even greater problem which remains to be solved for the application of the CW model is how to evaluate the cumulative fatigue (recovery) effect when off the job. The basic premise of this model is the assumption that when a sufficient amount of rest time is permitted during a work shift it will actually give train drivers the rest they need in a effort to reduce the fatigue level to 0 (or as close to zero as possible) for the next work shift. However, since this concerns the private lives of train drivers, it is difficult to evaluate and control.

As you can now see, there is a limit to how much a business can oversee workload. For the most part, this must be left up to the individual worker himself. To accomplish this, finding a useful and easy index which the worker himself can use is of great importance.

Yoneyama (1988), for example, takes a look at the train driver's awareness of his own "drowsiness" over a period of 30 days studying the times the train drivers (N=26) went to bed and got up and the time which they became drowsy on the job. The results of the analysis of these in contrast to the work schedule during this period showed that the self-awareness of drowsiness on the job could be used as an index of the amount of fatigue at that point in time. Furtheremore, a relatively simple relationship was found between the effect of rest off the job and drowsiness. Figure 2 plots the drowsiness rate per running hour during the whole period against the total score for the sleep effect grading for each individual illustrated in Table II for the same period. Subjects are clearly divided into two groups, both of which exhibit negative regression lines. In addition, each group reflects their

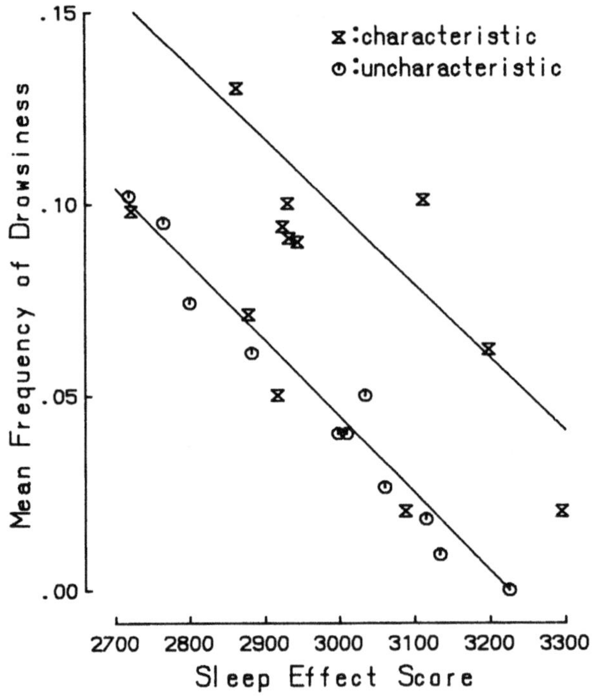

Figure 2. Mean frequency of feeling drowsiness per running hour as a function of 30-days score of Sleep Effect (Yoneyama 1988). "Characteristic" subjects were more active or in poorer condition than "uncharacteristic" ones during the period.

characteristics of the spare time off the job; drivers who were characterized by higher off-duty activities or by worse physical or mental conditions reported more drowsiness on duty than the others with the same sleep effect scores. This suggests that the rest effect is lower for the "characteristic" subjects than the "uncharacteristic" subjects. From this, a standard can be proposed requiring that "uncharacteristic" train drivers get sleep equivalent to 107 on the sleep effect rating scale to avoid the onset of drowsiness while on the job.

As a result this, the Sleep Effect Grading Table becomes adequate, and it indicates that train drivers themselves can find their own most appropriate individual activity and sleeping pattern by continuing work comparing their sleeping time data with their physical and mental condition the following day. This is useful knowledge for managing recuperation time by the worker himself.

REFERENCES

Ikeda, T. 1988. 'Workload Evaluation Scales for Train Drivers (1).' Railway Technical Research Institute Report, 2(3), 17-24. (in Japanese)

Ikeda, T. and Ohtake, H. 1986. 'Studies on Evaluation of Workload for Train Drivers.' Report of Railway Labour Science Research Institute, **86**. (in Japanese)

Ohshima, M. 1979. The Study of Fatigue, 2nd ed. Doubun-Shoin Publ., Tokyo. (in Japanese)

Yoneyama, S. 1988. 'Effects of Workload and Sleeping Hours on Railway Drivers.' Railway Technical Research Institute Report, 2(3), 25-30. (in Japanese)

A PHYSIOLOGICAL APPROACH TO THE EVALUATION OF WORKLOAD

O. Inomata
Labour Science Laboratory
Railway Technical Research Institute
2-8-38 Hikari-cho, Kokubunji-shi
Tokyo 185
Japan

ABSTRACT. The high-speed operation in keeping with a speedup ought to bring about an increased mental load on the driver, however the effect this has on the overall shortning of operating time far outweighs this shortcoming. Since operating fatigue becomes obvious when the time of a one-way trip is increased from three to four hours, this fatigue most probably occurs somewhere between these travel times.

In addition, when forced to maintain speed at a level approaching the maximum speed, it is a great load for the driver, and the long-range effects of this should be studied in greater detail.

1. INTRODUCTION

The characteristics of train operation have changed considerably along with the times. The work of operating the steam locomotive in its heyday called for advanced technology while at the same time creating a great physical load on the crew.

Consequently, the load was evaluated by measuring the Relative Metabolic Rate(RMR) through the analysis of the expired gas resulting from the work of stroking the boiler. These days, however, locomotives powered by electricity or diesel engines and electric trains have completely changed the picture. In short, the operation of such trains no longer requires the consumption of such a great degree of physical energy.

On the contrary, the improvement of train performance and operating equipment calls for increased train speed, greater transport capacity, and higher performance, and thus the nature of the crew's workload now depends on mental factors. This has also changed the items to be measured for the evaluation of workload with the wide adoption of such factors as the reflection of the autonomic nervous system (heart rate and blood pressure) and indices measuring the arousal level of the cerebral cortex (electroencephalograms and the critical flicker fusion frequency(CFF)).

Japan took the plunge into the age of high-speed mass transportation in October 1964 with the opening of the Shinkansen installed with its

A. Coblentz (ed.), Vigilance and Performance in Automatized Systems/ Vigilance et Performance de l'Homme dans les Systèmes Automatisés, 201–206.

advanced backup system.

The Shinkansen driver's cab was proposed at this laboratory in consideration of human engineering so that drivers encharged with this high-speed operation can work in the greatest comfort, and studies were carried out on what the driver should be put in charge of under never-before-experienced high speeds.

As a result, with regard to train operation, it was decided to put the driver in charge of acceleration but rely on automatic train control (ATC) for controlling deceleration. Existing surface signals would no longer be used, but rather, operations would be carried out with speed signals built in ATC resulting in considerably different driving conditions which created the need to study what effects this would have on drivers operating these trains.

Since the opening of the first Shinkansen, our institute has carried out a number of studies on the workload of Shinkansen drivers. The following is a report on a number of representative examples of this work.

2. THE PHYSIOLOGICAL EFFECTS ON THE SHINKANSEN DRIVERS OF CONTINUOUS OPERATION

When the first Shinkansen opened to service, it covered the Tokyo-Osaka distance of 515 km in four hours. Once the track bed stabilized, high-speed runs became possible and the time for the same section was reduced to three hours which has continued up to the present time. This led to experiments to study the effects of this one-hour speedup on the drivers.

Two groups of six drivers were used as experimental subjects for both conditions to compare two different driving schedules.

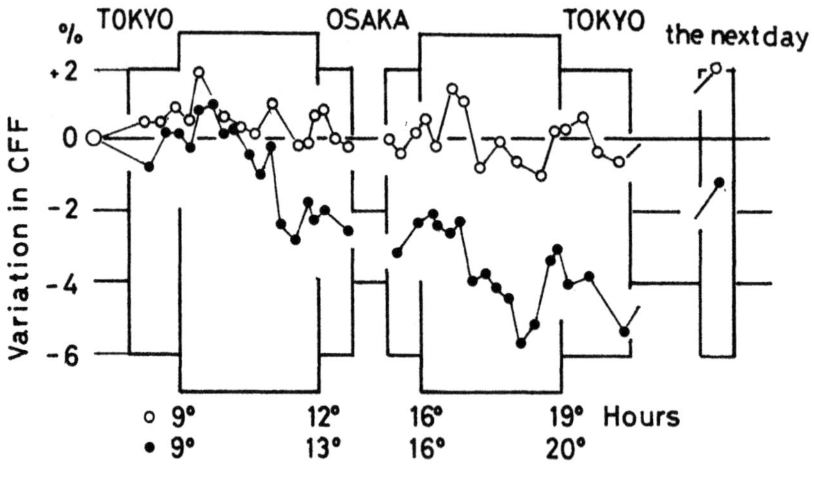

Figure 1. CFF variation curves in two different driving schedules: 3-hour(white) and 4-hour(black). CFF values measured every 20 min are shown as differences from those measured before work.

The experiment consisted of having each driver leave Tokyo for Osaka in the morning, take a short rest there, and then return to Tokyo the same day. Each driver was obligated to operate the train all alone for this section.

The critical flicker fusion frequency (CFF) results were referred to here as an index of the cortical activity level as shown in Figure 1. Figure 1 plots the rate of change in relation to value gotten for the drivers at rest prior to the round trip. The black circles indicate the results of the four-hour running time, while the white circles indicate the results of the three-hour running time.

For the four-hour trip the CFF begins to decrease about ninety minutes after departure with a 2.8% decline after three hours. After a short rest at the turning point the drivers went back on duty. The CFF value just prior to departure was still 2.8% lower than the standard value. Soon after departing on the return trip the CFF value began to decline, registering a decline of more than 5% after three hours.

On the other hand, for the three-hour trip there was only a slight decline in the latter half of the return trip, and a moderate maintenance of the driver's physical and mental functions was observed during the trip.

2.1. Discussion

It is only natural that in using the CFF to assess fatigue the standard will vary according to the nature of the work and job specifications. An operating error by a driver, for example, can cause a major accident, and in work which demands accuracy and precision, it is desirable that the worker's CFF when at work declines no more than 5% from the value at rest.

If this standard of judgment is applied, the four-hour return trip exceeds this value after three hours of operation, and thus, from the standpoint of the workload standard as well as the maintenance of safety, the one-day round trip should be limited to three hours for each leg of the journey.

The lessening of fatigue by reducing the one-way portion of the one-day round trip to three hours, was basically accomplished through the reduction of running time. High-speed operation increases mental load, and a related increase in fatigue can also expected. Thus, the shortening of time effectively ends the work before fatigue becomes noticeable.

In the case of continuous four-hour running, the CFF declines after about three hours, however, since this amount of fatigue does not interfere with the ability to operate the train, it is an acceptable running time. Nevertheless, it is difficult to get rid of this fatigue completely for the return trip. Therefore, rather than returning the same day, it is desirable that sufficient rest be gotten at the destination point and the return trip made the next morning.

In other words, the application of running time in combination with work shifts is crucial, and if this is properly carried out, fatigue will be light even if the workload is heavy.

3. THE PHYSIOLOGICAL LOAD OF INCREASED SPEED ON THE DRIVER

In setting ATC signals at 240km/h and as part of tests to improve the Shinkansen's speed, the effects on driver under two different sets of conditions were studied. Under Condition A, drivers had to maintain the speed as close as possible to the upper limit. Under Condition B, they were given relative leeway in controlling speed. There were six experimental subjects studied under Condition A and four subjects studied under Condition B. The section used for this study was the Tohoku Shin-

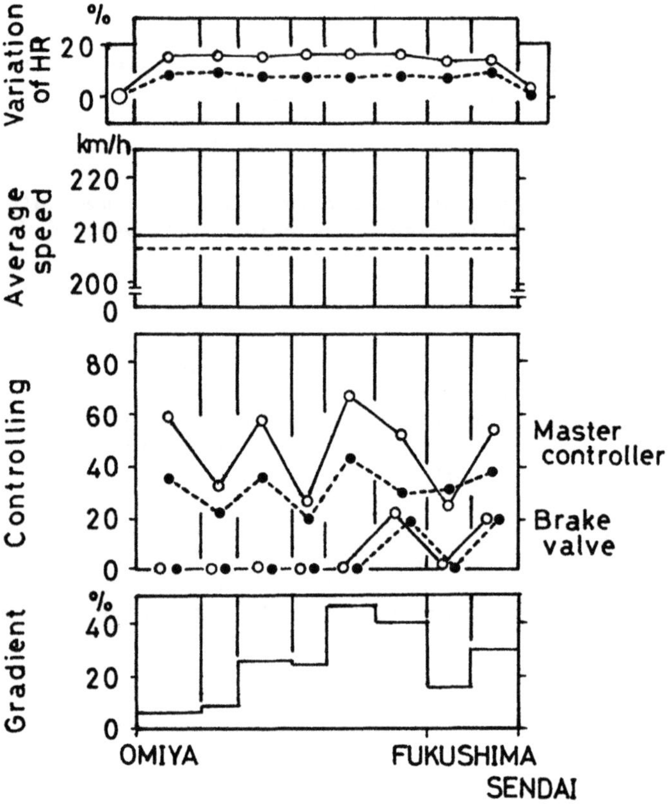

Figure 2. Relationship between variation of heart rate, average speed, driver's controlling action, and percentage of steep gradients between stops. Variation of heart rate between stops is shown as a deviation from that measured before work in the upper column. In the second column, average speed between Omiya and Sendai is shown. In the third column, frequency of operation of master controller and driver's brake handle between stops is shown, and in the lowest column, percentage of steep gradients over 10-15/1000 between stops is shown.
White circles and solid lines: Condition A. Black circles and dotted lines: Condition B.

kansen between Omiya and Sendai. Figure 2 shows the heart rate and average speed between Omiya and Sendai, the number of times the master controller and driver's brake handle were engaged, and the proportion of steep gradients between each stop.

Steep gradients are defined here as sections with gradients of more than 10-15/1000 which effect running speed. The white circles and solid lines in Figure 2 indicate the results under Condition A, while the black circles and dotted lines indicate the results under Condition B.

It is clear from Figure 2 that frequency of operation of the master controller corresponds with the proportion of gradients. The pattern of operation of the master controller is almost identical for both conditions A and B, but the amount of use was greater under Condition A. This is probably due to the need for greater use to maintain the speed as close as possible to the upper limit. This difference is also clearly illustrated by the effect on heart rates.

Figure 3 shows the relationship between the amount of the master controller and driver's brake handle and changes in heart rate. The white circles show the results under Condition A and the black circles Condition B. The heart rate shows that there was practically no recognizable effect of body movement related to the frequency of the

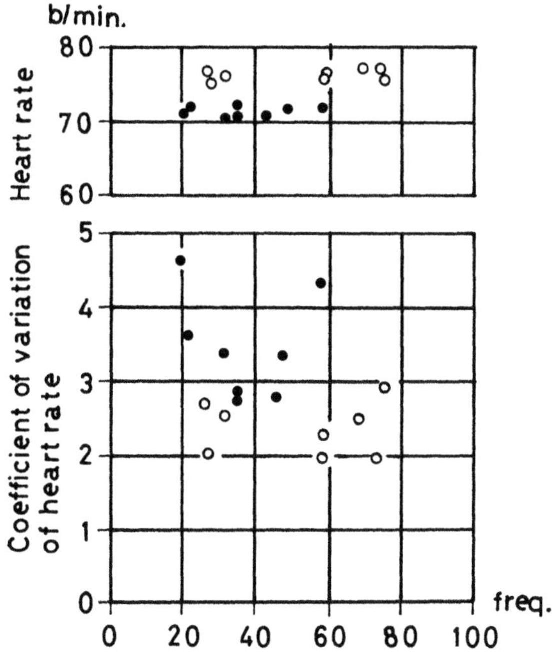

Figure 3. Relationship between heart rate and frequency of operation of the master controller and driver's brake handle.
White circles: Condition A, black circles: Condition B
There is a difference between the conditions, but there is no relation between heart rate and frequency of operation.

master controller and brake handle operation. From this it can be construed that the psychological tension of maintaining high speeds had a greater effect.

Compared to Condition B, the heart rate's coefficient of variation was small for Condition A, and this indicates that Condition A more obviously affected the maintenance of tension.

3.1. Discussion

In general, train operation in which an ATC safety backup system such as the Shinkansen's is fully considered, work requiring the driver's mental judgment decreases and the roll as mere observer increases. This tends to make the job monotonous.

In a situation such as Condition A which demands the maintenance of speeds at the very upper limit, it becomes a demanding rather than monotonous job.

When an ATC signal is tripped, the brakes are engaged automatically decelerating the train. Only after a specific amount of time passes can the train once more accelerate. Drivers don't like this situation so they try to operate trains in a way which avoids this as much as possible. This naturally demands the more delicate and frequent operation of the master controller and brake handle.

This kind of train operation is believed to raise the heart rate. The train schedule therefore should be drawn up to give drivers a little more leeway to control speed. This calls for further careful study of the long-term effects on drivers of such delicate and frequent train control.

REFERENCES

Endo, T. et al., 'Physiological Influences in High-Speed Driving.' Report of Railway Labour Science Research Institute. No.80-13, 1980. (in Japanese)

Hashimoto, K. et al., 'Physiological Strain of the Electric-Car Driving Schedules.' Bulletin of Railway Labour Science Research Institute. No.19, 1966. (in Japanese)

Ikeda, M. et al., 'A Researching Study on the Physiological Influence on the Tohoku Shinkansen Drivers During Their High-Speed Driving.' Bulletin of Railway Labour Science Research Institute. No.39, 1985. (in Japanese)

BEHAVIOURAL ADAPTATIONS TO POTENTIAL HAZARDS IN A SIMULATED DRIVING TASK

RAY FULLER
Trinity College
Department of Psychology
25-28 Westland Row
Dublin 2
Ireland

ABSTRACT. The task of vehicle driving typically involves a sustained approach to a goal destination, but from a safety viewpoint it is mainly characterised by avoidance responses. These responses may come under the control of stimuli early or late in the approach sequence to a potential hazard. Drivers' avoidance behaviour was studied in a simulated driving task comprising a 'journey' which included 15 potential hazards. Hazard probability and relative reward for delayed avoidance responses were systematically varied. High hazard probability had the effect of bringing avoidance responses forward in time. However no effects of relative reward for delayed avoidance were observed. The implications of these findings for driver safety and further research were discussed.

1. Introduction

In the exploration of the role of human factors in the genesis of road accidents, a typical view of the human agent has been as a component in an interactive system, whose main elements are the operator, the machine and the environment (Naatanen and Summala, 1976). Not surprisingly, given the traditions out of which this view arose, focus has to some extent been placed on what might be called the *ergonomics of error* (Singleton, 1976). Limitations of human performance and the conditions for its breakdown, whether in perception, attention, judgment, decision making or whatever, have been the subjects of considerable research effort and this effort has enabled significant improvement in the design of the driver's task. As a result, this task may be carried out today with greater efficiency, greater safety and greater comfort.

Nevertheless road use still continues to take an enormous

A. Coblentz (ed.), Vigilance and Performance in Automatized Systems/ Vigilance et Performance de l'Homme dans les Systèmes Automatisés, 207–217.

toll of human lives. In the European community and the United States alone, this amounts to about 100,000 deaths annually. Although this price is paid for the benefits of far greater human mobility than ever before, the *ergonomics of error* approach has clearly fallen short of a satisfactory solution to the problem of safety.

One reason for this may simply be that not enough attention has been given to the transactional nature of a large proportion of road accidents. Although drivers are occasionally passive victims of the system, accidents frequently arise out of an interaction between what the driver does at any particular moment and other features of the road and traffic environment. Thus an important question for safety is why a driver behaves in a particular way in a particular situation with the ultimate consequence of a road accident. This question has at least two obvious psychological dimensions, one to do with learning and the other with motivation. And there is one conspicuous theoretical framework which attempts to deal with both of these: behaviour theory. However there have been remarkably few attempts to utilise this conceptual framework in the theoretical exploration of the genesis of road accidents (see Parsons 1976). In a series of papers, I have begun to try to describe and analyse driver behaviour and the origins of accidents from the vantage point of a behavioural perspective (Fuller, 1984; 1988a; 1988b; 1988c; 1988d). It is to an outline description of that perspective that I shall now turn, before presenting preliminary results from some experimental work which arises directly from it.

From a behavioural perspective driver behaviour may be regarded as involving a sustained approach to some goal or destination, an approach which is characterised by a more or less continuous series of avoidance responses. These responses involve avoidance of obstacles in the driver's path, of leaving the roadway through inappropriate steering, of detection and indictment by traffic law authorities and so on. Avoidance responses which are successful in enabling drivers to avoid punishing consequences become reinforced. Such avoidance responses are typically under the control of discriminative stimuli ($S^D{}_s$) which signal the contingency between a particular response and its consequences. That is, it is only under certain conditions (discriminative stimuli) that particular responses are followed by reinforcing (rewarding) consequences. Thus the response of braking to avoid an obstacle on a dry road (S^D1) may be successful and therefore reinforced whereas the same response on an icy road (S^D2) may not. In its simplest form, then, the behavioural model has three terms: discriminative stimuli, responses and consequences.

A problem for the novice driver is that of discovering which S^Ds and which combinations of S^Ds signal the contingencies between particular responses and their consequences. Learning different contingencies can be especially difficult when it is hard to discriminate between the different stimuli which signal them: hence the widely experienced problems caused by intermittent patches of black ice and fog and sudden dash-outs from behind parked vehicles. It may be noted in passing that problems traditionally viewed as being of a more 'cognitive' nature, such as lapses in attention, may also be redefined in these terms - that is as failures to make appropriate discriminations.

Because it takes time, and therefore space, to alter the speed or direction of a moving vehicle, and because potential hazards ahead of the driver may be hidden or difficult to discriminate, designers of road systems have developed various ways of providing advance warnings of one kind or another. These warnings include stimuli such as road signs and markings, flashing lights and rumble strips. From a behavioural viewpoint, such stimuli have the potential to facilitate the driver's discrimination of potential hazards, should they arise, and enable the driver to make *anticipatory* as opposed to *delayed* avoidance responses (see Fuller, 1984 for a discussion of this distinction). Anticipatory avoidance responses may involve, for example, adjustments of speed or direction which effectively neutralise potential hazards ahead. The advance warning stimuli in such instances may be thought of as *higher-order* discriminative stimuli (see Figure 1). In general it is assumed that more risky driving may be characterised by delayed as opposed to anticipatory avoidance responding: a phenomenon described by others as 'defensive' driving.

A sequence of events remarkably similar to those represented in Figure 1 has received more than passing attention in the animal laboratory and has typically been described using classical rather than operant conditioning terminology. For our purposes this difference in terminology is of no practical or theoretical concern (see Zeiler, 1978, p. 51 for a discussion of this issue). In a review of this work (Fuller, 1988a) it was concluded that animals (such as rats, pigeons and Rhesus monkeys) tend to prefer delayed avoidance responding; to delay responding during the interval leading up to the aversive stimulus; and only make what is called here an anticipatory avoidance response if that response is made disproportionately more attractive (in terms of reward or reduced work demand). Although extrapolation from these results to humans is clearly fraught with pitfalls, the point was made that at present we simply do not have enough evidence about how humans behave under such conditions.

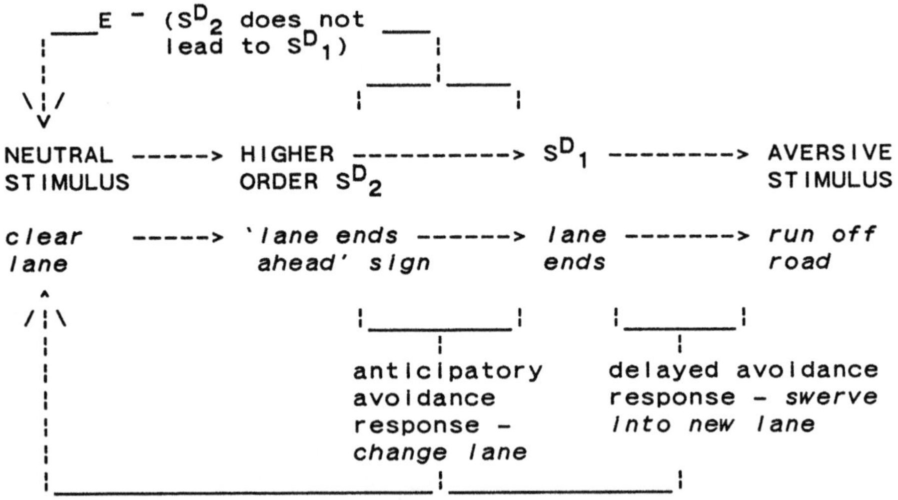

Figure 1. Representation of anticipatory and delayed avoidance responses, with an example.

A particular problem with extrapolation from the animal work is that in animal studies the discriminative stimulus is conventionally reliably followed by an aversive consequence, should the animal fail to make an avoidance response. This presumably facilitates learning of the contingency. However, in the driving situation, the learning of contingent relationships between signs and hazards is handicapped by the fact that particular higher-order S^Ds do not invariably or reliably signal the requirement for an avoidance response (represented by the upper path E in Figure 1). One example of this is the familiar 'school-ahead' sign which only provides an appropriate warning when children are *in transit* to and from the school premises. Because of this drivers get only occasional feedback that anticipatory avoidance responses, such as increased vigilance (in the sense of visual search behaviour) and decreased speed, are really necessary. The opportunity for such feedback, incidentally, has rarely been made available in animal studies. These variable contingencies should not only make learning more difficult but may also undermine the maintenance of safe driving behaviour. For this reason the effects of the *probabilistic* nature of hazardous contingencies was a major independent variable in the initial study reported below.

A further important difference between the conditions of typical animal studies and those of the driving situation is in the consequences of anticipatory and delayed avoidance responses. Unlike the conditions of most animal studies, drivers may experience some cost in making anticipatory avoidance responses over and above the sheer effort involved in the responses themselves, for example a loss of time during an urgent journey. In addition, delayed avoidance may be reinforced because an avoidance response may ultimately turn out to be unnecessary; because the experience of higher risk may be intrinsically rewarding; and because a more finely tuned response may be made to a hazard once that hazard has arisen. For these reasons the effects of variations in the *relative reward value* of anticipatory and delayed avoidance responses was the second major independent variable in our study.

To begin to investigate the effects on human behaviour of these variables a simulation of fundamental features of the driving task, as represented in Figure 1, was developed in the form of a computer game (TASCD-Version 1). On each play of the game, the subject embarks on a simulated journey which includes 15 potential hazards (trials). The sequence for each potential hazard involves presentation on a video display of a neutral stimulus, followed a predetermined number of seconds later by a higher-order S^D (S^D_2), equivalent to a road sign which is associated with a certain probability of hazard occurring. This probability is under full control of the experimenter who may determine whether or not to display the probability information to the subject. During the presentation of S^D_2, the subject may choose to make the equivalent of an anticipatory avoidance response which automatically returns him/her to a neutral stimulus condition.

If no response is made during the S^D_2 interval, after a number of seconds, again predetermined by the experimenter, the display changes, either back to the neutral stimulus, or to an S^D_1, that is a stimulus indicating that a hazard will definitely occur unless avoiding action is taken. At this point the subject *must* make a delayed avoidance response (by pressing a key) within a preselected time or a hazard (simulated collision) will result. It should be noted that if the subject responds during the S^D_2 interval (i.e. an anticipatory avoidance response) *and* this stimulus would otherwise have been followed by an S^D_1, the subject is informed of this. In this way the subject is routinely given feedback as to whether or not any anticipatory avoidance response was necessary.

During the game, selected points are awarded and displayed for responding to S^D_2 (an anticipatory avoidance response), S^D_1 (a delayed avoidance response once a hazard situation has occurred) and where no response is made and the displayed situation returns to neutral (i.e. the subject has opted for a delayed avoidance response but this response is not required because no hazard arises). Deductions of points are made for allowing 'collisions' to occur. Subjects are urged to score as many points as possible above a predetermined minimum, and to avoid 'collisions'.

With this simulation, which uses as discriminative stimuli a series of displayed colour names appearing within a rectangular frame, manipulation can readily be achieved of hazard probability, given an S^D_2, and relative rewards for anticipatory and delayed avoidance responses. Dependent variables which can be automatically measured are the frequency of each type of response (anticipatory and delayed avoidance; non-responses and 'collisions') and the latency and latency variability of each response. A record is also automatically made of each person's points score.

Thus with this simulation questions can be asked as to what happens to behaviour as a function of changes in hazard probability (either displayed or not) and as a function of changes in the relative reward value of anticipatory and delayed avoidance responses. For example, do avoidance responses tend to occur earlier in time as hazard probability increases or do human subjects behave more like laboratory animals, preferring to delay avoidance until the last few moments? Furthermore, is it possible to control the delay of avoidance responses by making such delay relatively more attractive or rewarding? The answers to such questions have obvious potential relevance to our understanding of the genesis of road accident situations.

2. Method

Subjects were three groups of 27 undergraduate students, aged between 18 and 25 years, with approximately equal numbers of both sexes. All groups completed 45 trials on the computer game (i.e. three complete 'journeys' each involving 15 potential hazards) after a preliminary period of task familiarisation and practice (15 trials). Each group experienced one of three displayed probabilities for an S^D_1 (hazard about to occur) given an S^D_2 (p=0.2, 0.5, 0.8), and also experienced 15 trials (i.e. one 'journey') at each of three levels of points for successfully opting for delayed avoidance responses - responses which were ultimately not required because no hazard occurred. These points were 100, 250 and 500 respectively. The order in which subjects were observed under levels of reward-for-delayed-avoidance-responses was randomised independently. Points for anticipatory avoidance responses were fixed at 50; for responding to a hazard occurring (S^D_1) at 0 and for a 'collision' at minus 2000. This system of allocating points was designed to simulate time-saved on a journey. Thus where the probability of a hazard occurring was high, it was much harder for the subject to accumulate points. In the same way it would be much more difficult to save time on a journey in high-density urban traffic.

The experimental design was therefore a 3x3 factorial experiment with hazard probability and reward-for-delayed-avoidance as main factors and repeated measures on the second factor.

Subjects were instructed to aim for the highest score possible above a minimum of 700 points. Trial number, last-trial score and the subject's accumulated total score were continuously displayed. Subjects were tested in groups of 2-4 individuals but at separate keyboard and monitor stations.

The software employed was written in BASIC and automatically presented each sequence of 15 trials or hazards. Intervals between all $S^D{}_s$ and between trials were set at 2 sec. The hardware used was a set of 5 Commodore C64 keyboards with 9 inch monochrome displays and a common disc drive.

3. Results

The number of anticipatory and delayed avoidance responses under conditions of varying hazard probability and reward-for-delayed-avoidance-responses were analysed separately using a 3x3 factorial AnoVa design, with hazard probability and reward-for-delayed-avoidance-response as main factors, repeated measures on the second factor and n=27 and n=26 respectively (data for 1 subject missing in second analysis).

For anticipatory avoidance responses there was a significant main effect for probability of hazard ($F_{2,78}$ = 259.4, p<.001) with progressively more anticipatory responses being given as hazard probability increased. The relevant means are presented in the first row of Table I. Although all comparisons are significantly different, there was a marked increase in the proportion of anticipatory avoidance responses when hazard probability was highest (where p[hazard] = 0.8). There were no significant effects for reward-for-delayed-avoidance-responses or interactions.

For delayed avoidance responses the relevant means are presented in the second row of Table I. Complementing the results for anticipatory avoidance responses, there was a significant and marked drop in delayed avoidance responses from the lower hazard probabilities to the high one of p[hazard] = 0.8 ($F_{2,75}$ = 193.8, p< .001). However the difference between the two lower probabilities was not significant. Again there were no significant effects for reward-for-delayed-avoidance-responses or interactions.

For the analysis of response latencies for anticipatory avoidance responses, data for level of reward-for-delayed-avoidance-responses were pooled (unweighted means) as inspection of the means revealed no differences. The latency results were then analysed using the *t* test for independent

TABLE I. Mean response frequency and latency as a function of hazard probability.

	hazard probability 0.2	0.5	0.8
anticipatory avoidance (AA) *(f)*	0.72[1]	2.79[1]	11.14
delayed avoidance (DA) *(f)*	13.76[2]	12.26[2]	3.96
response latency AA *(sec)*	0.90[2]	1.04[2]	0.74[2]
reponse latency DA *(sec)*	0.73	0.58[2]	0.52[2]

[1] p<.05 (all others p<.01)
[2] n/s (all others p<.01)

samples, comparing each level of hazard probability with each other. Although there was quite a drop in mean latency values from a hazard probability of p=0.5 to p=0.8 (see row three of Table I) this difference was not statistically reliable. There were also no other significant differences.

The mean response latencies of delayed avoidance responses which occurred to hazard onset are presented in row four of Table I. Responses were significantly slower where hazard probability was p=0.2 compared with other probability levels ($F_{2,78}$ = 103.0, p<.001). There were no other main effects or interactions.

The frequency of collisions as a function of hazard probability and reward-for-delayed-avoidance are presented in Table II. It may be seen that collisions increased with increases in reward for delayed avoidance responses and decreases in hazard probability. However overall numbers were relatively very small and observed differences were not reliable (*Chi-Square*, with *Yates Correction*, = 2.33, df=4, p>.75).

TABLE II. Numbers of 'collisions' at each level of hazard probability and of reward-for-delayed-avoidance-responses.

		hazard probability 0.2	0.5	0.8
reward for delayed avoidance response	100	2	1	1
	250	2	2	2
	500	5	3	1

4. Discussion and Conclusions

In this study, the occurrence of a hazard was associated with a points penalty (zero points compared with a *minimum* alternative of 50 points) and this penalty was described as being equivalent to the loss of time incurred by drivers in adjusting speed to a hazard in the road and traffic environment. The results reveal that manipulation of hazard probability can change behaviour, both in terms of the subjects propensity to make anticipatory as opposed to delayed avoidance responses and in terms of his or her response latency to the onset of a hazard. In particular, a high hazard probability was associated with an emphasis on anticipatory avoidance responses (a shift from about 5% to 73%) and also with shorter response latencies to hazard onset. These findings are consistent with those of two earlier studies by Wilde *et al.*, (1985) and Joly and Wilde (1987) using a rather different task from that employed here. Those studies also found that where the probability of an aversive stimulus arising was greater, subjects tended to bring their avoidance responses forward in time (see Fuller, 1988a for a brief discussion).

The corollary to the results obtained here is that where hazard probability was relatively low and there was something to be gained in gambling that the hazard *would not actually occur*, subjects appeared to adopt the more risky strategy and go for the gamble. Furthermore, under such low hazard probability conditions, response latencies to the actual occurrence of a hazard were reliably longer. These findings have obvious implications for performance safety.

Of course it is a moot point whether or not these preliminary results would generalise to experimental conditions with different but related values attached to the various parameters. Such parameters include not only hazard probability and rewards for delayed avoidance but also time intervals between the various stimuli and the rewards or penalties for anticipatory responses and collisions. It would also involve an enormous leap to generalise these results to the actual driving situation. Apart from the obvious facts that the game played by the subjects involved a rather abstract simulation of features of the driving task, and that it was a *game* that they were playing, some perhaps critical features were also absent. In particular in the real world of driving, the anticipatory avoidance response of, for example, reducing speed provides more time for the driver to discriminate potential hazards ahead; in other words it may have a partial rather than complete effect in enabling safe adaptation to hazards. Second, anticipatory and delayed avoidance responses frequently differ in both intensity and form whereas in the simulation used here the required responses were essentially the same.

A finding which highlights the problem of generalisation was that the relative reward value of delayed avoidance responses had no reliable effect on behaviour. Of course this may well be a valid general conclusion: but it is both highly implausible and inconsistent with the vast body of human operant literature. An alternative hypothesis would be that the exhortation to subjects to maximise their scores did not carry much motivational value for them. Future studies will have to take account of this possibility and employ a competitive monetary reward scheme, such as that used by Wilde *et al.*, 1985 or introduce more realistic penalties such as that described by Cownie, 1970. In Cownie's simulated driving task, subjects involved in an 'accident' had to spend two hours at a boring card sorting task!

Despite these reservations this study has shown that it is possible to modify simulated driving behaviour by manipulating hazard probability. In addition the study has also shown that the behaviour of human performers in the avoidance paradigm used is somewhat different from that of other animals. In particular there was very little evidence for the general disposition in pigeons, rats and monkeys to delay avoidance responses to the last possible moment. On the contrary such responses were extremely rare, even when the relative reward value of a delayed avoidance response was five times that of an anticipatory avoidance response. This difference may have arisen because of features of the avoidance paradigm used here or represent a true species difference. Only further research can resolve the issue.

Lastly it is worth noting another feature of behaviour in this driving simulation which merits attention and that is its sequential nature. As the subject proceeds along his or her simulated journey, the reward value of additional points may become a function not just of their absolute number but their number relative to the difference between the subject's current score and some target score. Where there is a large points discrepancy towards the end of a journey, subjects may prefer high risk/high reward response options to less risky but also less point-rewarding alternatives; that is they may well shift from anticipatory to delayed avoidance responding. Perhaps this is at the heart of many of those death and injury accidents described as being caused by 'excessive speed with regard to prevailing conditions' (Royal Ulster Constabulary, 1985).

5. Acknowledgements

The author is indebted to Colm Martin who tested the experimental subjects and to Rupert and Robin Fuller for their development of the software used in the simulated driving task.

6. References

Cownie, A.R., 1970, An operational game for the study of decision making in a hazardous activity. *Accident Analyis and Prevention*, **2**, 1-10.

Fuller, R., 1984, A conceptualisation of driving behaviour as threat avoidance. *Ergonomics*, **27**, 1139-1155.

Fuller, R., 1988a, On learning to make risky decisions. *Ergonomics*, **31**, 519-526.

Fuller, R., 1988b, Predicting what a driver will do: implications of the threat-avoidance model of driver behaviour. In *Road User Behaviour: Theory and Research* (Edited by J.A. Rothengatter and R.A. de Bruin) (Assen/Maastricht: Van Gorcum), pp. 93-105.

Fuller, R., 1988c, Psychological aspects of learning to drive. In *Road User Behaviour: Theory and Research* (Edited by J.A. Rothengatter and R.A. de Bruin) (Assen/Maastricht: Van Gorcum), pp. 527-537.

Fuller, R., 1988d, The application of behaviour theory to driver behaviour. *Paper presented at the International Symposium on Traffic Safety Theory and Research Methods*, Amsterdam, the Netherlands, April 26-28, 1988.

Joly, R. and Wilde, G.J.S., 1987, The effect of BAC at 0.8% by volume upon skill and risk-taking in a response-timing task. In *Alcohol, Drugs and Traffic Safety* (Edited by P.C. Noordzij and R. Roszbach) (Amsterdam: Elsevier).

Naatanen, R., and Summala, H., 1976, *Road User Behaviour and Traffic Accidents* (Amsterdam: North Holland).

Parsons, H.M., 1976, Caution behaviour and its conditioning in driving. *Human Factors*, **18**, 397-407.

Royal Ulster Constabulary, 1985, *Death and Injury Accidents in Northern Ireland 1985* (Belfast: Traffic Division Headquarters).

Singleton, T., 1971, Psychological aspects of man-machine systems. In *Psychology at work* (Edited by P.B. Warr) (Harmondsworth: Penguin Education), pp. 97-120.

Wilde, G.J.S., Claxton-Oldfield, S.P. and Platenius, P.H., 1985, Risk homeostasis in an experimental context. In *Human Behavior and Traffic Safety* (Edited by L. Evans and R.C. Schwing) (New York: Plenum), pp. 119-142.

Zeiler, M.D., 1978, Principles of behavioral control. In *Handbook of Applied Behaviour Analysis: Social and Instructional Processes* (Edited by A.C. Catania and T.A. Brigham) (New York: Irvington), pp. 17-60.

A BASIC STUDY ON A MULTIFUNCTION CONTROL PANEL

Kiyota KOREISHI and Kazuhiko YAMASAKI
T.R.D.I. Japan Defense Agency
2-2-1 Nakameguro Meguro-Ku Tokyo-To
153 JAPAN

ABSTRACT. From the viewpoint of ergonomics, properties between man and control system were studied. Various experiments regarding the simple reaction time of voice or touch input and accuracy of touch or lightpen input were performed. The effects of combining the waiting mode, speaking mode and words on voice reaction time could not be disregarded. The characteristics of touch input and lightpen input were discussed from the viewpoint of input time, input area and handiness for operators.

1. INTRODUCTION

In a CIC room of war vessel, vigilance task is performed for many hours with strain, consequently the control panel should be manageable for the operator.

A model of control system for a new type war vessel was made on an experimental basis. This model had multifaunction, for instance, voice recognition, synthesized voice output, keyboard, touch, lightpen or other input devices, graphic display, analogue data processing, etc.

The authers' duty was to examine the various functions and to study the effective application of this panel from the viewpoint of ergonomics.

2.METHOD

Present study consisted of two parts, that is, I) simple reaction time and II) man-machine characteristics of input devices. Furthermore the former and latter were divided into two experiments. The number of subjects was 3, 3, 4 and 5, respectively. All subjects were right-handedness and used the right hand in the whole of manual test mode.

For the present study some apparatus such as a small switch for the finger, voice switch, IO-box, timer board, software, and others were added newly. The graphic display and lightpen device (Sapiens MG-300) was delivered by Mutoh Inc. The total system and software were

A. Coblentz (ed.), Vigilance and Performance in Automatized Systems/ Vigilance et Performance de l'Homme dans les Systèmes Automatisés, 219–226.

prepared by NEC Inc.

All the process of experiment was controled by computer(NEC,PC-98). The outlines of each experiment were as follows.

2.1. Experiment I: Simple Reaction Time

2.1.1. Experiment Ia: Voice reaction time. Table 1 shows ten experimental conditions combined the waiting mode in lips, speaking mode and words. The optical and acoustic stimulus was given simultaneously. The optical stimulus was a character's appearance on the center of display, and acoustic stimulus was a beep sound.

Table 1. Conditions of the experiment Ia.

Code	Lips in Waiting	Speaking Mode	Word
1	compressed	with mouth wide open	"ah"
2		ordinarily	"ah"
3		rapidly	"alpha"
4		imagining voice input	"alpha"
5		ordinarily	"pa"
6		ordinarily	"rya"
7	parted	toungue-clicking	-
8		ordinarily	"ah"
9		ordinarily	"pa"
10		ordinarily	"pya"

2.1.2. Experiment Ib:Touch input reaction property in the different hand positions. A small switch was attached to the tip of right forefinger in order to exclude a systematic timelag of touch input devices. A plane swith which send a signal of motion of hand was attached to the waiting positions of hand.

When a target (optical stimulus) which was a small character was shown at the center of display accompanied by beep sound, a subject should do the touch input action as fast as possible.

Each distance between the center of display and the hand positions was 0.20, 0.35, and 0.55 m,that is,in the lower right corner of display, on the right side of space bar of keyboard, and on the edge of armrest of chair, respectively.

The latency (T1: stimulus - start of input motion), input time (T2: stimulus - input), moving time (T3: T2-T1) and mean velocity of hand motion (V) were measured.

2.2. Experiment II: Accuracy of Input

2.2.1. Experiment IIa: Accuracy in touch input. The effective size of display was 367×275 mm (20 in.), and the raster was 1024. The touch input device adopted a infrared method, the actual resolving power was 3.5 mm in X-axis and 3.0 mm in Y-axis.

Five grid conditions, that is, 5×5, 8×8, 12×12, 15×15 and 18×18 were prepared one by one within 350×250 mm display area. When some position of grid changed the color from null to sky-blue randomly accompanied by beep sound, subjects should search the target area and touch there with the right finger precisely in order to diminish the input errors. The target area changed the color from sky-blue to green for the right input or red for the error input.

The waiting hand position was on the side of space bar of keyboard. The interval from touch to appearance of next target was 1.5 sec. The distance from the eyes to display was about 50 cm. The inclination of display was 75 degrees.

2.2.2. Experiment IIb: Comparison between touch input and lightpen input. The logical address area of the lightpen input device was 5,600 in X-axis and 4,096 in Y-axis. Nine(3×3) target positions were prepared within a 300×200 mm display area, therefore each distance between the target positions was 150 mm in X-axis and 100 mm in Y-axis.

A small character was shown at some position randomly accompanied by beep sound. The experiment was performed in the same way as the experiment IIa.

2.3. Statistical Analysis

In the analysis of reaction time, each subject's data being exterior of mean±2s.d. were omitted. The two sided t-test was applied for statistical analysis.

3. RESULTS

3.1. Experiment I: Simple Reaction Time

3.1.1. Experiment Ia: Voice reaction time. The main results were shown in table 2. It was expected that the word "ah" would be easy to say when a subject waited the stimulus with his lips parted, contrary "pa" would be easy when the lips were compressed. Present data proved the above to be veritable.

The difference of mean voice reaction time between the fastest condition and the slowest condition was about 100 msec. It became clear that the effects of combining the waiting mode, i.e. the lips were parted or compressed, speaking mode and words on voice reaction time could not be disregarded.

3.1.2. Experiment Ib: Touch input reaction in the different hand positions. The results are shown in table 3. As a matter of course, in all three hand positions the latency (T1) did not change, and the input time (T2) and the motion time(T3) became longer by increasing a distance between the target and hand position. The mean velocity of hand motion (V) increased with the distance.

Table 2. Comparison of voice reaction time.

Compared Items	code	Subj.FY	Subj.KZ	Subj.YM
different waiting mode	2-8	>**	>**	>**
with same word	5-9	<**	<**	<**
same waiting mode	2-5		>*	<*
with different word	8-9	<***	<***	<***
articulation speed	3-4	<***		<***
number of syllables	2-3	>***	<***	
tongue effect	2-6			<***
mouth wide open effect	1-2	>***		>*
fastest condition	code	8	7	7
	mean	152	132	142
(msec,n=28-30)	s.d.	46	14	23
slowest condition	code	1	3	4
	mean	242	246	244
(msec,n=29)	s.d.	55	43	24

Two sided t-test was applied (*p<0.05, **p<0.1,***p<0.001). Meaning of each code is shown in table 1.

Table 3. Touch reaction time in three hand positions.

	Distance between the target and hand position 0.20m		0.35m		0.55m	
	mean	s.d.	mean	s.d.	mean	s.d.
T1(msec)	183	16	185	20	185	20
T2(msec)	457	63	549	62	619	60
T3(msec)	274	52	364	55	434	56
V (m/sec)	0.73	0.17	0.96	0.17	1.27	0.17

T1:latency (stimulus - start of input motion) T2: input time (stimulus - input), T3: motion time(T2-T1), V: mean velocity of the hand motion (distance /T3×1000). n=109-110

3.2. Experiment II: Accuracy of Input

3.2.1. Experiment IIa: Accuracy in touch input. The results show that the error would be within five percent when the total number of selective menu was less than fifty (figure 1). In the periphery of display many errors were generated (figure 2).

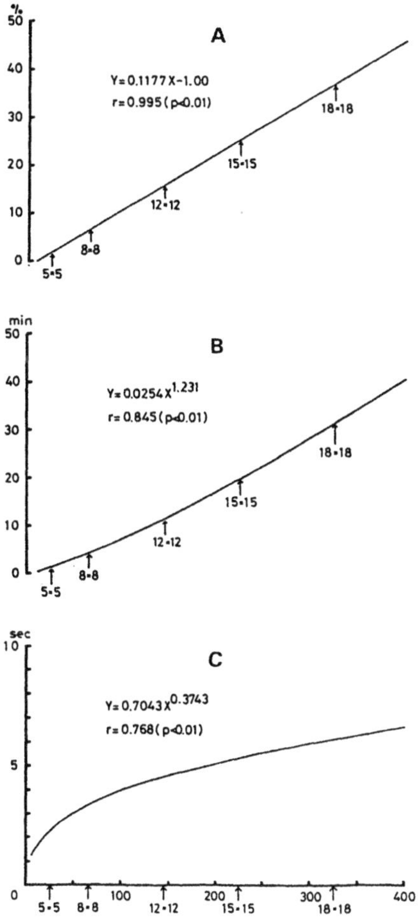

Figure 1. Touch input property in the different grid conditions. The axis of abscissa means the total number of selective menu. A: error percent, B: total input time, C: experimental input time per one area. The actual input time per one area is given by subtracting 1.5 sec from each case. n=20.

18×18

3	1	1	3	3	1	4	3	1	1	2	2	2	4	3	2	1	3
4	3	2	2	1	1	3	3	3	2	4	3	2	2	2		2	2
2	4	3	2	2	4	3	2	2	3	3	3	4	3	3	2	2	3
3	4	2	3	3	3	2	3	2	2	1	2	4	3	2	2	2	2
4	4	4	2	2	3	4	2	3	4	3	3	2	3	3	3	4	3
4	3	3	1	1	1			3	2	4	3			2	3	2	4
3	4	2	2			2		2	1	1	1	1		2	2	2	3
2	2	2	3	1	1	1	1	1			1		2	3	3	1	3
4	4	3	3		1	2		1	2	1	1		2	1	3	2	3
3	1	2	2	1	1	2	1	2	2			1	1	3	1	3	2
4	3	2	3	2		2		1	3		1	1				2	3
1	1		1		1	1		1		1							
3	3	1	1					1	1								1
3	1	1					2		1	1	1	1				1	1
1								1		2	1				2		
1	1		1	1	1	2	1	1	2	1	3		1	1	1		1
1	2	1			1				2	2	1	3	1				
	1			1		1			2	1	2	1	1	1	1	1	

Figure 2. Score in a case of 18×18 grid condition. The number of 4 means all subjects failed in touch input.

3.2.2. <u>Experiment IIb: Comparison between touch input and lightpen input</u>. Figure 3 shows the main result. In both touch input and lightpen input, the mean input position showed a tendency to close toward the center of display. The touch input device needed about from threefold to ninefold larger display area than lightpen input device.

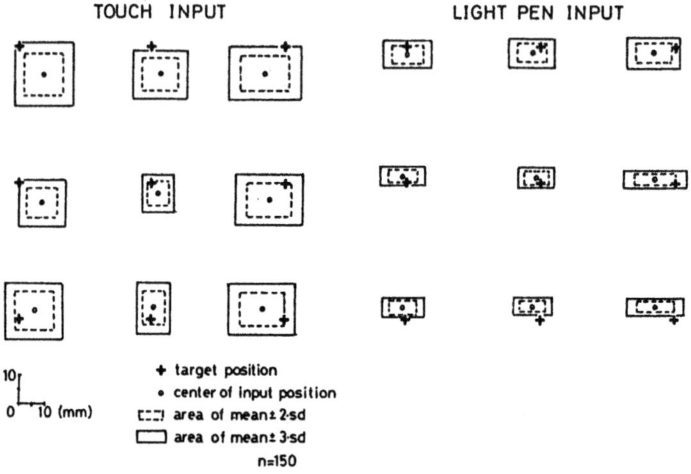

Figure 3. Comparison of scattering area between touch input and light pen input. The actual distances between target positions were 150 mm and 100 mm in horizontal and vertical direction, respectively.

4. DISCUSSION

Studies on the reaction time play an important part for design or evaluation of man-machine interface. A machine is diversified in hardware and software, however the properties of reaction time of mankind have inherent characteristics. For this reason, present study on the reaction time (experiment Ia,Ib) was performed disregarding the ability of machine side.

There were several studies about the voice reaction time. In almost cases voice reaction time was slower than manual reaction mode by about 40-70 msec in the simple reaction time (Oyama,1985), and about 150 msec in the selective attention task (Brownell,1976). Furthermore, property of voice input has attracted special interest from a viewpoint of interhemispheric transmission time(Bashore,1981), and a pilot's tasks(Wickens et al.,1984), recently. However, almost studies have not referred to the effects of waiting mode in lips, speaking mode or words.

It is definitely shown by the present study that the difference between the fastest condition and the slowest condition is about 100 msec in the simple voice reaction time.

Nowadays, the main theses of voice recognition are to eliminate the noise effect, minimizing the size of devices and improvement of algorithm, therefore the voice input reaction time has been seemed to be little account. However, it will be important for the simple on-off voice control such as emergency system, or in the near future when the computer processing time would be accelerated.

The data input using computer and display area rises in importance for the menu selection or decision of coordinates. Its management is connected with reaction time, positioning, search time, input accuracy, visualizing, etc. Regarding the above there are a lot of reports, however the authors should examine the control panel made on an experimental basis from an independent standpoint.

In the experiment Ib the properties of touch input reaction were measured. As the hands of operator are not always on a keyboard, touch input time would be dissimilar. Present data is considered to be available for the estimation of input properties in the various operator's situation.

Davies and Mebarki (1983) measured the velocity of forward hand motion,the mean movement time was about 180 msec in younger males, meanwhile the present data shows 274 msec at a distance of 0.2 m. In the motion of touch input, the tip of finger must decelerate near the display. That may be the reason there are some differences between the two.

The properties of search time and positioning are the main effector in touch input. There are various measurement methods in the studies of search time. Many authors have reported the increase of search time as the number of menu augments, similarly to the present study (e.g. Cahill and Carter, 1976; Drury and Clement, 1978). Present study shows that the input error was 5 percent, and actual input time per one area was about 1.5 second for 50 selective menu. That value would be a criterion of handiness for the touch input device similar to the present system.

As a matter of course, the accuracy of positioning of the lightpen input exceeds the touch input for the reasons of difference of resolving power and parallax reading error being inherent feature of IR method. However, the simplicity or suppleness for operators the former is inferior to the latter.

It is inevitable that the input accuracy decreases on the periphery of display similar to many other reports (e.g. Saito and Shimojo, 1979; Thomas and Schroeder, 1985). It is important the input area or position should be designed properly according to the frequency in use or degree of importance.

5. ACKNOWLEDGEMENTS

The authors would like to thank Mr.Y.Kasanami for his kind suggestions, and to Mr. K. Koizumi for his considerable assistance with the computer programs.

6. REFERENCES

Bashore, T. R., 1981, 'Vocal and manual reaction time estimates of interhemispheric transmission time', Psychological Bull., 89, 352-368.

Brownell, H. H., 1976, 'The effect of different response mode in a selective attention task', TR-83 (AD-A024-995).

Cahill, M. C. and Carter, R. C. Jr., 1976, 'Color code size for searching displays of different density', Human Factors, 18, 273-280.

Davies, B. T. and Mebarki, M., 1983, 'Speed of forward hand movement - the effect of age, sex, posture and hand' Ergonomics, 26, 1077-1079.

Drury, C. G. and Clement, M. R., 1978, 'The effect of area, density, and number of background characters on visual search', Human Factors, 20, 597-602.

Oyama, T., 1985, 'Historical background and the present status of reaction time studies (in japanese)', Japanese J. Ergonomics, 21, 57-64.

Saito, I. and Shimojo, M., 1979, 'A study on pointing task on the display board (in Japanese)', Japanese J. Ergonomics, 15, 65-71.

Thomas, B. W. and Schroeder, J. E., 1985, 'Final evaluation of the light pen as the key component in a microcomputer-based simulator', ARI RN 85-90 (AD-A160-461).

Wickens, C. D., Vidulich, M. and Sandry-Garza, D., 1984, 'Principles of S-C-R compatibility with spatial and verbal tasks: the role of display - control location and voice interactive display-control interfacing' Human Factors, 26, 533-543.

III

RYTHMES CIRCADIENS, SOMMEIL FRAGMENTE

CIRCADIAN RHYTHMS, FRAGMENTED SLEEP-WAKE SCHEDULES

TWENTY-FOUR HOUR PATTERNS OF SLEEP PROPENSITY

Peretz Lavie
Sleep Laboratory, Faculty of Medicine,
Technion - Israel Institute of Technology
Gutwirth Building, Haifa 32000, ISRAEL.

ABSTRACT: Recent years have seen a growing number of studies demonstrating the existence of ultradian cycles in various indices of arousal with periodicities centered at around 1.5h, 3-5h, and perhaps also at 12h. The present paper describes our studies on ultradian variations in arousal utilizing an ultrashort sleep-wake paradigm. The following picture of spontaneous variations in arousal emerged from the results of these studies: There are three distinct windows during the course of the 24 hrs characterized by spontaneous variations in sleep propensity: the midafternoon increase in sleepiness, an early evening period of decreased sleep propensity ("forbidden zone for sleep") and a gradual step-like nocturnal increase in sleepiness. This structure remained unchanged under sleep deprivation except for the nocturnal increase in sleepiness, which was modified into a sudden 'one-trial' increase in sleepiness. This latter phase was termed the 'primary' or nocturnal 'gate' for sleep. The timing of the nocturnal 'gate' was shown to be a stable individual characteristic. Implications of these findings to theories on sleep-wake regulation are discussed.

Introduction

Sleep propensity at different times of the day is a function of exogenous and endogenous factors. While environmental conditions may modulate sleep propensity, this modulation is bounded by the tonic level of arousal which is determined by cyclic components and homeostatic sleep regulatory processes. There is abundant evidence that endogenous cycles of arousal are comprised of both circadian and ultradian components. The periodicities of the ultradian cycles centered at 1.5 h, 3-4 h and possibly 12 hrs (Broughton, 1979; Zulley and Campbell, 1985; Lavie, 1982). The circadian and ultradian cycles interact with homeostatic regulatory processes to determine the tendency to fall asleep at different times of the day.

In recent years our research has focused on the 24h temporal structure of arousal. But instead of focusing on variations in the loosely defined "arousal level," we focused instead on 'sleepiness' or, more specifically, on the probability of initiating a sleep period at different

A. Coblentz (ed.), Vigilance and Performance in Automatized Systems/ Vigilance et Performance de l'Homme dans les Systèmes Automatisés, 229–239.

times during the 24 hours (p[w-s]). For this purpose we developed an experimental paradigm which allow a frequent sampling of sleep propensity across the day - the ultrashort sleep-wake cycle (i.e., Lavie 1986). The temporal course of sleepiness observed under this paradigm links together findings on ultradian variations in arousal showed previously in a variety of physiologic variables with the regulation of the circadian sleep-wake cycle. The present paper summarizes our findings with the ultrashort sleep-wake paradigm, and discusses their relevance to theories of sleep-wake regulation.

Experimental Procedure

The ultrashort sleep-wake cycle requires subjects either to fall asleep or to resist sleep repeatedly at frequent intervals throughout the day. Different studies varied from each other by the amount of prior wakefulness before the start of the experiment, by the starting time of the ultrashort cycle, and by the length of the ultrashort cycle. In the basic experiment, subjects came to the laboratory at 2100 after having a normal day without naps. They were fitted with electrodes to record EEG, EOG, and EMG, and spent the night asleep in the laboratory until approximately 0640; in the sleep deprivation condition, they spent the night awake under supervision. In both experimental conditions, at 0700, they began a schedule of 7-min sleep, 13-min awake, for 24 hours. Although in our preliminary studies (Lavie and Scherson, 1981; Lavie and Zomer, 1984) a 5/15 sleep-waking ratio was utilized, this was later modified to 7/13 to approximate more closely the 1:2 normal sleep-wake ratio. Electrophysiological recordings were carried out during the 7-min sleep attempts to determine sleep stages. At the end of each of the 7-min sleep attempts, whether asleep or awake, subjects were taken out of the bedroom for 13 min. In the middle of the 13-min subjects were tested on psychomotor tasks. Approximately equal size meals of light snacks and soft drinks were available every two hours throughout the experimental regimen.

The second part of the experiment investigated the temporal structure of subjects' ability to resist sleep. Since subjects who were not sleep-deprived could resist sleep without any difficulties, this part was done only after sleep deprivation. As in the first part, subjects came to the laboratory at 2100, spent the night awake, and at 0700 began an ultradian schedule of 7-min awake in bed with eyes closed, 13-min awake outside the bedroom, for 24 hrs. The specific instructions to the subjects were to lie in bed with eyes closed, and try to resist sleep for 7-min. Electrophysiological recordings were performed during the 7-min resisting sleep trials as before. Here, too, at the end of the 7-min trials, whether asleep or awake, subjects were taken out of the bedroom, and tested on the same psychomotor tests.

Sleepiness was investigated under both experimental conditions in order to determine whether the temporal structure of the ability to fall asleep is different from that of resisting-sleep. A similar temporal structure would support the interpretation that the 24-h variations in sleepiness are regulated by an active modulation of underlying hypnogenic structures. A different temporal structure, on the other hand, would weaken this interpretation and would suggest instead the possible influence of factors related to the subject's acquired sleep-wake habits.

To motivate subjects to conform to the specific experimental demands, monetary bonuses were paid to the best performing subjects in each condition.

Results

Fifty-seven normal young adults were investigated with the 7/13 ultrashort sleep-wake cycle under a variety of experimental conditions in 5 different experiments.

Amounts of sleep

The amount of sleep per 7 min trial obtained under the ultrashort sleep-wake cycle was found to be very stable. In all studies, total sleep per trial in both the AS and RS conditions was within the narrow range of 3.5-4.01 min. The differences between the amounts of sleep in the AS and RS conditions were not significant.

In addition to the between subjects consistency in the overall amount of sleep per trial, there was also evidence for a within-subject consistency in total amount of sleep. This is exemplified in Figure 1 which depicts the sleep histograms of two subjects tested under the AS and RS conditions from 1100 until 2300. It is evident that subjects showed an almost identical pattern of sleepiness under the two experimental conditions. In fact, based on their obtained amounts of sleep in the attempting- and resisting-sleep experiments, subjects could be divided into two types: "sleepy" and "alert". "Sleepy" subjects could easily fall asleep when instructed to do so, and also fell asleep when instructed to remain awake. "Alert" subjects, on the other hand, could not easily fall asleep but could easily resist sleep.

In the three experiments reported by Lavie (1986), the differences between the most sleepy and least sleepy subject for total sleep and stage 2 ranged from 189 to

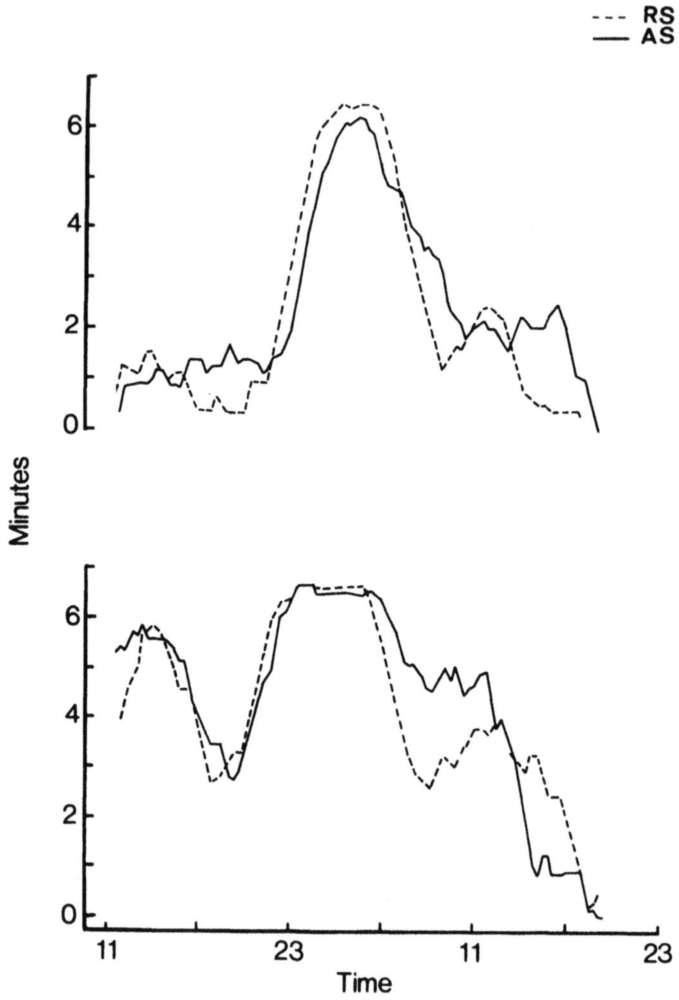

Figure 1: Total sleep of two subjects tested under the AS and RS conditions from 1100 until 2300 on the next day (36 h).

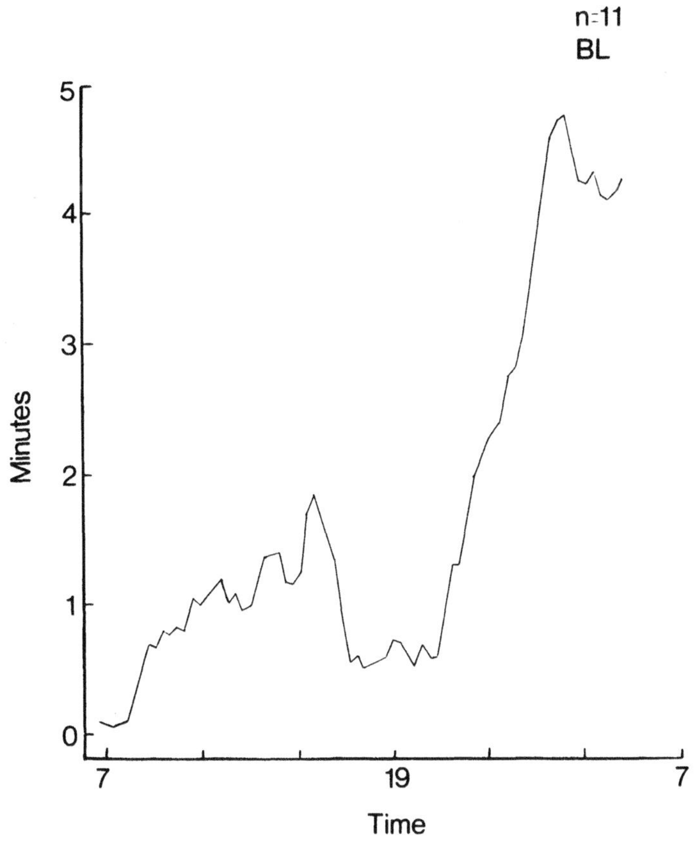

Figure 2: Mean of total sleep of 11 subjects tested from 0700 until 0700 on the next day (24 h) without sleep deprivation.

406%. The Pearson product moment correlation calculated across all subjects between total sleep in the resisting and attempting-sleep conditions was positive and statistically significant. The consistency was even higher when the stability was investigated within the same experimental condition of resisting sleep. (Lavie and Weler, in press).

'Gates' of sleep

The temporal structure of sleep propensity under normal sleep-wake cycle is depicted in Figure 2. This shows the sleep histogram of 11 subjects who were tested under the AS condition after having a 6-7h sleep in the laboratory. The probability of falling asleep varies in a systematic way across the 24h. It was low during the morning-midday period, increased during the afternoon, decreased again during the evening, and then gradually increased during the evening. It should be noted, however, that in individual subjects nocturnal sleepiness did not increase in a smooth, monotonous manner, but in a series of steps superimposed on the ascending slope.

This stepwise increase in sleepiness agrees with the subjective experience of evening sleep "spells" which are temporarily relieved by redundant "cat naps." It also agrees with the subjective experience of "missing the sleep gate," i.e., the subjective introspection of some patients that sleep must come at a certain precise time and therefore, if missed, can cause difficulties in initiating sleep.

The structure of sleep propensity was greatly modified by sleep deprivation. Here, there appear to be three pronounced peaks of sleepiness: a morning, a midafternoon and a nocturnal peak. In contrast with the gradual stepwise increase in nocturnal sleepiness under the basal condition, the increase in nocturnal sleepiness under the sleep deprivation conditions was rather steep, appearing for some subjects almost as an 'all or none' phenomenon. This is exemplified in Figure 3 which depicts the sleep histograms of the same subject, tested under the AS and RS conditions after 24h of sleep deprivation. Irrespectively of the experimental demand, nocturnal sleep steeply increased at 2300 and remained elevated for 4.5 hrs thereafter. This sharp transition period was termed the nocturnal sleep 'gate' (Lavie, 1986) and its timing was shown to be a stable individual characteristic. The correlation between the timing of the sleep 'gates' of subjects tested under the AS and RS conditions two weeks apart was 0.72 ($p<.01$).

We have preliminary evidence that this critical gating period seen under the sleep deprivation condition

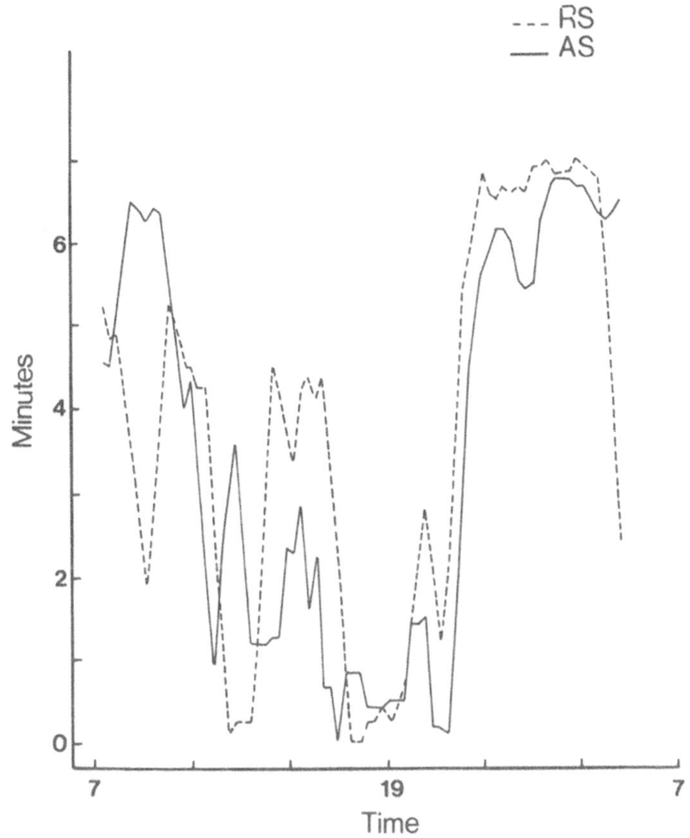

Figure 3: Total sleep of a subject tested under the AS RS conditions from 0700 until 0700 on the next day (24h) after 24h of sleep deprivation.

coincided with the first nocturnal increase in sleepiness under the baseline condition. Figure 4 presents the sleep histograms of the same subject tested under both baseline and sleep deprivation conditions. It can be seen that the first sleepiness peak under the baseline condition coincided with the timing of the sleep gate under the deprivation condition. The following explanation may be offered for the stepwise increase in sleepiness under the baseline conditions. Without prior sleep deprivation the pressure to sleep experienced around the time of the nocturnal sleep gate was relieved by the small amount of sleep accumulated during the 7 min trials. Consequently, sleepiness during subsequent trials decreased. This process of sleep pressure building up and then dissipitating because of sleep accumulation persisted

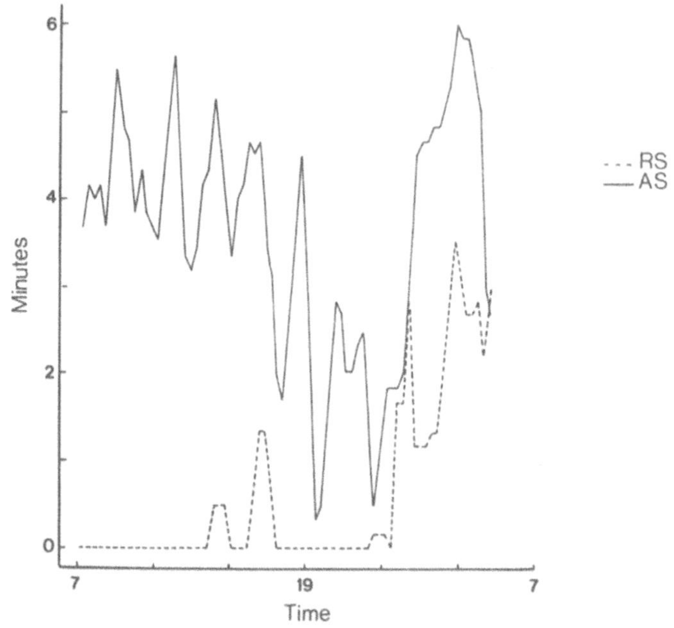

Figure 4: Total sleep of a subject tested under baseline and sleep deprivation conditions. Note the increase in sleepiness under the baseline condition at the time of the nocturnal gate.

until sleep pressure was high enough up to ensure sleep in each successive 7 min trial.

Further results from our laboratory demonstrated that the nocturnal "sleep gates" remained stable even when 2-h naps were interjected at different times during the 7/13 resisting-sleep schedule (Lavie and Weler, in press). In that study, nine subjects were tested twice on the 7/13 resisting-sleep paradigm, after one night of total sleep deprivation. In contrast to the other experiments, here, the ultradian sleep wake cycle was interrupted at either 15:00, or at 19:00, when subjects were allowed uninterrupted 2-h sleep periods. At the end of these periods, i.e. at 17:13 and 21:13, the ultradian schedule was resumed until 04:00. Subjects did not know in advance the timing of the naps nor their planned length.

Although the 2-h sleep periods had differential effects on the post-sleep sleepiness levels, the early sleep period being much more efficient than the late sleep period, the mean times of the gates were very close: 23:40 and 24:00 for the early and late sleep period conditions, respectively. The Spearman rank order correlation between the times of the gates in the two conditions was 0.68 ($p<.025$).

The "forbidden" zone for sleep

The period just before the sleep 'gate' was characterized by a large decline in the tendency to fall asleep, or alternatively, by a large increase in the ability to resist sleep. This may indicate that during this period there was spontaneous activation of arousal related neural networks which interfered with the process of falling asleep, or greatly facilitated the ability to resist sleep. Therefore we termed this period the 'forbidden zone' for sleep (Lavie, 1986). The occurrence of the 'forbidden zone for sleep', which was also reported by Strogatz et al in subjects isolated from time cues, was found to be a very robust phenomenon. Even in studies continued for 36 hrs, from 1100 to 2300 on the next day, the 'forbidden zone' for sleep was clearly visible during the first and second day of the study.

Discussion

Subjective experience shows that the level of sleepiness at any given time is a function of the length of prior wakefulness and the time of day. At 0400 most people feel sleepy regardless of prior sleep deprivation. Sleepiness is considerably augmented at this time if there is an accumulated sleep loss. These principles have been elegantly incorporated in a sleep-wake regulation theory by Daan et al (1984). Combining homeostatic principles and a circadian oscillator, Daan et al. assumed that a sleep-promoting substance S is built up during active waking, and decreases during sleep. Sleep onset is triggered when S approaches an upper threshold, and waking occurs when S reaches a lower threshold. These thresholds are suggested to be controlled by a circadian oscillator.

Some aspects of the temporal structure of sleepiness revealed by the ultrashort sleep-wake cycle pose difficulties to theories combining only homeostatic principles and circadian oscillations in order to explain sleep-wake behaviour. The present findings show that the increase in sleepiness during the day is not a linear or exponential process of the length of prior wakefulness. At two distinct time periods, during the midafternoon and early evening, there are pronounced changes in sleepiness which appear independent of prior wakefulness. The midafternoon increase in sleepiness and the early evening forbidden zone for sleep occurred whether or not subjects were sleep deprived. Furthermore, in some subjects the transition from the 'forbidden zone' to the nocturnal sleep gate was rather rapid within 1 or 2 trials which is even more difficult to reconcile with a linear process.

It should be added, however, that some aspects of the present findings can be accommodated by Daan et al's model. The secondary midafternoon 'gate' for sleep, and the ultradian cycles in sleepiness can be simulated by the model by reducing the threshold for sleep. The abruptness of the nocturnal sleep gate under the sleep deprivation condition can be also accommodated by such a model by assuming a non-linear relationship between the hypothetical sleep factor S and the sleep gate. Sleep is therefore triggered only after a critical accumulation of the sleep factor. Other models also cannot easily account for the present findings (Kronauer et al 1982). After a detailed examination of all the prevailing models of sleep–wake behavior, Strogatz (1986) summarized by saying: "The existence of the nap phase, the evening wake-maintenance zone and anomalous circadian phase during entrainment, are all beyond the reach of current models." (p.210)

Is there a need for a different sleep-wake model? While the answer to this question should be no, the present data indicate that some modifications of the prevailing models are certainly needed. A third principle, that of ultradian variations, in addition to circadian and homeostatic regulating processes should be taken into account. The stability of the temporal structure of sleepiness under all experimental conditions, and the close similarity between the AS and RS conditions, implicates an active participation of a 'clock-like' ultradian mechanism. This, at certain privileged circadian phases, overrides temporarily the activity of homeostatic processes, and increases or decreases sleep propensity independently. It is not clear at this stage if this mechanism is operating as an independent ultradian oscillator, or its principle of activity is under feedback control mechanisms.

Although more information is needed to elaborate these points, including ultradian elements in sleep-wake regulation, models will certainly improve their predictability.

References

Broughton, R. Biorhythmic variations in consciousness and psychological functions. Can. Psychol. Rev. 16: 217-239, 1979.

Daan, S., Beersma, D.G.M. and Borbely, A.A. Timing of human sleep: recovery process gated by a circadian pacemaker. Am. J. Physiol. 246: R161-R178, 1984.

Kronauer, R.E., Czeisler, C.A., Pilato, S.F., Moore-Ede, M.C., and Weitzman, E.D. Mathematical model of the human circadian system with two interacting oscillators. Am. J. Physiol. 242:R3-R17, 1982.

Lavie, P. Ultradian Rhythms in Human Sleep and Wakefulness. In: W. B. Webb (Ed.), Biological Rhythms, Sleep and Performance. Wiley: Chichester, 1982.

Lavie, P. Ultrashort sleep-waking schedule. III. Gates and "forbidden zones" for sleep. EEG. & Clin. Neurophysiol. 63, 414-425, 1986.

Lavie, P. and Scherson, A. Ultrashort sleep-waking schedule. I. Evidence of ultradian rhythmicity in "sleepability". EEG. & Clin. Neurophysiol. 52, 163-174, 1981.

Lavie, P., Zomer, J. Ultrashort sleep-waking schedule. II. Relationship between ultradian rhythms in sleepability and the REM-NONREM cycles and effects of circadian phase. EEG and Clinical Neurophysiology, 57, 35-42, 1984.

Lavie, P., and Veller, B. The timing of naps: relationship with post-nap sleepiness. EEG and Clinical Neurophysiology (in press).

Strogatz, S.H. The mathematical structure of the human sleep-wake cycle. In: S. Levin (Ed.), Lecture notes in Biomathematics vol. 69, Springer: Berlin, 1986.

Zulley, J., and Campbell, S.S. Napping behavior during "spontaneous internal desynchronization": Sleep remains in synchrony with body temperature. Human Neurobiol. 4:123-126, 1985.

SLEEP LOSS AND SUSTAINED PERFORMANCE

A.W.K. Gaillard and F.J.J.M. Steyvers
TNO Institute for Perception/Tilburg University
P.O. Box 23
3769 ZG Soesterberg
The Netherlands

ABSTRACT. Deterioration of human performance due to sleep loss is largely dependent on the type of task and the environment in which the task has to be performed. In general, the effects are most prominent in boring tasks that require continuous attention during a prolonged period. These effects are enhanced when the task has to be performed in a monotonous environment that is not very stimulating.

A review is given of the task characteristics and the environmental features that are critical in producing the detrimental effects of sleep loss. A selective approach is proposed to investigate these effects in reaction-time tasks.

1 In search of selective effects

Recent research has provided considerable insight into the mechanisms involved in the impairment of human performance due to stressors, such as sleep loss, drugs and mental fatigue (see Hockey, 1986, and Johnson, 1982, for reviews). Some tasks or task situations appear to be more vulnerable to sleep loss than others. From an applied point of view it is important to be able to predict the type of job that is likely to be vulnerable to sleep loss. One way to do this is to investigate in laboratory tasks which aspects of a task or which task demands are critical in producing the negative effects of sleep loss. In other words we have to know 1. Which psychological processes are sensitive to the effects of sleep loss, 2. Which aspects of the task configuration (e.g., task duration, feedback) or the environment (e.g., isolation) facilitate or inhibit these effects, 3. The extent to which these factors are involved in a particular job or practical situation.

The effects of stressors have often been investigated in test batteries. Results obtained in this way are often difficult to interpret and their validity is to be questioned seriously. As is argued elsewhere (Gaillard, 1988) a test battery approach has the following problems: Firstly, an effect of sleep deprivation obtained in a memory task not necessarily implies that sleep loss affects memory; there are different

A. Coblentz (ed.), Vigilance and Performance in Automatized Systems/ Vigilance et Performance de l'Homme dans les Systèmes Automatisés, 241–250.

types of memory (e.g., echoic, iconic, working memory) and different subprocesses (e.g., acquisition, retrieval); some of these may be sensitive to sleep loss, whereas others may be hardly affected. Secondly, the effects may be due to other processes than memory (e.g., speed of processing) which also determine the performance in the memory task, and therefore may cause the sleep loss effect. For example, tests, such as the classic Digit Symbol Substitution Test, involve perceptual and motor processes, central processing, memory, but also attention, concentration and eye-hand coordination. Therefore it is hard to say which psychological process is responsible for an observed decrement in performance. Thirdly, it may be the case that the effects are not specific for memory tasks, that is, the effect would also be obtained in tasks where memory does not play a role at all.

Since Broadbent (1971) several researchers have advocated the use of laboratory tasks to examine the effects of stressors. In the literature two basic approaches can be found: In the first approach the effects of several stressors, administrated alone or in combination, are investigated in one task (Wilkinson, 1969; Broadbent, 1971; Hockey, 1986); The second approach, based on the additive factors method concentrates on the interactions between task variables with and without a particular stressor (Frowein, 1981; Sanders, 1983). In both approaches conclusions are based on the pattern of the effects on human performance of the independent variables (i.e. stressors and task variables). When the effect of one variable depends on the another variable it is assumed that both influence the same psychological process; when their effects are additive they are assumed to affect different processes. In both approaches tasks are used that have a solid theoretical basis, that is, the task should have been investigated extensively in the laboratory and the psychological processes involved in the task should be largely known and described in a theoretical framework. As is outlined elsewhere (Gaillard, 1988), several experimental designs can be used to show selective effects of stressors. Basically two tasks or two conditions of the same task are compared that differ only in the extent to which one particular psychological process is involved in the task. For example in a memory search task the memory load is varied by increasing the size of the memory set; two conditions are compared: one with a high memory load and the other with a low memory load. As a function of sleep loss three types of effect may occur: 1. <u>No effect</u>: sleep loss does not influence performance in either condition; 2. <u>General effect</u>: the impairment of performance is the same in each condition, which implies that memory is not the decisive factor in generating the sleep loss effect. 3. <u>Specific effect</u>: the effect of sleep loss is significantly larger in the high memory than in the low memory condition; this suggests that memory is selectively affected by sleep loss and that the impairment in performance due to sleep loss will be larger, the more the performance in the task depends on that type of memory processes.

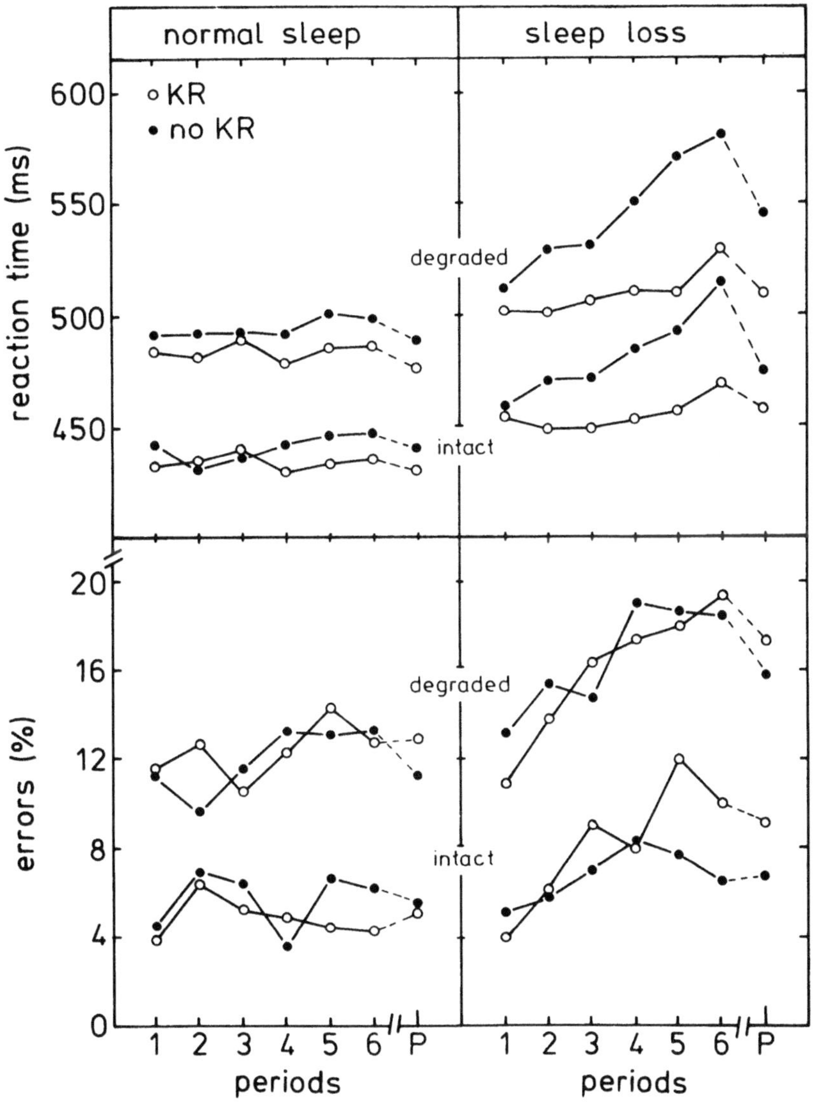

Fig. 1. Reaction-time and percentage errors after normal sleep and after one night of sleep loss (30 hrs.) during a 30 min working period (6 periods of 5 min.) After a short interruption a posttest (P) of 5 min was given (Results are obtained from Steyvers & Gaillard, unpublished manuscript).

2 Incentives and sleep loss

The effects of sleep loss and drugs have been investigated with the additive factors method in choice reaction tasks (Frowein, 1981; Sanders, 1983). In a more recent series of experiments (Steyvers, 1987; Steyvers & Gaillard, unpublished manuscript) it was examined whether the negative effects of sleep loss could be counteracted with incentives. In these experiments subjects always performed a choice reaction task during a prolonged working period (30-35 min.). Besides one night of sleep loss, knowledge of results (KR) was manipulated; KR was given immediately after each response. The stimuli were either degraded or intact, as was done in earlier experiments (Frowein, 1981; Sanders et al., 1982). The results of one experiment are presented in fig. 1. The experimental findings of these series of experiments can be summarized as follows:

Amount of sleep loss: The effects were larger in the after-noon than in the morning, that is the larger the period of sleep deprivation (25 hrs versus 30 hrs.) the larger the effect of sleep loss. Since the subjets were also tested in the morning this effect is confounded with the effects of general fatigue and time-of-day.

Task variables: The effects of sleep loss are larger with degraded stimuli than with intact stimuli which suggests that the perceptual encoding process is affected by sleep loss. The effects of sleep loss are also larger, when the stimuli are presented with irregular intervals than with a fixed interval; this suggests that sleep loss affects the preparation for a motor response (See also fig. 2 and Frowein et al., 1981). In contrast, the effects of sleep loss do not interact with stimulus-response compatibility that reflects the load on central processing (i.e., the selection of an appropriate response given a particular stimulus category).

Time-on-task: The effects of sleep loss increase towards the end of the half hour working period, in particular when the stimuli are degraded or presented irregularly (see also fig. 1).

Incentives: KR reduces the effects of sleep loss, in particular the increase in reaction-time during the working period; in contrast KR did not affect the increase in errors (see also fig. 1). KR did, however, reduce the increase in omissions after sleep loss from 4 to less than 1%; after normal sleep the omission rate was always less than 1%. KR is assumed to have both specific and aspecific properties: in the former case it provides subjects with information on the quality of their performance, which may prevent an impairment of performance. KR has, of course, also motivating properties which may prevent a reduction in task performance during a prolonged task period.

To examine further the role of incentives, the task was shortly interrupted after 30 min.; the experimenter opened the door of the cabin and announced that "the task only lasted another 5 min". As can be seen in fig. 1 this social stressor (see Hockey, 1986) induces a considerable reduction in the impairment due to sleep loss, in particular when no KR is given. In other experiments it was shown that a short pause of 1 min had a similar, but smaller effect, whereas a visual time cue, which announced the last 5 min, had only marginal effects on performance. As with KR these interruptions mainly affected the reaction-time, and not errors or omissions.

The effects of sleep loss may be modulated by other stressors, both in a positive and in a negative way. A barbiturate appears to have the same effects as the above described effects of sleep loss. As is shown in fig. 2 administration of an amphetamine can compensate for the impairment of sleep loss (Frowein et al., 1981). These findings suggest that barbiturates, amphetamines and sleep loss influence the same psychological process. Similar results have been found in a series of experiments with the five choice reaction task (see the reviews of Wilkinson, 1969; Hockey, 1986). In general, alcohol, sedative drugs, and fatigue enhance the effects of sleep loss; noise, incentives and stimulating drugs may compensate for the impairment. These results have been shown in choice reaction tasks, but may be generalized to other tasks. However, more research is necessary to examine whether the selective approach also can be applied to other tasks.

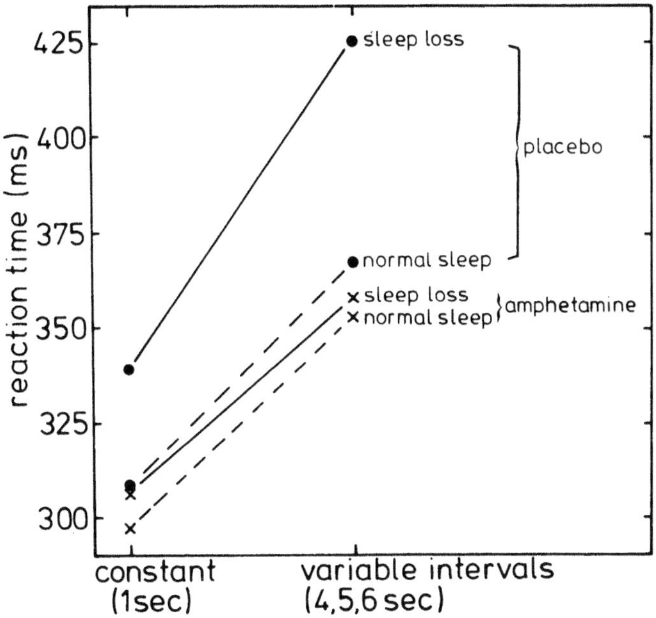

Fig. 2. Reaction-time performance in the afternoon after normal sleep and after one night of sleep loss (30 hrs.) with and without the intake of an amphetamine, as a function of the irregularity of the interval between stimuli (Results are obtained from Frowein et al., 1981).

3 Type of task

The following factors should be taken into consideration when the effects of sleep loss on task performance are evaluated:

Task structure. This refers to the psychological processes involved in the task. Results based on the selective approach in reaction time studies suggest that lower and more peripheral processes (either perceptual or motor) are more sensitive to sleep loss than central, cognitive or memory processes. Also Haslam (1981) observed that cognitive tests are not very sensitive to the effects of sleep loss. According to Wilkinson these tasks are less sensitive, because they are more interesting for the subject (see below under motivating properties).

Time structure. Effects of sleep loss have been found in continuous reaction tasks, but also in tracking and vigilance tasks, where lapses in attention immediately result in performance failure (Johnson, 1982). The performance on these tasks mostly deteriorates with task durations longer than 20 min. This impairment generally increases with sleep loss. Thus, tasks which are known for their tendency to induce performance impairment as a function of time on task, are likely to be vulnerable to the effects of sleep loss.

Self-paced, rather than experimenter-paced, tasks are less likely to deteriorate during sleep loss (Haslam, 1981). This result can also be interpreted in terms of motivating properties or active control (see below).

Feedback. Most tasks that show performance decrement do not implicitly or explicitly provide feedback about the quality of the performance. However, when knowledge of results is provided the effects may disappear at least partially. KR influences particularly the speed of processing, leaving accuracy unaffected (See also fig. 1 and Hockey, 1986).

When operators receive feedback about their performance, they can prevent an impairment by mobilizing more energy. Besides this specific effect, feedback has also a general motivating effect. Thus, the effects may also disappear when "fake" feedback or monetary incentives are given, even when they are not based on performance efficiency.

Motivating properties. Some tasks are intrinsically more motivating or stimulating than other tasks (e.g., Wilkinson, 1969). On the one hand, subjects have to be paid to do reaction tasks in a laboratory setting; the same subjects, on the other hand, may want to pay to do video games. The psychological processes involved in both tasks are rather similar, but the motivating properties are quite different. Although little research has been done on this issue, it is clear that tasks may vary enormously in their stimulating properties and therefore in their potential vulnerability to sleep loss.

Active versus passive control. Tasks differ in the amount of control an operator has over the way in which the task may be performed. In monitoring tasks the operator generally has no control over the type of events that will occur nor the timing of these events. This is also the case in watchkeeping and failure detection. In general, task performance appears to be more resistent to the effects of stress when the task or the situation gives the operator some freedom to cope actively. When operators are at liberty to work at their own pace under sleep loss, they perform relatively slowly but at a high level of accuracy. At a macro level a similar result may be obtained when operators can schedule their own activities and determine their own work/rest scheme (lunch and coffee pauses, short periods of sleep) or decide how the work is divided between operators (job rotation).

4 Environmental features

An important factor that has so far been neglected is the environment in which the task has to be performed. An operator, for example, may have to work in an isolated environment for either technical or security reasons. Isolation can take place at different levels:

Physical. Operators often have to work in a confined environment shielded from the outside world. The apparatus to be controlled or monitored may be placed in a different room or building. This often implies a lack of stimulation by sounds or voices or by changes in day light and temperature.

Psychological. Increased technology and computerization often increases the distance between the operator and the apparatus to be controlled. Given the complexity of automatized systems an operator may only know procedures and have no insight into the system and therefore he is not able to evaluate the importance of the task. Often he may not be able to perceive the effects of his actions or he may receive only indirect or delayed feedback about his performance. In combination with a low probability of detecting targets or failures, this may lead to reduced motivation and alertness.

Social. Operators may have little social contact with their coworkers. Mutual stimulation to remain fit may be an important determinant to reduce the effects of sleep loss and fatigue. Moreover, it is generally acknowledged that social support reduces the harmful effects of stress-related factors.

Thus, working in confined or isolated environments may lead to sensory deprivation, boredom, feelings of unreality and disruption of diurnal rhythm. These aspects may facilitate the deterioration in performance due to sleep loss. The most important aspect may be the monotony of the total situation (task and environment) as perceived by the operator. Monotony is characterized by an environment which does not change or only changes in a repetitive or highly predictable manner and an environment over which the operator's control is perceived to be minimal.

5 Energy management

The effects of sleep loss and other stressors are most commonly described in terms of an inverted U-curve relationship between performance efficiency and the level of activation. Each task or psychological activity is assumed to have an optimal activation level (see fig. 3).

With a too high activation level one may become restless and anxious; anxiety is assumed to have distracting effects and to reduce the available capacity which may deteriorate performance. In extreme stress situations this may lead to overstimulation and finally to panic in which state 'rational' information-processing is hardly possible. When the activation level is too low, performance may be impaired because too few resources are available to execute the task.

The extent to which the activation level can vary without affecting the performance dependens on the type of task and the environment. It has been maintained that the optimal activation level can be higher for easy than for difficult tasks. One could also maintain that for easy tasks optimal performance is less dependent on variations in the activation level.

Although it is generally assumed that the impairment in performance due to sleep loss are caused by a decreased activation level, little is known about how this takes place or which physiological mechanisms are involved. This issue is related to the fundamental question of mental fatigue. We still do not know why performance deteriorates in boring, monotonous and prolonged task situations. In general, the effects of sleep loss are most prominent in tasks that tend to impair performance even under normal sleep conditions. Therefore the mechanisms causing performance decrements in vigilance tasks are likely to be the same as those mediating the effects of sleep loss.

The effects of sleep loss can very often be compensated by an active compensatory process. Fig. 3 illustrates that outside the range where performance is optimal there is an area where the person may adapt his activation level to the optimal range by mobilizing extra effort. Thus, sleep-deprived persons may perform a task as well as non-deprived persons but at greater costs in terms of processing resources. Consequently, the effects of sleep loss will be most serious in those situations where the activation is already in the lower adaptation area, even without sleep loss. In this case the effects of sleep loss are more difficult to compensate for, because the person has already invested more effort to maintain the required performance level under conditions of no sleep deprivation.

Mobilizing extra effort to execute the task is only possible for a limited time span, depending on the person's motivation; motivation is of course an important factor in counteracting the effects of sleep loss, but there is a point beyond which the sheer need for sleep will overwhelm even a highly motivated subject.

There are large individual differences in the vulnerability to sleep loss and fatigue. Physical fitness, personality and prior experience in similar situations are important. Individuals seem to differ in their capacity to mobilize effort and to work efficiently at the limit of their capacity.

In general, introverts are superior in sustained performance, in particular when no feedback or incentives are provided. Introverts seem to be able to focus longer on the task and to be more resistent to boredom and monotony. Introverts have more control over their behavior and have chronically a higher activation level. For this reason their activation level is likely to remain optimal in situations of under-stimulation; therefore they do not have to mobilize extra effort (see fig. 3) and are less quickly tired in this type of situation.

In the series of experiments described earlier (Steyvers & Gaillard, unpublished manuscript) extraverts also tended to be more vulnerable to the effects of sleep loss; the increase in reaction-time during the 30 min working period (see also fig.1) was larger for extraverts than for introverts. Similar results have been found by May & Kline (1987; see also Eysenck, 1982 and Hockey, 1986, for reviews).

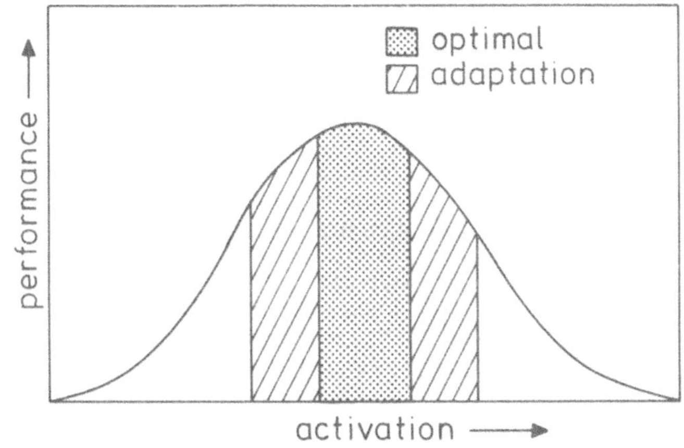

Fig. 3. The inverted U-relationship between performance efficiency and the level of activation. Indicated is an area where performance is optimal and an area where performance can only be maintained at a satisfactory level, when activation is actively manipulated by mobilizing more energy either to increase the level of activation or to suppress the effects of anxiety.

6 Concluding remarks

Detrimental effects of sleep loss are most likely to occur when the environment is relatively monotonous and isolated; sleep deprivation appears to potentiate the de-arousing influence of this situation. The effects tend to become larger the longer the working period and when no feedback is provided. Tasks that are intrinsically motivating (i.e., video-game features) are less sensitive to sleep loss. Administration of amphetamine, noise or incentives often can compensate for the negative effects of sleep loss. The results on the complexity of the task are equivocal. The critical factor appears to be how interesting, important or attractive the task is for the subject. When the task is not motivating (e.g. adding digits), the effects of sleep loss increase with complexity, whereas the reverse may be found when the task is stimulating (video game). Cognitive tasks (e.g., intelligence, memory tests), which are often of short duration and self-paced, are not or hardly affected by the effects of sleep deprivation (See also Haslam, 1981; Hockey, 1986; May & Kline, 1987). There is a need for more research on the interaction between the motivating properties of the task on the one hand and the difficulty or processing demands on the other hand. Moreover, the characteristics of the task should be examined in relation to the more general features of the task situation.

REFERENCES.

Broadbent., D.E. (1971) Decision and stress. New York, Academic Press.

Eysenck, M.W. (1982) Attention and arousal. Berlin, Springer Verlag.

Frowein, H.W. (1981) Selective drug effects on human information processing. Monograph of the Institute for Perception TNO, Soesterberg.

Frowein, H.W., Reitsma, D., and Aquarius, C. (1981) 'Effects of two counteracting stresses on the reaction process'. In: Baddeley, A.D. & Long, J.L. (Eds.), Attention and Performance IX. Hillsdale (N.J.), Erlbaum.

Gaillard, A.W.K. (1988) 'The evaluation of drug effects in laboratory tasks'. In: Hindmarch, I., Aufdembrinke, B. & Ott, H. (Eds.), Reaction time and drug effects. New York, Academic Press.

Haslam, D.R. (1981) 'The military performance of soldiers in continuous operations'. In: Johnson, L.C., Tepas, D.I., Colquhoun, & Colligan, M.J. (Eds.), Biological rhythms, sleep, and shift work. New York, Spectrum Publications.

Hockey, G.R.J. (1986) 'Changes in operator efficiency as a function of environmental stress, fatigue, and circadian rhythms'. In: Boff, K.R., Kaufman, L., & Thomas, J.P. (Eds.), Handbook of perception and human performance. New York, Wiley.

Johnson, L.C. (1982) 'Sleep deprivation and performance'. In: Webb, W.B. (Ed.), Biological rhythms, sleep, and performance. New York, Wiley.

May, J. & Kline, P. (1987) 'Measuring the effects upon cognitive abilities of sleep loss during continuous operations'. British Journal of Psychology, 78, 443-455.

Sanders, A.F. (1983) 'Towards a model of stress and human performance'. Acta Psychologica, 53, 61-97.

Sanders, A.F., Wijnen, J.L.C., Van Arkel, A.E. (1982) 'An additive factor analysis of the effects of sleep loss on reaction processes'. Acta Psychologica, 51, 41-59.

Steyvers, F.J.J.M. (1987) 'The influence of sleep deprivation and knowledge of results on perceptual encoding'. Acta Psychologica, 66, 173-187.

Steyvers, F.J.J.M. & Gaillard, A.W.K. 'The combined effects of sleep loss and incentives on reaction-time performance (unpublished manuscript).

Wilkinson, R.T. (1969) 'Some factors influencing the effect of environmental stresses upon performance'. Psychological Bulletin, 72, 260-272.

VIGILANCE AND SLEEPINESS: A LABORATORY ANALYSIS

Roger J. Broughton
Division of Neurology
Ottawa General Hospital and University of Ottawa
Ottawa, Canada K1H 8L6

ABSTRACT: This paper reviews the contribution of the author's laboratory to aspects of vigilance. A proposed 2/day endogenous rhythm of sleep propensity has been confirmed. Ultradian fluctuations in vigilance tasks and in EEG measures of alertness related to sleep quality were documented. Subjective measures have been found to be entirely unreliable in sub-chronic (5 days sleep deprivation) and chronic (narcoleptics') sleepiness. Performance deficits were seen both during episodes of drowsiness and after arousals from them (i.e., as 'sleepiness inertia'). Slow waxing and waning of alertness rather than punctate 'microsleeps' determined performance deficits in sleepy patients. Evoked potentials were a sensitive measure of sleepiness which helped differentiate qualitatively different states of 'REM sleepiness' and 'NREM sleepiness'.

1. INTRODUCTION

The British neurologist Henry Head (1923) defined vigilance as the "efficiency" of an entire organism or one of its parts. Head (1926) later stated: "When vigilance is high, mind and body are poised in readiness to any event external or internal". Vigilance theory and the study of vigilance were markedly accelerated during and immediately after World War II. At this time it became obvious that greater knowledge of human performance and "fatigue" was needed to better understand ability to sustain efficiency, particularly during relatively boring tasks such as radar detection of hostile from friendly ships, submarines and planes. Unpublished classified studies of this type were done by Sir Frederick Bartlett of the Psychology Laboratories, Cambridge University; and the topic was subsequently pioneered in laboratory studies mainly in Cambridge by N. Mackworth, Broadbent, Wilkinson and others.

Vigilance tasks came to be defined as those involving detection of slightly different (by intensity, duration, frequency, shape or other characteristics) "target" signals from a series of mainly non-target signals. Signal detection theory (Tanner and Swets 1954) distinguishing ability to respond (d') from willingness to respond (beta) was soon applied (Broadbent and Gregory 1963).

A. Coblentz (ed.), Vigilance and Performance in Automatized Systems/ Vigilance et Performance de l'Homme dans les Systèmes Automatisés, 251–261.

Vigilance research was subsequently rapidly broadened to include data on reaction times to regular and easily distinguished signals as a more-or-less direct return to Head's original concept of overall efficiency. Buck (1966) concluded that such RT tasks represented similar functions as did signal detection tasks. Further subsequent extensions included a number of measures of efficiency of "higher cognitive functions" such as tests of attention, memory and decision making. Over the last decade it has become increasingly apparent that sleep-wake states and circadian rhythms of body core temperature variables are significant modulators of performance efficiency.

This paper reviews my laboratory's contribution to the topic of vigilance in respect to chronobiological aspects, performance measures, subjective estimates and physiological indices. Much of this research was inspired by a 1975 sabbatical period with Robert Wilkinson, MRC Applied Psychology Unit, Cambridge University.

2. CHRONOBIOLOGICAL ASPECTS OF VIGILANCE

There is strong current evidence for endogenous sleep-wake biorhythms with periods of around 24 hr (circadian), 12 hrs (circasemidian), 3-4 hr (slow ultradian), and 90-120 min (ultradian) i.e., at the periodicity of the NREM/REM cycle. Each exerts effects on vigilance-type and other performance tasks. This topic has been recently reviewed in detail elsewhere (Broughton, in press) and only highlights will be mentioned here.

Circadian aspects include, of course, the endogenous rhythmic expression of the major sleep period (whose free-running period is some 24.5-25.5h) and the consequent predictable circadian acrophases of such individual sleep components as sleep onset, slow wave sleep, REM sleep and final awakening. It is now established that a major peak of alertness usually exists in the early evening. Most measures of performance efficiency (as well as body core temperature) are maximum at this time. Exceptions are those heavily involving memory, which have a separate mid-morning peak around 10-11 a.m. (Laird 1925; Baddeley et al. 1970; Folkard et al. 1976). It has been proposed that the peak of evening alertness is due to synchrony of minimum pressure for both SWS and REM sleep associated with maximum body core temperature (all optimizing vigilance functions), whereas the a.m. memory peak may be due to facilitation of REM sleep mechanisms which in extended sleep persists well across the morning (Broughton 1988; Broughton, in press).

Existence of a circasemidian (approximately 2/day) endogenous sleep rhythm in man was first proposed by Broughton (1975) and Webb (1978). It remains uncertain whether this rhythm is simply normally entrained to the circadian one (Broughton 1985) or is an intregal component of the latter (Aschoff and Gerkema 1985). The second pole of this rhythm causes the "post-lunch dip" in performance (Blake 1947) whose existence is unrelated to food intake (Hockey and Colquhoun 1972). It has been postulated that the fundamental circasemidian rhythm may not be of sleep but rather of wakefulness with two daily peaks in the usual entrained state around 1200 and 2000 hrs with gates for sleep in between (Broughton, in press). In any event, this 2/day rhythm with pressure for sleep at about 0100-0300h and again around 1500-1600h is a prominent modulator of vigilance functions across the day.

Although several studies have shown evidence for circa-4 hr fluctuations in sleep or alertness (cf. Broughton, in press), to my knowledge there are no reports as yet of such a periodicity in vigilance-type performance measures.

There is evidence for ultradian approximately 90-120 min fluctuations in daytime alertness at the periodicity of the within sleep REM cycle, as postulated by Kleitman's Basic Rest-Activity Cycle (Kleitman 1963, 1982). As extensively reviewed elsewhere (Broughton 1975, in press; Lavie 1982), such waking rhythms have been shown for EEG measures of alertness, amount of sleep in appropriate ultrashort sleep schedules, pupillometry, and in some performance tasks of vigilance type. Orr, Hoffman and Hegge (1976) and Busby and Broughton (1983) have documented such performance rhythms in some normal subjects. In research paradigms maximizing motivation and alertness, however, these rhythms are generally minimal or not demonstrable (Kripke et al. 1985). It appears that drowsiness enhances the daytime ultradian rhythm of vigilance, whereas high alertness suppresses it (Broughton 1985).

Evidence for seasonal, monthly and weekly rhythms in vigilance tasks remains weak.

It is important to keep in mind these time-dependent aspects of vigilance, when considering its other features.

3. PERFORMANCE MEASURES

Studies with Glenville, Wing and Wilkinson found that one night of total sleep deprivation led to significant deterioration on brief (10 min) portable cassette-recorded simple RT (in mean RT) and 4-choice reaction time (in mean RTs and "gaps", i.e., reponses over 1 sec, but not errors) tests, as well as on a standard 1 hr auditory vigilance task (fewer hits and lower d′, without any change in beta or in false positive rate), but not in short-term memory or hand writing variables (Glenville et al. 1978). It therefore appeared that, to be sensitive, tasks need not be long and boring, as Wilkinson (1970) had previously maintained.

Joel Herscovitch and I subsequently compared those 3 tasks plus the anagram and Wisconsin Card Sort Tests in a study of 5 days of partial sleep deprivation (- 40% from baseline) and recovery oversleep (+ 40% on first and + 20% on second recovery night) using a 16 day protocol (Herscovitch and Broughton 1981). Time of awakening was held constant. We found that partial sleep deprivation led to: fewer hits on the 1 hr auditory vigilance task, increased mean RT and increased RT variability in the simple RT task; an overall learning curve obscuring any changes in both the 4-choice serial RT and anagram measures; and more perseverative and fewer non-perseverative errors on the Wisconsin Card Sort Test

(Herscovitch et al. 1980, Herscovitch and Broughton 1981. By comparison, during oversleeping there was generally recovery of performance deficits with an inversion of Wisconsin Card sort effects which showed fewer overall perseverative errors and more non-perseverative errors.

The latter findings gave some evidence for qualitative differences in vigilance measures under conditions of undersleeping versus oversleeping, as found by Taub and Berger (1969, 1973) and reported in a case study of a 2 hour ultrashort natural sleeper (Stuss and Broughton 1978). As well, time-of-day (a.m.-p.m.) differences in test results were maximum on day 1 of partial sleep deprivation and day 1 of oversleeping with inversion of some measures such as auditory vigilance detections. These findings supported the results of Taub and Berger (1976) indicating that misalignment of circadian rhythms underlies some of the effects of sleep deprivation and oversleeping on vigilance type tasks.

Further understanding of performance efficiency in sleepy individuals has come from studies of untreated patients with narcolepsy-cataplexy compared to matched controls. As would be anticipated from studies in sleep deprived normals, we found that performance on the 1 hour auditory vigilance task was very poor with narcoleptics detecting fewer than one half the target signals than normals (without any change in false positive rate) associated with poorer performance on the 10 min 4-choice serial RT task (Valley and Broughton 1981, 1983). There were, however, no deficits in narcoleptics in very challenging short tasks including the Paced Auditory Serial Addition Task (PASAT) and the Digit Span test. These differences were due to the ability of narcoleptics to rally to (recorded) full alertness during the short challenging tasks, whereas they became drowsy in the more boring tasks. Microanalysis of periods of "sustained wakefulness" (13 sec or more of W) prior to target signals in the 1 hr vigilance task showed normal levels of detection (Valley and Broughton, 1981, 1983). Narcoleptics, therefore, were found to have no inherent irreversible organic impairment of vigilance.

Ability to rally and perform normally has also been found for memory tasks. Subjective memory problems, mainly for recent events, were encountered in approximately one half of 180 narcoleptics (Broughton et al. 1981a). However, 10 patients chosen for having major memory problems did not differ from controls, either when treated or withdrawn from treatment, in a battery of tests of immediate recall, short term memory and long term memory using both visual and auditory forms of presentation (Aguirre et al. 1985). Again, it appeared that the deficit was entirely due to sleepiness which could be temporarily overcome by motivational effects in a challenging and structured laboratory situation.

It seems probable that "fighting off sleepiness" to sustain full vigilance is not done without penalty. Sleepy patients have described different levels of effort required to accomplish different tasks. In the study of Valley and Broughton (1981) there was increasing subjective effort for the 4-choice serial RT, digit span, auditory vigilance and PASAT tasks. The high effort to accomplish the PASAT was followed in two instances by cataplexy, which was never encountered with other tests (Valley and Broughton 1983). In studies of seated narcoleptics over many hours of sustained performance Volk et al. (1984) reported higher levels of polygraphically defined "active wakefulness" (EMG and movement artifacts) compared to normals, suggesting motoric struggle against sleepiness. However, the opposite was found in ambulatory home monitorings in which greater amounts of active wakefulness occur in normal subjects (Broughton et al., in press-a). It is evident that sleepy normals and patients often must fight off their sleep pressure; and the effects upon performance after the period of effort warrants careful study.

4. SUBJECTIVE ASSESSMENT

The ability to self-assess level of alertness and related performance efficiency is an important issue. Can one depend on self estimations to know reliably whether one can fulfill an important task? The two most widely used subjective alertness measures are the 10 cm analog scale and the 7-level Stanford Sleepiness Scale of Hoddes et al. (1973). Our experience is limited to the latter.

After one night of total sleep deprivation, normal subjects could accurately predict their performance on sensitive measures, the overall correlation coefficients ranging from +0.50 - +0.69 (Glenville and Broughton 1979). This result confirmed the findings of Hoddes et al. (1973) who, however, had employed an imbalanced design and found only a non-significant trend. However, after only 5 days of partial sleep deprivation in normals, reliability of self assessment was totally lost for all sensitive vigilance tasks with the correlations varying randomly for individual tests and in individual subjects between +.58 and -.61 (Herscovitch and Broughton 1981). And in patients with chronic sleepiness, such as narcoletics, there was total inability to subjectively assess the degree of impairment measured by performance (Valley and Broughton 1981), sleep latency in the MSLT (Aguirre 1985) and evoked potentials (Aguirre and Broughton 1987).

5. PHYSIOLOGICAL MEASURES OF VIGILANCE

Physiological measures are relevant for several reasons. First, vigilance can be defined in terms of physiological (as well as performance) efficiency. Second, as stated earlier, changes in sleepiness-alertness are major modulators of performance efficiency. Third, such measures have promise to help elucidate the underlying mechanisms of performance inefficiency.

EEG measures have, of course, long been employed. Authors have at times distinguished a large number of levels of vigilance between full alertness and stage 1 sleep by the resting or stimulus-related EEG features (e.g., Roth 1964). Others have normals (Oswald 1962; Morrell 1966) and narcoleptics (Pond 1952; Daly and Yoss 1957). More recent emphasis has been upon brief 5-15 sec "microsleep" patterns associated with temporary arrest of continuous performance tasks, i.e. "lapses" (Guilleminault et al. 1975). Indeed many authors have stated that microsleeps are the major cause of the performance deficit in sleepy subjects.

Fine-grained analysis of the EEG-polygraphic patterns during Wilkinson's 1 hour auditory vigilance task in untreated narcoleptics and controls have clarified a number of these issues and have shown three main features. Valley and Broughton (1981, 1983).

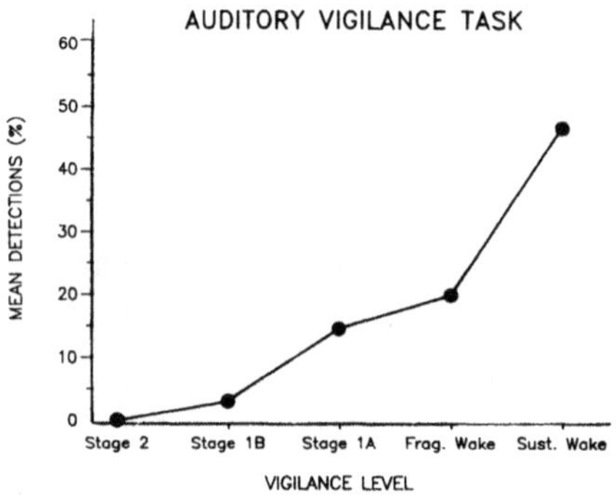

Fig 1. Mean number of detections of target stimuli in the Wilkinson 1 hour auditory vigilance task in sustained and fragmented wakefulness and stages 1A, 1B and 2 sleep.

First, detection ability was exquisitely correlated with EEG level of alertness showing normal detection rates (47%) during sustained wakefulness, a marked drop during fragmented wakefulness (18%), further decrease (14%) in stage 1A (anterior diffusion and mild slowing of alpha rhythm), very few detections (4%) in stage 1B (medium voltage theta activity), and no detections during stage 2 (Fig. 1).

Secondly, the performance deficit was not limited to fewer hits (potentially explicable by lapses). False positives and slower RTs also occurred during drowsiness. This result and the essential absence of polysomnographically defined microsleeps in the presence of marked EEG "waxing and waning" negate any major role for the "microsleep-lapse" mechanism; they indicate that slow fluctuations of alertness are much more significant. The results support the early conclusion of Oswald (1962, p. 184) that: "Lapses represent only the more extreme end of a recurrent downward sliding of cerebral vigilance... General disorganization and inappropriateness of response are also consistent with decline".

Thirdly, detection ability was much more impaired during fragmented wakefulness than sustained wakefulness. This indicates that recent somnolence has a carry-over effect and expresses a "sleepiness inertia" comparable in nature, although less marked, than the sleep-inertia or full-blown "sleep drunkenness" effects of awakening from sleep, especially SWS (Gastaut and Broughton, 1965).

It is therefore evident that it is sustained alertness over time which is important for optimum vigilance function. Transitions from alertness to drowsiness and from drowsiness to alertness are both associated with deterioration of performance. State stability over time is therefore essential to fulfill the functions of wakefulness, just as sustained unfragmented sleep appears paramount in accomplishing the latter's recuperative function (Bonnet 1985).

Cerebral evoked potentials averaged by computers were introduced by our laboratories to as sess impaired vigilance. In our first study, the auditory evoked potential triggered by tone stimuli during periods of sustained wakefulness in the 1 hour auditory vigilance test in which narcoleptic patients performed as well as controls was found to be significantly altered with lower amplitude and longer latency of late components in the narcoleptics (Broughton et al. 1981b, 1982). Variability was sufficiently high, however, that the simple AEP could not be used to actually diagnose sleepiness.

The amplitude of component P3 of the P300 odd-ball paradigm, which involves a vigilance-type detection of higher from lower frequency sound stimuli (Picton et al. 1978), was later found to be a much more sensitive measure of sleepiness (Broughton and Aguirre 1986; Aguirre and Broughton 1987). This result has been confirmed by Ollo et al. (1987). The component amplitude reduction occurs in the absence of any increased latency as is seen in aging and in dementia. Using discriminant analysis, P3 amplitude was almost as sensitive as the MSLT in separating narcoleptic from normal subjects (Broughton et al., in press-b). It is considerably less reliable, however, on an individual nap basis.

Evoked potentials give insight into the underlying sleepy cerebral state and its consequent poor efficiency on vigilance-type tasks. Component P3 amplitude has been found in various studies to index a number of higher cognitive functions including the ability to match signals with a template, confidence in decision making and equivocation, all of which appear impaired in pathological sleepiness.

6. QUALITATIVELY DIFFERENT SLEEPINESS STATES

Sleepiness has traditionally been considered as a unidimensional state varying only in magnitude. However, just as there are two fundamentally different sleep states (so that one can talk of three basic mammalian and avian states of wakefulness, NREM sleep, REM sleep), it can be postulated that qualitatively different sleepy states exist (Broughton 1982). Pathophysiologically, these would represent relative pressure for NREM sleep ("NREM sleepiness"), REM sleep ("REM sleepiness") and impaired waking mechanisms (dearousal sleepiness, or so-called subvigilance syndrome).

We have documented evidence that the sleepy brain state is in fact different immediately prior to REM versus NREM sleep, at least in narcoleptics who fall asleep directly into either state. Specifically, REM sleepiness is perceived as subjectively greater, has shorter sleep latencies on MSLT, and correlates with different event-related potential changes. The latter include a higher amplitude component P2, a strong trend for a lower amplitude component P3

(P300 odd-ball paradigm), and almost complete suppression of the so-called contingent negative variation (Broughton and Aguirre 1985, 1987). Others have found similar differences immediately prior to REM versus NREM sleep in click evoked potentials (Pressman et al. 1982) and in memory functions (Scrima 1982). It therefore appears that sleepiness and its concomitant reduction of vigilance functions may be a far more complex state than is currently envisaged.

ACKNOWLEDGEMENTS

The author would like to thank his many colleagues cited below for their collaboration in these studies, as well as the Medical Research of Canada for its support, and Barbara Reynolds for typing the manuscript.

REFERENCES

Aguirre, M. "Behavioral and physiological indices of normal and pathological sleepiness". Ph.D. thesis, University of Ottawa, Canada, 1985.

Aguirre, M. and Broughton, R.J. "Complex event-related potentials (P300 and CNV) and MSLT in the assessment of excessive daytime sleepiness in narcolepsy-cataplexy". Electroenceph. clin. Neurophysiol., 1987, **67** : 298-316.

Aguirre, M, Broughton, R. and Stuss, D. " Does memory dysfunction exist in narcolepsy-cataplexy." J. Clin. Exp. Psychol., 1985, **7** : 14-24.

Aschoff, J. and Gerkema, M. "On diversity and unity of ultradian rhythms." In: H., Schulz and P. Lavie (Eds.), Ultradian Rhythms in Physiology and Behavior. Springer-Verlag, Berlin, 1985: 321-334.

Baddeley, A.D., Hatter, J., Scott, D. and Snashall, A. "Memory and time of day." Brit. J. Psychol., 1970, **22** : 605-609.

Blake, M.J.F. "Time of day effects on performance in a range of tasks." Psychonom. Med., 1947, **9** : 349-350.

Bonnet, M.H. "Effect on sleep disruption on sleep, performance and mood." Sleep, 1985, **8** 11-19.

Broadbent, D.E. and Gregory, M. "Vigilance considered as a statistical decision." Brit. J. Psychol., 1963, **54** : 309-323.

Broughton, R.J. "Biorhythmic variations in consciousness and psychological functions." Can. Psychol. Rev., 1975, **16** : 217=230.

Broughton, R. "Performance and evoked potential measures of various states of daytime sleepiness". Sleep, 1982, **5** : S125-S146.

Broughton, R. "Three central issues concerning ultradian rhythms. In: H. Schulz and P. Lavie (Eds.), Ultradian Rhythms in Physiology and Behavior. Springer-Verlag, Berlin, 1985: 217-223.

Broughton, R. "The circasemidian sleep rhythm and its relationship to the circadian and ultradian sleep-wake rhythms." In: W. Koella, F. Obal, H. Schulz and P. Visser (Eds.), Sleep '86, Fischer-Verlag, Stuttgart/New York, 1988: 41-43.

Broughton, R. "Chronobiological aspects and models of sleep and napping." In: D. Dinges and R. Broughton (Eds.), Sleep and Alertness. Raven, New York, in press.

Broughton, R. and Aguirre, M. "Further evidence for qualitatively different types of excessive daytime sleepiness." In: W.P. Koella, E. Ruther and H. Schulz (Eds.), Sleep '84. Fischer-Verlag, Stuttgart, 1985: 115-116.

Broughton, R. and Aguirre, M. "Differences between REM and NREM sleepiness measured by event-related potentials (P300, CNV), MSLT and subjective estimate in narcolepsy-cataplexy." Electroenceph. clin. Neurophysiol., 1987, **67** : 317-326.

Broughton, R., Ghanem, Q., Hishikawa, Y., Sugita, Y., Nevsimalova, S. and Roth, B. Life-effects of narcolepsy in 180 patients from North America, Asia and Europe compared to matched controls. Can. J. Neurol. Sci., 1981a, **8** : 299-304.

Broughton, R., Low, R., Valley, V., da Costa, B. and Liddiard, S. "Auditory evoked potentials compared to EEG and performance measures in the assessment of impaired vigilance in narcolepsy-cataplexy." Sleep Res., 1981b, **10** : 184.

Broughton, R., Low, R., Valley, V., da Costa, B. and Liddiard, S. "Auditory evoked potentials compared to performance measures in assessing excessive daytime sleepiness in narcolepsy-cataplexy". Electroenceph. clin. Neurophysiol., 1982, **54** : 579-582.

Broughton, R., Dunman, W., Lutley, K., Duschesne, P. and Rivers, M. Ambulatory 24-hour sleep-wake monitoring in narcolepsy-cataplexy compared to controls. Electroenceph. clin. Neurophysiol., in press.

Broughton, R., Aguirre, M. and Dunham, W. "A comparison of multiple and single latency and cerebral evoked potential measures in the assessment of narcolepsy-cataplexy." Sleep, in press-b.

Buck. L. "Reaction time as a measure of perceptual vigilance." Psychol. Bull., 1966, **65** : 291-304.

Busby, K. and Broughton, R. "Waking ultradian rhythms of performance and motility in hyperkinetic and normal children". J. Abn. Child Psychol., 1983, **11** : 431-442.

Daly, D.D. and Yoss, R.E. "The electroencephalogram in narcolepsy." Electroenceph. clin. Neurophysiol., 1987, **9** : 109-120.

Folkard, S., Knauth, P., Monk, T.H. and Rutenfranz, J. "The effect of memory load on the circadian variation in performance efficiency under a rapidly rotating shift system." Ergonomics, 1976, **19** : 479-488.

Gastaut, H. and Broughton, R. "A clinical and polygraphic study of episodic behavior during sleep." Rec. Adv. Biol. Psychiat., 1965, **7** : 197-221.

Glenville, M. and Broughton, R. "Reliability of the Stanford Sleepiness Scale compared to short duration performance tests and the Wilkinson auditory vigilance task." In: P. Passouant and I. Oswald (Eds.), Pharmacology and the States of Alertness. Pergamon, Oxford, 1979: 235-244.

Glenville, M., Broughton, R., Wing, A.M. and Wilkinson, R.T. "Effects of sleep deprivation on short-duration performance measures compared to the Wilkinson auditory vigilance task."

Sleep, 1978, **1** : 169-176.
Guilleminault, C., Billiard, M., Montplaisir, J. and Dement, W.C. "Altered state of consciousness in disorders of daytime sleepiness." J. neurol. Sci.,1975, **26** : 377-393.
Head, H. "The concept of nervous and mental energy II. Vigilance: a physiological state of the nervous system." Brit. J. Psychol., 1923, **14** : 126-147.
Head, H. Aphasia, Cambridge University Press, Cambridge, 1926.
Herscovitch, J. and Broughton, R. "Performance deficits following short-term partial sleep deprivation and subsequent recovery oversleeping." Can. J. Psychol., 1981, **35** : 136-158.
Herscovitch, J., Stuss, D. and Broughton, R. "Changes in cognitive processing following short term partial partial sleep deprivation and subsequent recovery oversleeping." J. Clin. Neuropsychol., 1980, **2** : 301-319.
Hockey, G.R. and Colquhoun, W.P. "Diurnal variations in human performance." In: W.P. Colquhoun (Ed.), Aspects of Human Efficiency. English Universities Press, London, 1972: 1-24.
Hoddes, E., Zarcone, V., Smythe, H.R. and Dement, W.J. "Quantification of sleepiness: a new approach." Psychophysiology, 1973, **10** : 431-436.
Kleitman, N. Sleep and Wakefulness. University of Chicago Press, Chicago, 1963.
Kleitman, N. "Basic rest-activity cycle - 22 years later." Sleep, 1982, **5** : 311-317.
Kripke, D.F., Mullaney, D.J. and Fleck, P.A. "Ultradian rhythms during sustiained performance." In: H. Schulz and P. Lavie (Eds.), Ultradian rhythms in physiology and behavior. Springer-Verlag, Berlin, 1985: 201-216.
Laird, D.A. "Relative performance of college students as conditioned by time of day and day of week." J. exp. Psychol. 1925, **8** : 50-63.
Lavie, P. "Ultradian rhythms in human sleep and wakefulness." In: W.B. Webb (Ed.), Biological Rhythms, Sleep and Performance. Wiley, Chichester/New York, 1982: 239-272.
Morrell, L. "EEG frequency and reaction time - a sequential analysis." Neuropsychologia, 1966, **4** , 41-48.
Ollo, C., Squires, N., Pass, H., Walsleben, J., Baker, T. and Gujaverty, K. "Electrophysiological and neuropsychological assessment of cognitive function in narcolepsy." Sleep Res., 1987, **16** : 402.
Orr, W.C., Hoffman, H.J. and Hegge, F.W. "The assessment of time dependent changes in human performance." Chronobiologica, 1976, **3** , 293-305.
Oswald, I. Sleep and Waking. Elsevier, Amsterdam, 1962.
Picton, T.W., Campbell, K.B., Baribeau-Braun, J. and Proulx, G.B. "The neurophysiology of human attention: a tutorial review." In: J. Requin (Ed.), Attention and Performance, VII. Lawrence Erlbaum,, Hillsdale, NJ, 1978: 429-467.
Pond, D.A. "Narcolepsy: a brief critical review of eight cases." J. ment. Sci., 1952, **98** : 595-604.

Pressman, M.R., Spielman, A.J., Pollak, C.P. and Weitzman, E.D. "Long latency auditory evoked responses during sleep deprivation and in narcolepsy." Sleep, 1982, **5** (Suppl. 2): 147-175.

Roth, B. "L'EEG dans la narcolepsie-cataplexie." Electroenceph. clin. Neurophysiol., 1964, **16** : 170-190.

Scrima, L. "Isolated REM sleep facilitates recall of complex associative information." Psychophysiology, 1982, **19** : 252-259.

Stuss, D. and Broughton, R. "Extreme short sleep: personality profiles and minimum performance." Waking and Sleeping, 1978, **2** : 101-105.

Tanner, W.P. and Swets, J.A. "A decision-making theory of visual detection." Psychol. Rev., 1954, **61** : 401-409.

Taub, J.M. and Berger, R.J. "Extended sleep and performance: The Rip Van Winkle effect." Psychonomic Science, 1969, **16** : 204-205.

Taub, J.M. and Berger, R.J. "Performance and mood following variations in the timing and duration of sleep." Psychophysiology, 1973, **10** : 559-570.

Taub, J.M. and Berger, R.J. The effects of changing the phase and duration of sleep." Journal of Experimental Psychology: Human Perception and Performance, 1976, **2** : 29-33.

Valley, V. and Broughton, R. "Day-time performance deficits and and physiological vigilance in untreated patients with narcolepsy-cataplexy compared to controls." Rev. EEG Neurophysiol. (Paris), 1981, **11** : 133-139.

Valley, V. and Broughton, R. "The physiological (EEG) nature of drowsiness and its relation to performance deficits in narco-narcoleptics." Electroenceph. clin. Neurophysiol., 1983, **55** : 243-251.

Volk, S., Simon, O., Schulz, H., Hansert, E. and Wilde-Frenz, J. "The structure of wakefulness and its relationship to daytime sleep in narcolelptic patients." Electroenceph. clin. Neurophysiol., 1984, **57** : 119-128.

Webb. W.B. "Sleep and Naps." Speculations in science and technology, 1978, **1** : 313-318.

Wilkinson, R.T. "Methods for research on sleep deprivation and sleep function." Int. Psychiat. Clin., 1970 **7** 169-176.

POLYPHASIC AND ULTRASHORT SLEEP-WAKE SCHEDULES: RELEVANCE TO PERFORMNACE IN EXTENDED WORK SITUATIONS

Claudio Stampi
Human Neurosciences Research Unit
University of Ottawa
451 Smyth Rd., Ottawa, Ont. K1H 8M5, Canada

ABSTRACT. In those situations where continuous and prolonged work is experienced, the usual human monophasic sleep pattern can rarely be accomplished, producing a sometimes severe decrease of performance efficiency. It is postulated that adult humans may have an endogenous ability to adapt to polyphasic sleep patterns, and that these may represent feasible and useful strategies for management of sleep in continuous work. Several phylogenetic, ontogenetic and chronobiological issues are discussed relative to this hypothesis, which is also supported by the findings of a number of laboratory and field studies in which ultrashort sleep-wake schedules have been experimentally imposed or spontaneously adopted. The factors relevant to the design and experimentation of such schedules are discussed by presenting a laboratory based polyphasic sleep study that is being currently undertaken.

SLEEP STRATEGIES DURING CONTINUOUS WORK

The usual monophasic human sleep pattern (one 6-8 hr long sleep episode per 24 hrs) can rarely be accomplished by those professionals who may occasionally be exposed to relatively extended periods of Continuous Work (CW). These include fire fighters and police officers; health care providers; air, rail, or sea transportation personnel; nuclear, petrochemical and steel industry workers; expeditionary groups (e.g., astronauts, support crews); and rescue and emergency personnel dealing with natural (e.g., forest fire, earthquake, volcanic eruption) and manmade disasters (e.g., hazardous materials catastrophes).

Sleep need is the most important limitation during CW, since without its correct management, sleep simply tends to be partially or totally avoided, resulting in accumulated sleep debt and consequently in a sometimes serious decrease of efficiency of performance.

Which could be the best strategies to satisfy sleep need, in order to maintain high quality task performance for prolonged periods? The four theoretical ways to deal with the problem are to: (1) permit total sleep loss, (2) lessen the requirements for sleep, (3) combine partial sleep loss and recovery napping, and (4) adopt polyphasic and ultrashort sleep-wake patterns. They will be briefly discussed.

A. Coblentz (ed.), Vigilance and Performance in Automatized Systems/ Vigilance et Performance de l'Homme dans les Systèmes Automatisés, 263–273.

1. Permit Total Sleep Loss

The detrimental effects on performance, mood and efficiency of total sleep deprivation are well known (Naitoh, 1969; Johnson and Naitoh, 1974) and it is generally accepted that the upper limit of human performance for working intensively and continuously are 2 to 3 days, even though the initial effects of sleep loss may appear within the first 24 hrs of continuous work (Naitoh et al., 1983).

2. Lessen the Requirements for Sleep

If sleep could be "stored" for later use, then the individual could attempt to sleep longer before the start of a SUSOP; however, no study has yet demonstrated that sleep might be a cumulative process beyond minimal amounts, while it is well established that sleep in excess may produce performance and mood decrements similar to those experienced with equivalent amounts of sleep loss (Taub and Berger, 1973). Similarly, no pharmachologic or natural agents have yet proved to increase sleep "efficiency" per unit time. In brief, while the idea of enhancing the efficient utilization of monophasic sleep is conceptually appealing, it is well clear that there are limitations to such an approach.

3. Combine Partial Sleep Loss and Recovery Napping

Research has demonstrated that if sleep is reduced in amount, most performance can be mainatined with 60 to 70% of the usual amount of daily sleep [i.e., approximately 4.5 to 5.5 hrs of sleep per day (Johnson et al., 1977)]. Interestingly enough, it has also been shown that naps appear to have a very robust restorative power under CW and after partial or total sleep loss: when some sleep is permitted, a significant improvement of performance is reported as compared to no sleep (Naitoh et al., 1983).

The schematic analysis presented shows that there is no way to restore performance effectiveness without permitting individuals to sleep; it also suggests that short periods of sleep (naps), are disproportionately effective in recovering functioning. However, under such conditions, the effects of sleep loss are still evident and performance may be very far from 100% of baseline levels. High performance levels and maintained capacity for decision making are prerequisites for individuals dealing with such high responsibility CW tasks.

4. Adopt Polyphasic and Ultrashort Sleep-Wake Patterns

One of the main limitations of the 3 sleep strategies discussed so far is that sleep is allowed only after a more or less pronounced amount of sleep debt has been accumulated. Individuals who are already sleep deprived are certainly more sensitive to further sleep loss. Conversely, the main point of the polyphasic sleep or multiple napping strategies is to systematically sleep for short periods before the accumulation of severe sleep debt, hence minimazing its accompanying performance deficits. By adopting a polyphasic sleep strategy, the individual would in theory have a higher probability of being awake and alert when demand appears, and could also take advantage of any, even brief, interval between tasks to recuperate by voluntary napping.

The latter appears therefore as the most promising and feasible strategy to maintain high levels of efficiency during prolonged CW situations. However, typical adult human sleep is monophasic or biphasic and is apparently an exception to the general rule of the animal kingdom: although sleep is generally preferred at certain phases of the 24-hr light-dark cycle, most mammalian species show polyphasic sleep-wake patterns (Campbell and Tobler, 1984). The question arises: would adult humans easily adapt to such unusual polyphasic sleep-wake patterns?

It is postulated that humans may have an endogenous, although usually masked, ability to adapt to polyphasic sleep patterns. This hypothesis is strongly suggested by the mentioned phylogenetic, and by several ontogenetic and chronobiological considerations. Sleep is a typically poliphasic phenomenon in the human neonate and in the first months of life. This is then accompained by a triphasic pattern of one long nocturnal sleep episode and two daytime naps (one in the morning and one in the afternoon) in the first year, shifting into a single afternoon nap from the second to the fifth year. Nap tendency is diminished in the post puberty period, although it is widely recognized that particular settings may behaviorally facilitate naps, e.g., in college populations and "siesta" cultures (Webb, in press).

Chronobiological studies have well established that, besides the major nocturnal peak in sleep propensity observed around the major nocturnal core body temperature trough, humans show at least a second mid-afternoon sleep pressure pulse located close to the temperature peak (a circasemidian cycle as proposed by Broughton, 1975). However, ultradian fluctuations in sleep tendency have also been observed in different situations and experimental designs. During constant bed-rest routines, Zulley and Bailer (in press) observed a 4-hr cycle in sleep tendency, while shorter (90-120 min) rhythms of "sleepability" were described by Lavie and Scherson (1981) during ultrashort sleep experiments.

When adult human sleep was studied under similar experimental conditions as those employed in the sleep studies of other species (social and time-free environment with few alternative activities to sleep - a situation termed disentrainment), Campbell and Zulley (1985) observed polyphasic sleep-wake patterns with multiple sleep episodes dispersed throughout the nychtemeron: the median sleep episode duration was of only 2.4 hrs.

The hypothesis that adult humans can adapt to polyphasic sleep - at least for limited periods of time - is also partially supported by the findings of a number of laboratory and field studies in which ultrashort sleep-wake schedules have been experimentally imposed or spontaneously adopted. The main results of these studies, which elucidated several important characteristics of the human sleep system, will be briefly discussed.

SIMULATED ULTRASHORT SLEEP STUDIES

The first study to evaluate a polyphasic pattern of several short sleep episodes relative to monophasic sleep was conducted by Husband (1934). Forty years later, following the discovery of NREM/REM intrasleep cyclicity, several researchers were stimulated to study ultrashort sleep/wake schedules (Table 1). The sleep episode durations (SEDs) most adopted in these studies were of 30 and 60 min, with waking intervals generally being of respectively 60 and 120 min (1:2 ratio). Most experimental durations adopted were probably not long enough to allow full subjective adaptation, i.e., entrainment, to such unusual schedules; and performance measures were reported only for a few studies.

Despite these two major limitations, these pioneer studies represent an invaluable source of information towards the description of the physiology of the human polyphasic sleep system. Their main findings will be briefly summarised.

Sleep Structure

Sleep efficiency was reduced to around 50%, while maintaining a clear and somewhat shifted circadian rhythm (maximun Total Sleep Time -TST- between 0600-1300). Strogatz's (1986) re-analysis of Carskadon & Dement's (1975) data interestingly showed that subjects' TST and core body temperature demonstrated a free running rhythm (period: 25.25 hrs). Slow Wave Sleep (SWS) amounts were not reduced, while Rapid-eye-movement (REM) sleep amounts were significantly penalized.REM amounts maintained a circadian rhythm with peaks during the rising phase of core body temperature. Mean REM latencies were significantly reduced, with REM tending to occur on alternate nap periods (Carskadon & Dement 1975, Weitzman et al. 1974). "Compressed Sleep" re-analysis (Moses et al. 1978) of the two latter and of the Moses et al. (1975) studies revealed a sleep-dependent ultradian rhythm (60-100 min) in the REM cycles.

The overall response of the sleep system to these polyphasic schedules was therefore remarkably similar to what has been found in gradual sleep reduction studies (Johnson et al. 1977). This implies that these changes may be related more to reduction in TST than to polyphasic sleep *per se*.

Behavioral Functioning

Sleepiness ratings showed a shifted circadian rhythm parallel to TST. Naps containing SWS were generally followed by increased subjective sleepiness, whereas greater pre-nap sleepiness predicted

Table 1: LABORATORY ULTRASHORT SLEEP STUDIES

AUTHORS		SED min	WI min	TBT hrs	N	DURN. days
HUSBAND	1934	180	180-900	6	1	30
CURTIS & FOGEL	72	120	random	8	6	14
KELLEY et al.	73	30	60	8	1	6
HARTLEY	74	80	300-480	4	12	4
WEITZMAN et al.	74	60	120	8	7	10
CARSKADON & DEMENT	75,77	30	60	8	10	6
MOSES et al. LUBIN et al.	75 76	60	160	*6.5	8	1.7
WEBB & AGNEW	75	180	360	8	2	6
LAVIE & SCHERSON	81	5	15	*6	9	0.5
LAVIE	85,86	7	20	8.4	8	1.5
MULLANEY et al.	83	60	360	*3.4	10	1.7

SED: Sleep Episode Duration WI: Wake Interval
N: Number of subjects DURN: Duration of Experiment
TBT: Total Bed-Time per 24hrs *: + previous night sleep

greater amounts of REM in the subsequent nap. The generally high daily mean sleepiness found at the beginning of the schedule tended to decrease with adaptation (towards the end of the study) to close to baseline levels (Carskadon & Dement 1975). Curtis & Fogel (1972) found that ability to sleep at irregular times may be determined in part by personality factors. Significant ameliorative effects on performance were produced by 1-hr naps every 220 min (Lubin et al. 1976) or every 7 hrs (Mullaney et al. 1983), as compared to no sleep. Interestingly enough, Hartley (1974) has shown that a polyphasic schedule of three 80-min naps lead to significantly higher perfomance levels compared to a monophasic pattern of the same amount (4 hrs) of sleep per day.

From the perspective of sleep management in the face of demands for prolonged functioning, the most interesting findings of these studies were that the sleep system seems to have a high plasticity in terms of scheduling and duration, and that adult humans appeared to adapt surprisingly well to schedules of multiple naps.

REAL-LIFE ULTRASHORT SLEEP STUDIES

Before any attempt to design napping schedules, it was hypothesyzed that putative human polyphasic sleep propensity may be best expressed and revealed under extreme situations where the individual cannot follow the habitual monophasic sleep pattern. Which are the spontaneous reactions of an individual to prolonged continuous work when no sleep-wake schedule is imposed? Will the natural tendency be to sleep less, to totally avoid sleep, or will a napping regime be adopted? If a polyphasic sleep pattern is chosen, will it be beneficial to the subject's performance?

In order to address these questions, a peculiar real-life model of extreme and prolonged work situation was selected as experimental setting: solo sailing races across the oceans. During a solo race, the yachtsman is exposed, over periods ranging from 2 up to 6 weeks, to recurrent, almost unpredictable and sometimes extenuating demands to be awake, to steer and adjust his yacht to varying and occasionally prohibitive conditions of sea and wind, and to carefully study race strategies that will permit to optimize boat performance and obtain maximun speed.

Behavioral studies (through questionnaires, sleep logs and interviews) were conducted during 3 races: the Observer Single-handed Transatlantic Race of 1980 (OSTAR, 3000 nautical miles), the 1982 Round Britain Race (RBR - 1900 miles) and the 1983 Mini-Transat (MT - 4300 miles), for a total of 99 (54 + 29 + 16) subjects studied (Stampi, 1985; in press). Competitors took an average of respectively 33, 36 and 14 days to finish the races, while their ages ranged between 20 and 67 years, with means of respectively 38.7, 29.7 and 35.6 years.

As anticipated, ultrashort sleep patterns with mean sleep episode durations (SEDs) ranging between 10-20 min to 2 hr were widely adopted (66.5% of the sailors. The overall mean SEDs per each race were quite short and respectively of 2.06 hrs, 2.7 hrs and 37 min for OSTAR, RBR and MT. Competitors therefore tried to meet the need of staying awake and on watch as much as they could without greatly reducing TST [overall mean: 6.3 (sd 1.7) hrs/24h], but mostly by fragmenting sleep into short naps.

In order to verify if the greatly reduced SEDs were capable of maintaining competitors' performance at reasonable levels, a linear regression analysis was performed between individual mean SEDs and mean speed of yachts. We found that in all 3 competitions race performance was significantly negatively correlated to sleep episode duration: the shorter the sailor's mean SED, the better his race standing. The best results were achieved by competitors whose sleep episodes ranged from 10-20 min to 1 hr, while obtaining a total of approximately 4.5 to 5.5 hr of sleep per day.

The surprising and successful use of polyphasic sleep strategies by solo sailors has been recently reconfirmed and studied in much greater detail by adoption of ambulatory wrist movement recorders (Actigraph, Ambulatory Monitoring, Inc., NY). Twenty-eight wrist actigraphs have been distributed amongst the 95 competitors of the last edition of C-STAR races (Plymouth, UK, to Newporth, RI, 5 June 1988). Preliminary results obtained from one competitor (Fig. 1)show that frequent and very short sleep episodes were adopted. Further analysis of the data collected will permit to comprehensively describe the sleep strategies adopted, incuding important aspects that were not documented in the first behavioral study, such as the temporal placement (preferred phases) of naps.

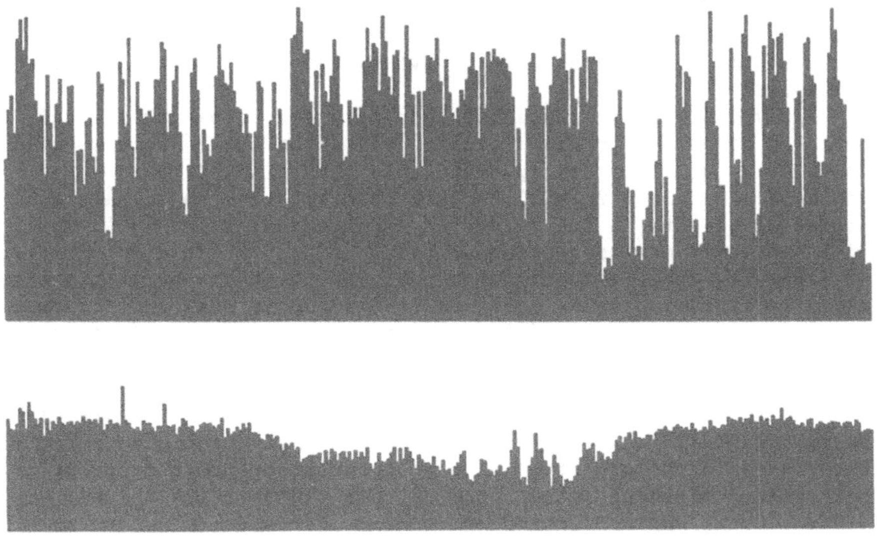

Figure 1. Twenty-four hour period actigraphic recordings of a solo sailor's wrist movement combined to his yacht motion (top). Sailor's movement is obtained by subtracting yacht motion (bottom) from combined (top) values per each (5 min) epoch. Note the multiple and ultrashort sleep episodes adopted. Mean sleep episode duration was of aproximately 20 min. Data refers to the seventh day after transatlantic race start.

DESIGN AND EXPERIMENTATION OF POLYPHASIC SLEEP STRATEGIES

It not only appears that adult humans adapt reasonably well to artificially (laboratory) imposed ultrashort sleep-wake schedules, but that when motivated individuals are free to self-administer sleep under prolonged and sustained performance conditions (up to 40 days in the present studies), most tend to adopt ultrashort sleep patterns, and that this procedure improves performance as compared to longer sleep episodes.

It appears therefore that humans may not be an exception to the general polyphasic sleep rule observed across phylogeny: when obstructions are removed (social, occupational or other pressures), or when specific stimuli appear, the endogenous polyphasic sleep ability tends to be revealed. All these issues represent evidence strongly suggesting that, indeed, polyphasic sleep-wake schedules may represent feasible and useful strategies to manage sleep requirements during sustained operations.

This, however, does not necessarily imply that all individuals are capable of adopting polyphasic sleep schedules, nor that these are the ideal strategies for every CW situation. In future research one may find considerable differences among individuals in their constitutional ability to adapt to polyphasic sleep, analogously to what is commonly observed in various sleep related characteristics. There are, in fact, long and short sleepers, natural nappers and non-nappers, morning and evening types, adapters and non-adapters to shift-work.

Polyphasic sleep schedules should therefore be studied and tailored according to each specific situation and to each individual. If a number of easily measurable physiological parametres could be identified related to such ability, they could have a potential utility in the selection of personnel expected to deal with emergency or sustained operations.

Despite their potential benefits for workload, performance and sleep optimization under continuous work conditions, and for emergency or disaster preparedness training for various professional categories, no systematic study of optimal polyphasic sleep schedule designs has been published to date.

It is expected that the results of a study that is currently being undertaken will help to elucidate these issues.One purpose of this research project is in fact to investigate the changes of performance measures, sleep inertia, nap recuperative values, biological rhythms, as a function of several ultrashort sleep episode durations, sleep structures and time of day. In other words, it is expected to clarify if there is any individual "ideal" duration (and timing) of naps that permits one to maximize sleep recuperative value, while minimizing sleep-inertia effects immediately after awakening.

Subjects are exposed to 2 different paradigms of polyphasic sleep:

1. <u>Short-term combined "anchor" sleep plus ultrashort sleep.</u> Three

different sleep episode durations of 80, 50 and 20 min each are tested independently. During four consecutive 24 hr periods (4 days), subjects are allowed 4 hours of "anchor" (Minors and Waterhouse, 1980) sleep (01:00 to 05:00), plus three 80 min naps equally distributed throughout the remaining 20-hr span; after 2 weeks of "wash-out", the paradigm is repeated adopting respectively five 50 min and twelve 20 min naps.

2. Long-term ultrashort sleep only. Accurate analysis of the data in paradigm (1) will indicate, per each subject, which is the nap episode duration, and which periods of the 24-hr span, that provided best recuperative values, or higher performance levels, relative to baseline conditions. Each subject will then be exposed, for 12 consecutive days, to a schedule of ultrashort sleep only (no anchor sleep), with multiple naps in the "ideal" duration range.

A total of eight hour of sleep per 24-hr period is allowed in each paradigm. Continuous EEG, EOG, EMG and tympanic temperature recordings are performed by means of ambulatory Medilog units (Oxford Medical Instruments). During all paradigms performance levels on several tasks (including Memory and Search Task, Descending Subtraction, Simple Reaction Time, Grip Strength, self assessed Sleepiness and Mood ratings) for a total of 121 different variables per each Session are measured immediately after arousal from each nap, 20 min and 54 min after arousal respectively, and are compared to the referential performance level measured 20 min before the start of each nap. The purpose of this paradigm is to obtain a precise description of the sleep-inertia behaviour as a funtion of several factors including: time of day, sleep structure, sleep stage at arousal, nap and previous wakefulness durations. The average duration of each Test Session is of aproximately 11 min, for a total of 30 sessions (workload: 5.5 hrs) per 24-hr period.

A preliminary analysis of the data from one female subject, age 30, shows that non significant differences in performance on the Memory and Search Task were observed between the baseline (BL), the 80 min (POLY 80) and the 50 min (POLY 50) napping conditions. Performance immediately after arousal from sleep episodes showed a decrease, as expected, relative to performance before sleep. Mean performance decrements were more pronounced after the 80 min (-23%) rather than after 50 min (-13%) naps. Conversely, recuperation speed was somewhat faster in POLY 80, although recuperative global amounts measured 54 min after arousal were quite similar in both conditions.

Time-of-day effects were observed in both POLY 80 and 50 conditions. A bimodal circadian rhythm of sleep inertia produced greater performance decrements after the anchor sleep and the mid-afternoon naps (around 15:00 h). Recuperative speeds were greater in the mid-morning hours (between 09:30 and 11:30) in both conditions, while recuperative global amounts peaked at 11:30 h only in the POLY 50 condition.

These preliminary results, presented as an example of a new methodology for analyzing in great detail sleep inertia and recuperative effects of napping schedules, represent only a minor fraction of the data that has been so far collected. It is anticipated that the results of the studies and the methodology described will elucidate the influence of circadian, circasemidian or ultradian sleep pressure on performance, and provide much needed data on the efficacy and value of polyphasic sleep regimes for decreasing performance deficits. The results will assist in the development of strategies to enhance performance during situations involving sustained operations.

REFERENCES

Broughton, R. Biorhythmic Variations in Consciousness and Psychological Functions. Canadian Psychological Review 16:217-239, 1975.

Campbell S.S., & Tobler I. Animal Sleep: a Review of Sleep Duration Across Phylogeny. Neuroscience and Behavioral Reviews 8:269-300, 1984.

Campbell, S.S., & Zulley, J. Ultradian Components of Human Sleep-/Wake Patterns During Disentrainment. Experimental Brain Research, Suppl. 12:234-255, 1985.

Carskadon, M.A., & Dement, W.C. Sleep Studies on a 90-minute day. EEG. & Clin. Neurophysiol. 39:145-155, 1975.

Carskadon, M.A., & Dement, W.C. Sleepiness and Sleep State on a 90-min Schedule. Psychophysiology 14:127-133, 1977.

Curtis, G.C., & Fogel, M.L. Random Living Schedule: Psychological Effects in Man. J. of Psychiatric Research 9:315-323, 1972.

Hartley, L.R. A comparison of continuous and distributed reduced sleep schedules. Quarterly J. of Experimental Psychology 26:8-14, 1974.

Husband, R.W. The Comparative Value of Continuous Versus Interrupted Sleep. J. of Experimental Psychology 17:792-796, 1934

Johnson, L.C., & Naitoh, P. The Operational consequences of sleep deprivation and sleep deficit. (AGARDograph n. 193). London: Technical Editing and Reproduction Ltd., 1974.

Johnson, L.C., Naitoh, P., & Moses, J.M. Variation in Sleep Schedules. Waking and Sleeping 1:133-137, 1977.

Kelley, J., Laughlin, E., Carpenter S., Simmons, J., Sidoric, K., & Lentz, R. A study of ninety minute sleep cycles. Stanford Review 3:1-5, 1973.

Lavie, P. Ultradian Rhythms: Gates of Sleep and Wakefulness. In: H. Schulz and P. Lavie (Eds.), Ultradian Rhythms in Physiology and Behavior. Springer-Verlag, Berlin, 1985.

Lavie, P. Ultrashort sleep-waking schedule. III. Gates and "forbidden zones" for sleep. EEG. & Clin. Neurophysiol. 63:414-425, 1986.

Lavie, P., & Scherson, A. Ultrashort Sleep-Waking Schedule. I: Evidence of Ultradian Rhythmicity in "Sleepability". EEG. &

Clin. Neurophysiol. 52: 163-174, 1981.
Lubin, A., Hord D., Tracy M.L., & Johnson L.C. Effects of exercise, bedrest and napping on performance decrement during 40 hours. Psychophysiology 13: 334-339, 1976.
Minors, D.S. and Waterhouse, J.M. Do "anchor sleep" entrain the oscillator that controls circadian rhythms? Journal of Physiology 308:92-93, 1980.
Moses, J.M., Hord, D.J., & Lubin A. Dynamics of Nap Sleep During a 40 Hour period. EEG. & Clin. Neurophysiol. 39:627-633, 1975.
Moses, J.M, Naitoh, P., & Johnson, L.C. The REM cycle in altered sleep-wake schedules. Psychophysiology 15:569-575, 1978.
Mullaney, D.J., Kripke, D.F, & Fleck P.A. Sleep loss and nap effects on sustained continuous performance. Psychophysiology 20:643-651, 1983.
Naitoh, P. Sleep Loss and its Effects on Performance. Report n. 68-3. US Naval Medical Neuropsychiatric Research Unit, San Diego, California, 1969.
Naitoh, P., Englund, C.E & Ryman, D.H. Extending Human Effectiveness During Sustained Operations Through Sleep Management. Report n. 83-13, Naval Health Research Center, San Diego, California, 1983.
Stampi, C. Ultrashort Sleep-Wake cycles Improve Performance During One-man Transatlantic Races. In: W.P. Koella, E. Ruther & H Schulz (Eds.) Sleep 1984. Gustav Fischer Verlag, Stuttgart, pp.271-272, 1985.
Stampi, C. Sleep Management in Sustained Operations: a Field Study on Single-and Double-handed yacht races. Work & Stress, (in press).
Strogatz, S.H. The mathematical Structure of the Human Sleep-Wake Cycle. Lecture notes in Mathematics 69. Springer Verlag, Berlin-Heidelberg-New York-London-Paris-Tokyo. pp.239, 1986.
Taub, J.M., & Berger, R.J. Performance and Mood Following Variations in Length and Timing of Sleep. Psychophysiology 10:559-570, 1973.
Webb W.B.. Napping across the life-span. In: D. Dinges & R. Broughton (eds.): Sleep and Alertness: the Nature of Napping (in press).
Webb, W.B., & Agnew, Sleep efficiency for Sleep-Wake Cycles of Varied Length. Psychophysiology 12:637-641, 1975.
Weitzman, E., Nogeire, C., Perlow, M., Fukushima, D., Sassin, J., McGregor, P., Gallagher, T., & Hellman, L. Effects of a prolonged 3-hour sleep-wakefulness cycle on sleep stages, plasma cortisol, growth hormone and body temperature in man. J. Clin. Endocrin. 38:1018-1030, 1974.
Zulley J., Bailer J.. Polyphasic sleep-wake patterns and their significance to vigilance. In: J.P. Leonhard (ed.): Vigilance: Methods, Models and Regulation. (in press).

IRREGULARITY OF WORK AND CIRCADIAN RHYTHMICITY: IMPLICATIONS FOR AIRLINE OPERATIONS

A. N. Nicholson and M. B. Spencer
Royal Air Force Institute of Aviation Medicine
Farnborough, Hampshire, GU14 6SZ
United Kingdom

ABSTRACT. Sleep, rectal temperature, electrolyte excretion and performance were studied for 9 days during an irregular schedule of work and rest. Sleep structure varied with the circadian rhythm and the length of the preceding activity period. Circadian rhythms no longer showed a period of exactly 24h, and the acrophase of the temperature rhythm varied throughout the experiment under the influence of the sleep-wake cycle and the time of day cues. A model which included time of day and time since sleep was used to predict variations in performance for work periods starting at different times of day. Poorest performance occurred when a long period of duty terminated close to the nadir of the circadian rhythm, and it is considered that this has important implications for the scheduling of aircrew.

1. INTRODUCTION

Irregular work schedules are a feature of many occupations, including that of the airline pilot operating world-wide. Duty can occur at almost any time of the day or night as the schedule meets operating constraints, commercial requirements and the needs of the passengers, so that the normal relationship between circadian rhythm and sleep and performance is no longer maintained. Fatigue induced by long periods of work interacts with the circadian rhythm of performance (1) and the nature of this interaction is of particular interest to the safety of world-wide air operations.

While adaptation to conventional shiftwork has been the subject of considerable research, less information is available on the effect of irregularity of work and rest. The patterns of work in irregular schedules differ in two important respects from those of conventional shiftwork. Firstly the amount of work done between successive rest periods may vary considerably. Secondly, since sleep is not taken at regular 24h intervals, the circadian rhythm of the individual may not resynchronize with the sleep-wake cycle.

To investigate those factors that may influence performance during these irregular schedules, we have undertaken laboratory

A. Coblentz (ed.), Vigilance and Performance in Automatized Systems/ Vigilance et Performance de l'Homme dans les Systèmes Automatisés, 275–286.

studies over a 9 day period.

2. METHODS

The same schedule was used in 9 individual experiments. The first experiment was carried out using three subjects at the sleep laboratory at Farnborough. The remaining 8 experiments, involving a total of 30 subjects, were carried out in an isolation unit (2), under constant conditions of temperature and humidity.

2.1. The schedule of work and rest

The schedule of work and rest (Figure 1) involved 24 6h periods of work and 12 6h periods of rest over 9 days, and it was preceded by at least two days of normal nycthemeral activity. Since it was the primary intention to study irregularity of sleep and to avoid serious sleep deprivation, an average of 8h rest was provided every 24h.

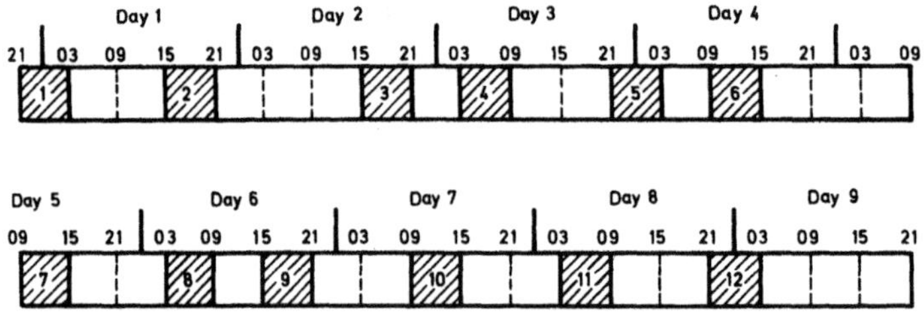

Figure 1. Schedule of sleep periods (hatched) during the experiment.

2.2. Subjects

33 healthy volunteers aged 18 - 32 years were studied, and their details are given in Table I. Subjects abstained from alcohol and refrained

TABLE I Details of subjects and experiments

Group	No	Age	Sex
1	3	26-32	Male
2	4	18-20	Male
3	4	18-19	Male
4	4	18-22	Male
5	3	18-19	Male
6	4	18-22	Male
7	4	18-20	Female
8	3	19-21	Female
9	4	18-19	Male

from napping throughout the study and were trained on the various tasks to reach steady levels of performance.

2.3. Sleep electroencephalography

Sleep was recorded electroencephalographically, during the first experiment only, for two nights preceding (control) and for two nights after (recovery) the 9 day period, and throughout eachof the 6h rest sessions. On control and recovery nights, subjects retired at their normal bedtime (2300 - 2330h), and were woken at 0730h. During the 9 day period the subjects were woken, if necessary, exactly 6h after "lights out". The sleep recordings were scored manually according to conventional criteria.

2.4. Temperature and Urinary Constituents

Temperature was measured using a rectal probe and values at 2h intervals were retained for analysis. Urine was collected at 2h intervals during waking hours, and immediately before and after the rest periods, and the analysis was carried out as described by Minors et al (3).

2.5. Performance and Mood

During each work period performance was tested every 2h beginning 1h after rising. In all experiments, digit symbol substitution was measured and mood was assessed subjectively. Subjects in groups 4 - 6 also carried out mental arithmetic and six-letter cancellation tasks, and those of groups 7 - 9 carried out symbol copying, logic, and two- and six-letter cancellation tasks. The details of these tasks are given elsewhere (3).

3. STATISTICAL ANALYSIS

The analysis of the sleep variables was based on the data from the first experiment. The analysis of performance, temperature, mood and the urinary constituents was based on experiments 2 - 9.

3.1. Sleep

Comparisons between sleep obtained on the second control night and during the experiment were made using a two factor analysis of variance. In order to avoid heterogeneity of variance, it was necessary to transform some variables prior to the analysis.

3.2. The Basic Model

The following basic model was used to analyse the performance data, mood assessments, temperature data and urinary constituents:

$$y_{ij}(t) = \Sigma\theta_k t^k + a_1\cos(2\pi t/\lambda) + a_2\sin(2\pi t/\lambda) + \gamma_j + \varepsilon_i$$

where t is the time in h relative to an arbitrary origin, and i and j index the measurements in sequence through the experiment, and from the preceding sleep period respectively. The data are therefore represented as the sum of a trend term and circadian rhythm and time-since-sleep (t.s.s.) effects.

3.3. Assumptions of the Model

3.3.1. Additivity of time since sleep and time of day. One of the assumptions of the basic model was that the t.s.s. effect was independent of circadian rhythmicity. To check whether this assumption was reasonable, an analysis was carried out on the data from the digit symbol substitution task from two experiments in which 8 subjects remained in the isolation laboratory for a total of 18 days, repeating the schedule of the first 9 days. The model was extended to include a circadian rhythm by t.s.s. interaction. For this analysis the period of the circadian rhythm was fixed (the period chosen being that of the temperature rhythm for the same subject) and the interaction term was constructed from the product of the circadian terms with linear and quadratic contrasts in t.s.s.

3.3.2. Sinusoidal circadian variation. A second assumption of the model was that the circadian variation in performance was sinusoidal. To test this hypothesis the basic model was extended to include the first five harmonics of the fundamental circadian frequency. The circadian trend was reconstructed from the estimates of the harmonic coefficients and values at ten equally spaced points during the circadian period were analysed by analysis of variance.

3.3.3. Constant circadian period. Another assumption of the basic model was that the period of the underlying circadian rhythm was constant for each subject throughout the experiment. To test this hypothesis on the temperature data, the circadian term in the model was replaced by one in which the period was set to 24h and the phase allowed to vary as a cubic polynomial in time.

4. RESULTS

4.1. Sleep

The means of sleep measures for the sleep periods during the schedule are compared with the values for the first 6h of the second control night (Table II). Sleep onset latency did not differ from that of the control night throughout the schedule. However, total sleep time varied with the length of prior wakefulness for each subject ($p < 0.05$) and for the group ($p < 0.001$). Reductions in total sleep time were only observed after the daytime rest periods (those commencing at 0900h or 1500h) that did not follow a waking period of 18h.

The latency to stage 3 increased compared with control nights after short periods of prior wakefulness. On the other hand the latency to

TABLE II. Various sleep measures for each sleep period (means for 3 subjects).

Sleep period	Time of start	Total sleep time (min)	Latency to stage 3 (min)	Stages 3 & 4 (min)	REM (min)	Awake (min)
CN2	2300	331.3	15.0	58.0	72.0	7.2
1	2100	285.2	14.7	47.0	58.0	19.3
2	1500	174.8†	25.0	40.4	34.3**	5.2
3	1500	311.0	12.0	83.4	69.5	19.0
4	0300	313.7	103.0†	21.1**	121.8	12.7
5	2100	303.7	16.7	41.2	61.8	10.8
6	0900	169.2†	116.5†	24.5**	57.0	14.8
7	0900	304.8	24.0	50.3	71.5	12.5
8	0300	316.0	32.0	29.8*	98.5	7.8
9	1500	118.3†	51.5*	13.5†	25.8	30.5
10	0900	256.5*	21.8	18.5†	66.7	26.5
11	0300	326.3	19.2	46.5	94.3	6.0
12	2100	304.2	17.3	30.7*	74.0	14.8

Significance levels: * $p < 0.05$; ** $p < 0.01$; † $p < 0.001$
Significant differences are from control night (CN)2

rapid eye movement sleep decreased when sleep began at 0300h or 0990h, except after a long period of prior wakefulness. Significant reductions in the duration of REM sleep occurred only between 1500h and 2100h, while stages 3 and 4 decreased after all 6h activity perods and after three of the six 12h activity periods.

4.2. The Basic Model

Periods and acrophases for variables with significant circadian rhythms are given in Table III. Acrophases based on 24h periods are presented for those variables whose circadian period did not differ significantly from 24h; otherwise acrophases are based on the best fitting rhythms and are correct for the first day of the experiment only. It was not possible to identify a significant circadian rhythm in those tasks that were included in only three experiments.

Two mood assessments (alert-dulled and efficient-useless) exhibited circadian rhythms with periods in excess of 24h ($p < 0.05$). Circadian rhythms were also present in body temperature and in all urinary constituents, and, with the exception of calcium excretion, the best-fitting periods were significantly greater than 24h.

Changes in mean levels of performance and mood assessments with t.s.s. (corrected for circadian rhythm and trend) are given in Table IV. Decrements in performance and changes in mood were observed from 9h onwards.

Temperature was lower during sleep than during the waking periods ($p < 0.01$). A t.s.s. effect was established, and temperature 7h after

TABLE III. Mean circadian periods (h) and acrophases

Variable	No. of experiments	Period & 95% confidence level	Acrophase
Performance			
Digit symbol substitution	8	**24.51 ± 0.31	17.05
Cancellation of six letters	6	24.43 ± 1.01	19.38
Subjective assessments			
Awake	8	24.21 ± 0.43	18.33
Energetic	8	24.22 ± 0.87	18.45
Alert	8	**24.68 ± 0.46	16.25
Concentration	8	24.51 ± 0.53	17.47
Efficient	8	*24.47 ± 0.47	15.38
Temperature	8	†24.43 ± 0.16	17.65
Urinary constituents			
Sodium	8	*24.59 ± 0.47	13.07
Potassium	8	†24.66 ± 0.20	11.92
Calcium	8	24.15 ± 0.52	10.23
Phosphate	8	*24.64 ± 0.57	17.85
Chloride	8	†24.61 ± 0.25	11.58
Urate	8	**24.61 ± 0.44	10.93

Significance levels * $p < 0.05$; ** $p < 0.01$; † $p < 0.001$ refer to deviation from 24h.

TABLE IV. Time-since-sleep (t.s.s.) effects on performance and subjective assessments of mood

t.s.s. (h)	digit symbol substitution	6-letter cancellation	awake	concentration	efficient
1	234.3	838.2	56.2	57.1	60.4
3	<u>236.1</u>	<u>839.9</u>	<u>58.2</u>	<u>59.0</u>	<u>61.7</u>
5	234.4	836.8	56.6	57.3	61.2
7	233.4	824.9	58.0	57.0	61.3
9	228.9**	819.4	50.3*	51.8**	58.0
11	228.8**	820.4	45.6 †	50.2**	55.6*
13-17	224.7 †	796.5*	45.5 †	46.1 †	51.9 †

Significance levels (* $p < 0.05$; ** $p < 0.01$; † $p < 0.001$) refers to differences from peak value (underlined) using the Newman-Keuls test.

awakening was higher than that after 3 and 5h (Table V). The excretion rates of electrolytes (Table VI) were influenced by time awake.

($p < 0.001$). The time taken to reach the peak rate after sleep varied from 4h with calcium to 10h with sodium, potassium and phosphate.

TABLE V. Temperature during sleep and wakefulness.

Time (h)	Temperature (°C)	
	Asleep	Awake
1	36.91	37.12
3	36.88	37.07**
5	36.83**	37.09*
7		37.16
9		37.12
11		37.13
13-17		37.11

Significance levels (* $p < 0.05$; ** $p<0.01$) using Newman-Keuls test refer to differences from the peak value (underlined).

TABLE VI. Urinary excretion rates related to time-since-sleep (t.s.s.)

t.s.s. (h)	Urinary constituent (μmol min^{-1})				
	Sodium	Potassium	Calcium	Phosphate	Chloride
0	114.6†	42.5†	3.06†	30.3†	104.3†
2	96.6†	37.2†	3.74†	24.6†	101.1†
4	116.6†	41.3†	6.05	15.8†	129.1†
6	135.2**	56.2	5.34	19.4	151.4
8	160.1	62.2	5.32	28.0	167.6
10	168.7	62.9	5.01*	31.3	161.8
12	157.9	58.7	4.82**	31.0	148.1
14-18	163.4	55.3	5.10*	30.3	151.0

Significance levels (* $p < 0.05$; ** $p < 0.01$; † $p < 0.001$) refer to differences from peak value (underlined) using the Newman-Keuls test.

4.3. Assumptions of the model

The investigation of the additivity of the two components revealed a tendency for the circadian amplitude to increase with increasing time awake, but this increase fell below even the 10% level of significance. Nevertheless two individuals did exhibit a significant interaction. In one, the circadian amplitude was greatest immediately after waking while in the other it was greatest at the other extreme, after a long period of wakefulness.

The circadian variation in digit symbol substitution, as a function of the circadian phase of temperature, is shown in Figure 2, with the

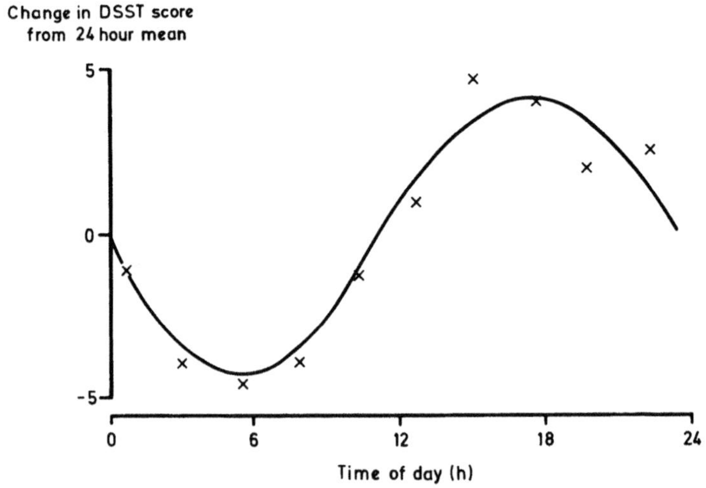

Figure 2. The circadian variation in performance.

time axis rescaled to the 24h clock. There was no evidence, even at the 10% level, of any departure from the sinusoidal fit, and in particular there was no evidence of a midday or afternoon dip in performance.

The mean acrophase of the temperature rhythm during the experiment is shown in Figure 3. The phase variation was non-linear ($p < 0.01$):

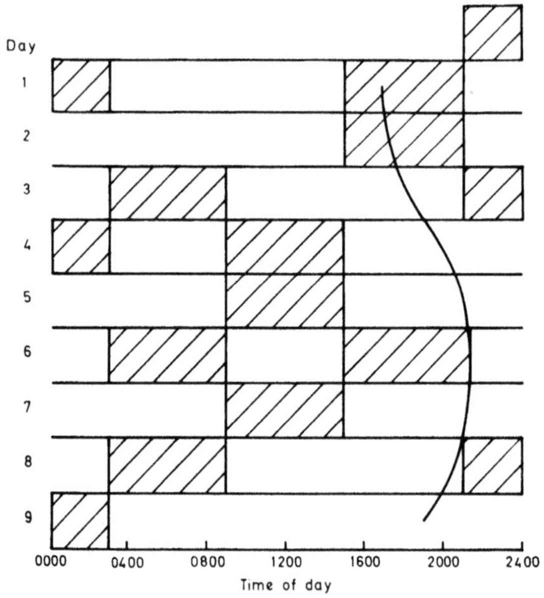

Figure 3. Trend in the temperature acrophase through the experiment. The hatched areas represent the rest periods.

during the first half of the experiment the phase lengthened, but during the last two days it began to shorten.

5. DISCUSSION

5.1. Sleep

During this schedule involving irregular periods of activity, the duration and internal structure of sleep was found to be related to the time of day as well as to the preceding pattern of work and rest. The relationships between slow wave sleep and length of prior wakefulness, and between REM sleep and circadian rhythmicity have been observed in other studies in which the amount of prior wakefulness and timing of sleep were varied (4,5,6). In this study the lack of a significant increase in slow wave sleep during any of the individual sleep periods suggests that the subjects did not experience problems due to lack of sleep in the course of the experiment.

5.2. Circadian rhythms

The present study provides further evidence that rhythms no longer remain synchronized to the 24h day during a schedule which involves irregularity of work and rest. Almost all fitted rhythms exceeded 24h, and in many cases the difference was significant. However, the more detailed analysis of the temperature rhythm revealed that the delay of the acrophase was not uniform through the experiment (Figure 3).

This change in phase partially reproduced the pattern of the sleep periods, suggesting that the sleep-wake cycle was acting as a zeitgeber for the temperature rhythm. Simulations using models based on the Van der Pol equation confirmed this impression and suggested that both the sleep-wake cycle and the time of day (which was known to the subjects) had separate zeitgeber influences (7).

5.3. Time-since-sleep effects

Sleep exerts a direct 'masking effect' upon circadian rhythms of body temperature and urinary excretion, the size of which depends upon the variable under consideration (8). In addition, t.s.s. effects are marked for urinary constituents (Table VI). The exact explanation of these is unknown but effects of dietary intake after fasting during sleep are likely to be involved. The t.s.s. effect for temperature (Table V) may be due to the activity pattern enforced by the schedule.

There was evidence (Table IV) that performance deteriorated after a 9h period of wakefulness. The fall was approximately 5 - 10% of the mean value after 9h or more of wakefulness, and this is in accord with a previous study (9).

5.4. A model for performance

These results suggest that changes in performance during an irregular schedule of work and rest arise from both the endogenous circadian rhythm and the time that has elapsed since the end of the preceding rest period. In these experiments the score for digit symbol substitution, corrected for the long-term trend, could be expressed as the sum of a sinusoidal component in the time of day (T) and a cubic trend in t.s.s. (t):

$$DSST(T,t)=233.3+1.54t-0.304t^2+0.0108t^3+4.97\cos(2\pi(T-17.05)/24)$$

where both T and t were measured in hours. This model assumes that the two components are additive which, though true for most subjects, may not be true for all.

This model can be used to predict changes in performance at different starting times (figure 4). Starting at 0200h, performance

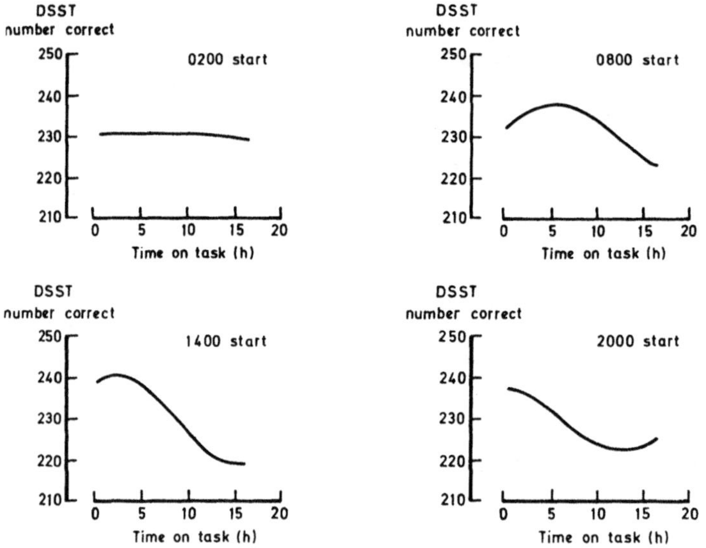

Figure 4. Predicted trends in performance corresponding to different starting times

remains relatively constant for the first 16h, since the fall in performance towards the end due to time since sleep coincides with the acrophase of the circadian rhythm, so that the factors approximately cancel out. Starting at 1400h the opposite effect is seen and there is a sharp fall in performance after 7h when the t.s.s. effect is magnified by the effect of the circadian rhythm, which is then well past its peak. The 0800h start corresponds most closely to the normal working day and performance is maintained at a relatively high level for over 10h.

During work periods that begin in the early hours of the morning, the model predicts that performance, though never very high, will be maintained above the lowest levels for at least 16h. This presupposes the conditions of the experiment, in particular that an individual is fully rested when work commences, and in many practical situations this may be an unrealistic assumption. However, for the individual who starts work in the evening, even if he is initially fully rested, the downward slope of the contour lines suggests that his performance will rapidly fall to very low levels due to the combination of time of day and t.s.s.

5.5. Concluding remarks

The performance task upon which this model has been based is limited in nature and the maximum predicted decrements do not exceed 10%. It is also possible that this task is not representative of tasks with a high memory load which some authors have suggested are controlled by a separate oscillator (10) with an acrophase in the early morning (11). Nevertheless it is believed that this approach has important implications for those who work on irregular schedules including aircrew on long-haul civilian operations.

The model has been derived from an experiment which avoided overall sleep deprivation and in which the workload was low, so it may have identified only the immediate effects of the work schedule. Indeed, other factors need to be considered when the effects of a particular schedule are evaluated. These include the nature of the task as well as the availability of adequate time for the individual to rest and to avoid cumulative sleep-deprivation.

However, even in the absence of problems associated with sleep, it would appear that the ability of an individual to maintain the highest levels of performance is restricted both by his circadian rhythm and by the length of the duty period. The most severe problems occur when a long period of duty terminates near the lowest point of the circadian rhythm of body temperature, which occurs at approximately 0500h or 0600h. At greatest risk, therefore, are those long duty periods commencing in the early evening, during which performance is not maintained at the same level as during a duty period of a similar length that commences in the morning.

REFERENCES

1. COLQUHOUN W.P., HAMILTON P. and EDWARDS R.S. (1980) Effects of circadian rhythm, sleep deprivation, and fatigue on watchkeeping performance during the night hours. In: Studies of Shiftwork (W.P. Colquhoun and J. Rutenfranz eds.) Taylor & Francis, London, 225-233.
2. ELLIOTT A.L., MILLS J.N., MINORS D. and WATERHOUSE J. (1972) The effect of real and simulated time zone shifts upon the circadian rhythms of body temperature, plasma 11-hydroxycorticosteroids and renal excretion in man. Journal of Physiology, **221**, 227-257.
3. MINORS D.S., NICHOLSON A.N., SPENCER M.B., STONE B.M. and WATERHOUSE J.M. (1986) Irregularity of rest and activity: Studies on circadian rhythmicity in man. Journal of Physiology, **381**, 279-296.
4. HUME K.I. and MILLS J.N. (1977) Rhythms of REM and slow-wave sleep in subjects living on abnormal time schedules. Waking and Sleeping, **1**, 291-296.

5. TAUB J.M. and BERGER R.J. (1973) Sleep stage patterns associated with acute shifts in the slee-wakefulness cycle. Electro-encephalography and Clinical Neurophysiology, **35**, 613-619.
6. WEBB W.B. and AGNEW H.W. (1971) Variables associated with split-period sleep regimes. Aerospace Medicine, **42**, 847-850.
7. SPENCER M.B. (1987) The influence of irregularity of rest and activity on performance: a model based on time since sleep and time of day. Ergonomics, **30**, 1275-1286.
8. MILLS J.N., MINORS D.S. and WATERHOUSE J.M. (1978) The effect of sleep upon human circadian rhythms. Chronobiologia, **5**, 14-27.
9. ALLUSI E.A. (1972) Influence of work-rest scheduling and sleep loss on sustained performance. In: Aspects of Human Efficiency. Diurnal Rhythm and Loss of Sleep (Colquhoun W.P. ed.) English Universities Press, London, 199-215.
10. FOLKARD S., WEVER R.A. and WILDGRUBER C.M. (1983) Multi-oscillatory control of circadian rhythms in human performance. Nature, **305**, 223-226.
11. FOLKARD S., KNAUTH P., MONK T.H. and RUTENFRANZ J. (1976) The effect of memory load on the circadian variations in performance efficiency under a rapidly rotating shift system. Ergonomics, **19**, 479-488.

TOWARDS THE PREDICTION OF ALERTNESS ON ABNORMAL SLEEP/WAKE SCHEDULES

Simon Folkard[1] and Torbjorn Akerstedt[2]
[1]MRC/ESRC Social and Applied Psychology Unit, University of Sheffield, Sheffield, S10 2TN, England.
[2]Karolinska Institute, Box 60205, S-104 01, Stockholm, Sweden.

ABSTRACT. This paper describes our recent progress in developing a mathematical model that may eventually allow the prediction of alertness on any given sleep/wake schedule. In its present form it comprises an endogenous sinusoidal function, and two exogenous logarithmic functions that are "set" on awakening. These functions are shown to simulate published time-series of alertness ratings with considerable accuracy.

1. INTRODUCTION

One of the unfortunate characteristics of automatized systems is that they tend to be expensive. Thus there is considerable pressure to maximise their use, if only for purely economic reasons, and this frequently involves their operators being required to work on some form of shift, or irregular, work system. Such systems require people to sleep at unusual times of day, and often result in disturbances of their circadian rhythms. Both these factors may affect their ability to maintain their vigilance and performance levels. The present paper is concerned with the further development of a mathematical model which may eventually allow the prediction of an operator's alertness or vigilance (and hence, hopefully, performance) on such systems.

Psychologists have long recognised that vigilance or alertness levels are influenced by both the time of day and by how long people have been awake. Michelson (1897) recognised that there are variations in people's level of sleepiness over the day that were to some extent independent of when they had slept. He proposed a hypothetical function relating the level of sleepiness to time of day and argued that sleepiness was lowest, or wakefulness highest, at about 17:00. Other early psychologists, such as Thorndike (1900), were more concerned with the effects of work on the accumulation of "mental fatigue" over a period of wakefulness, and viewed sleep as having a recuperative or restorative function. However, it was also recognised that despite this restorative nature of sleep, people take some time to wake up "properly" in the morning. Thus, for example, Gates (1916) combined

A. Coblentz (ed.), Vigilance and Performance in Automatized Systems/ Vigilance et Performance de l'Homme dans les Systèmes Automatisés, 287–296.

Michelson's (1897) sleepiness function with Howell's (1897) "anaemia" theory of sleep and argued that "although the return to consciousness (on awakening in the morning) is sudden, the complete re-adjustment of the anaemic condition takes place more slowly, since it is dependent on the cumulative effects of sensory experience and the increasing activity of the brain" (Gates, 1916, p11).

Between them, these early psychologists had thus identified three factors affecting variations in sleepiness or alertness at different times of day. Namely (i) an endogenous circadian rhythm, (ii) a build up of fatigue, or reduction in alertness, over time awake, and (iii) a short-lived "waking up" or recovery from sleep effect. The first two of these factors correspond to the hypothetical processes "C" and "S" respectively in Borbely's (1982) model of sleep regulation. Our concern in this paper is to show how we have attempted to isolate all three factors, to describe them in terms of mathematical functions, and to then illustrate how these functions may enable us to predict variations in alertness on abnormal sleep/wake schedules.

2. ISOLATING THE FACTORS

The starting point for isolating these factors was the results obtained in a fractional desynchronization study (Folkard et al. 1985) in which three groups of four students each were isolated from environmental time cues (zeitgebers) in a temporal isolation unit. They were provided with artificial time cues that ran progressively faster over the successive days of the three week study, such that the "day" length reduced from 24h to 22h. At regular intervals throughout each day the subjects performed a 20-minute battery of performance tests and rated their alertness on a visual-analogue scale (VAS) labelled from "Drowsy" to "Alert". The subjects were instructed to regard the extremes of this scale as representing the most drowsy, or alert, that they ever <u>normally</u> felt. These measures were recorded on-line on a microcomputer, with the four subjects in each group logging on one at a time in a constant order such that the complete session lasted 1 hour 20 minutes.

The alertness ratings (scored on a 1-21 scale) exhibited a marked circadian rhythmicity for the duration of the study. For about the first half of the study this rhythm was entrained by the artificial zeitgebers, i.e. it phase advanced in line with the timing of sleep, etc. However, it then "broke out" from the artificial zeitgebers when they reached a period of 23.0h (s.e.=0.06h). This break out was followed by a "free-run" of the alertness rhythm through approximately 360 degrees with respect to the artificial zeitgebers and the sleep/wake cycle (see Folkard et al., 1985, for details). The "time of local day effect" in the alertness ratings up to the point at which the rhythm broke out will clearly reflect both endogenous and exogenous influences. Conversely, it is arguable that from this break-out point to the end of the study the "time of local day effect" will reflect only the exogenous influences averaged over all possible phases of the endogenous component. By subtracting this "time of local day effect" following break-out from that prior to it we can thus estimate the endogenous component (see Folkard and Akerstedt, 1987, for further details).

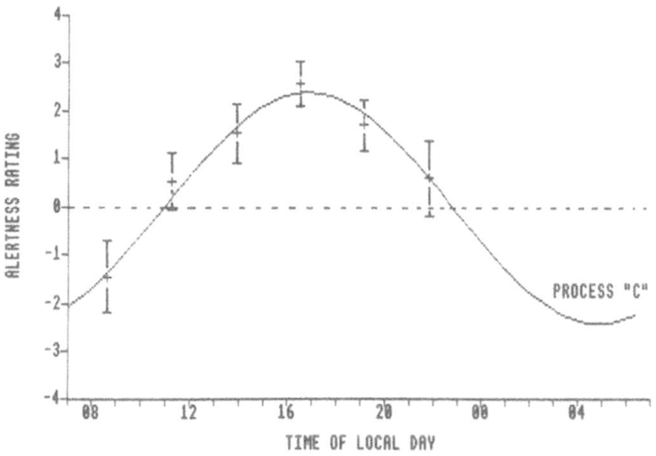

Figure 1. The estimated endogenous component of the alertness rhythm together with the best fitting 24h sine curve (process C).

The resultant estimate of the endogenous component of the alertness rhythm is shown in Figure 1. A 24-hour sine curve provided an extremely good fit to this data, and is shown in Figure 1 extrapolated over a complete 24 hours. It accounted for 97.9% of the variance and estimated the peak (or acrophase) to have occurred at 16.835h. We shall refer to this sine curve as "process C" since it represents our estimate of the endogenous, circadian rhythm component of variations in alertness over the day (cf Borbely, 1982).

The "time of local day effect" following break-out, i.e. our estimate of the exogenous component, is shown in Figure 2. The readings obtained from the first testing session are shown separately for the 1st, 2nd, 3rd, and 4th sub-groups of subjects to perform the tests at each session. With the exception of this first session data, the points were closely fitted (r = -0.983) by the logarithmic, process "S" function shown as a solid line in Figure 2. We refer to this function as "process S" since it reflects a decrease in alertness, or increase in drowsiness, over time awake, and is thus very similar to Borbely's (1982) hypothetical increase in sleep need. This has been extrapolated back to the point of awakening in Figure 2 since we assume sleep need to start increasing from this point.

Finally, the more detailed plotting of the readings from the first session of the day suggests a third, relatively short-lived, logarithmic deviation in alertness from this extrapolation of process "S" over the first three hours after awakening (see Folkard and Akertstedt, 1987, for further details). This is labelled "process W" in Figure 2 since we assume it to reflect a "waking-up" or "sleep inertia" effect similar to that proposed by Howell (1897). This logarithmic regression accounted for the data shown very well (r = -0.994), but is based on a "between-subjects" comparison of extremely small (N=3) groups.

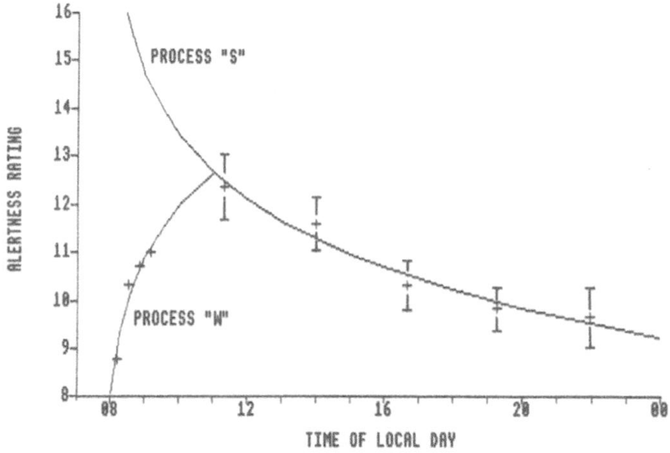

Figure 2. The estimated exogenous component of the alertness rhythm and the two logarithmic functions used to model it (processes S and W).

A further group of four subjects were thus run in a similar "fractional desynchronization" study (in collaboration with Drs D. S. Minors and J. M. Waterhouse) to check on our estimate of process "W". This study was identical to that described above except that (a) the period of the zeitgebers was only reduced to 22.8h, and then kept constant, and (b) in addition to their normal on-line sessions, the subjects rated their alertness on "paper and pencil" 10cm VASs (scored on a 0-100 scale) 5 minutes after awakening (i.e. at 07:50), at 20 minute intervals from 08:00 to 10:00, at 2 hourly intervals from 10:00 to 22:00, and finally on retiring at 23:45 (all on local time).

Our estimate of the exogenous component of the alertness rhythm for this study (i.e. from the days when the zeitgeber period was less than 23.0 hours) was then cross-correlated with the model values from the first study, i.e. those derived from processes "S" and "W". The fit for the mean data was extremely good (r = 0.985), and indeed it was reliable for each individual subject (r = 0.577 to 0.921; $p<0.05$ to <0.001). Thus it would appear that our original estimate of process "W" was reasonably accurate despite it being based on impoverished data.

In sum, our model of the variations in alertness over the day comprises three, highly reliable, mathematical functions. The sinusoidal process "C" is thought to reflect the endogenous circadian component of variations in alertness over the day, and is thus likely to be relatively unaffected by minor variations in sleep timing, and to adjust only slowly to longer term changes in the timing of sleep. In contrast, the logarithmic processes "S" and "W" are thought to be more exogenous in origin, and to be "set" by the time of awakening. They will thus be crucially dependent on the precise day-to-day timing of sleep.

3. VALIDATING THE MODEL

Testing the predictive validity of this model involves the derivation of " predictions" from these three mathematical functions, and their subsequent cross-correlation with "real" data. This derivation of predictions requires only the "wake-up" time on a given day to be known, and an estimation to be made of the probable phase of process C if subjects are on an abnormal sleep/wake routine. A "predicted" or hypothetical alertness curve can then be obtained by simply adding the values derived from these three functions. Any differences in the nature of the alertness scale used will affect the slope and intercept of the linear regression between the "predicted" and "actual" time series, but not the magnitude of the correlation. Thus, for example, when modelling fatigue ratings both the correlation and the regression slope are negative.

Our first concern was whether we could successfully "predict" variations in alertness over the day in people living under normal conditions, outside a temporal isolation unit. Folkard (1983) collated data from five such studies that had used a variety of measures, ranging from 10 cm VASs to Mood Adjective Check Lists (MACLs). The scores from these studies were expressed as z scores, and then averaged to produce a "normative" time of day effect for alertness. Since, on average, the individuals taking part in these studies habitually awoke at 06:50, we assumed a wake-up time of 07:00, and an acrophase of process C of 16:00, in deriving our predicted trend. The cross-correlation between the predicted trend and this "normative" trend was extremely high ($r = +0.994$) (see Folkard and Akerstedt, 1987), confirming the generality of the results obtained in the fractional desynchronization study of Folkard et al. (1985).

The second type of data that we have modelled concerns the effect of sleep deprivation on fatigue ratings. The "prediction" of this data is critical since it requires the extrapolation of process S over a prolonged period of wakefulness. Three similar studies conducted by Akerstedt and Froberg (e.g. Froberg, 1977) have obtained three-hourly fatigue ratings over 75h of sleep deprivation. Figure 3 shows the averaged actual data from these three studies together with the predicted trend from the model (solid line). Since the subjects awoke at about 07:30, we assumed a normal 17:00 acrophase for process C. The cross-correlation between the actual and predicted trends was again high ($r = -0.938$). Indeed the mean correlation (across the three studies) between the predicted and actual trends was as high ($r = -0.898$) as the mean intercorrelation between the results from the three studies (r = +0.885). This implies that the predictive power of the model is as high as the test-retest reliability of the data. Nevertheless, close examination of Figure 3 suggests that there may be a systematic discrepancy at about 15:00 on the second and, more markedly, on the third days of these studies (see discussion).

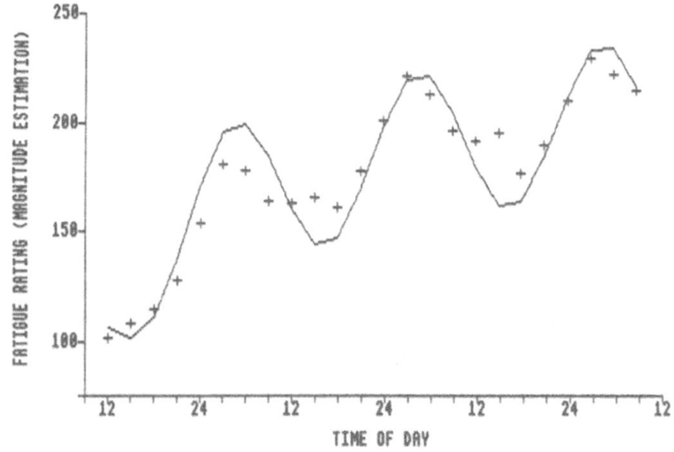

Figure 3. The effect of prolonged sleep deprivation on subjective ratings of fatigue together with the "predicted" trend from the model.

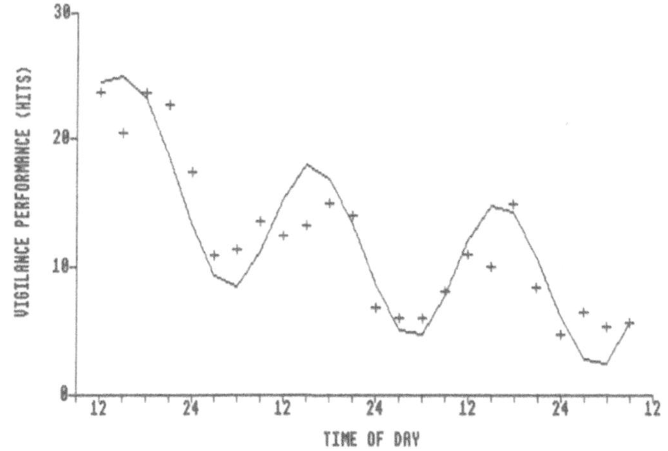

Figure 4. The effect of prolonged sleep deprivation on vigilance performance together with the "predicted" trend from the model.

The study of Froberg (1977) also obtained objective measures of vigilance, namely performance on a 30-minute, Bakan, vigilance task (Figure 4). The predicted alertness trend described the number of correct detections (hits) rather well (r = 0.910), suggesting that we may be able to generalise our model to more objective measures of vigilance. Again, however, there appeared to be a systematic error at about 15:00, but in this case on all three days of the study.

The third type of data that we chose to validate this model concerned the alertness ratings of nurses over a week of night shifts. The cross-correlation between the predicted and actual trends was again high (r = +0.951, see Folkard and Akerstedt, 1987, for details). However, the day sleep duration of these nurses between night shifts was as long as normal night sleeps (about 8 hours), and it was thus relatively easy to model this data. The day sleeps of shiftworkers, and those on irregular routines, are usually considerably shorter than normal sleeps. The question thus arises as to how process "S" recovers during sleep, and thus what the effect of a truncated sleep might be.

4. PROCESS "S" DURING SLEEP

We assume that process "S" reflects a build up of sleep need (or more strictly its inverse) over a period of wakefulness, that is subsequently dissipated by sleep. This implies that sleep need is zero on the point of waking spontaneously from a normal duration night sleep. In this context, it is of interest to note that backward extrapolation of process "S" reaches the scale maximum (i.e. 21) some 3.8 minutes after waking. Unfortunately, in view of its logarithmic nature, it also produces infinite values at the point of awakening! Under normal circumstances these are "compensated" for by an infinite deviation due to process "W", and the model thus produces "sensible" predictions.

However, if we are to model the recuperation of "S" during sleep, these infinite values create problems. For the purpose of the present paper, we have assumed a process "S" value of 21 at the point of awakening from a normal sleep, and that recuperation is a linear function of the logarithm of "hours asleep plus one". This produces a very similar curve to the exponential function fitted by Borbely (1982) to EEG sleep measures, and indeed the use of exponential functions for processes "S" and "W" makes virtually no difference to the results.

This hypothetical recuperation of process "S" is shown in Figure 5, and will be referred to as process "S'". In deriving this function, we have assumed sixteen hours of wakefulness prior to sleep starting, and that process "S" achieves complete recuperation over an eight hour sleep. This hypothetical process "S'" can be validated against alertness ratings made during brief intervals of wakefulness during a normal night's sleep, assuming "C" and "W" to operate normally. The mean three-hourly alertness ratings of eight subjects averaged over 6 complete 24hour cycles are shown in Figure 6 (Wegmann et al., unpublished data), together with the trend predicted by this extended model. The overall cross-correlation was again high (r = 0.986). Of particular note is the fact that this extended model successfully predicted the

the values obtained at 02:00 and 05:00 which were taken during the sleep period, and only 10 minutes after the subjects were woken. Thus it would appear that process "S'" is a reasonable representation of the recovery of process "S" during sleep.

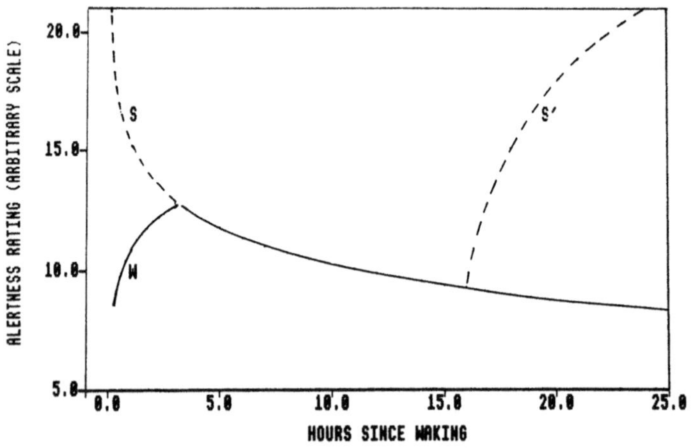

Figure 5. The hypothetical recuperation of process "S" during sleep.

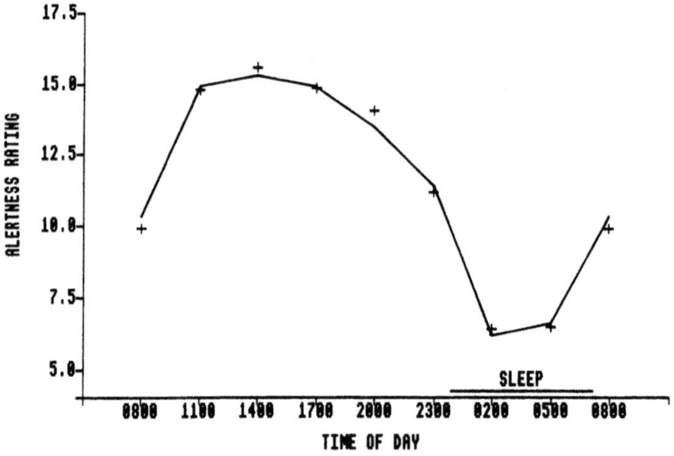

Figure 6. The fit to alertness ratings over a complete 24h period including those from brief intervals of wakefulness during sleep.

Finally, if we assume the shortened day sleeps of nightworkers to truncate the recuperation due to process "S'", we can predict what might happen to the process "S" component of variations in alertness over successive night shifts. This is illustrated for three successive night shifts in Figure 7, in which we have assumed a normal 8-hour night sleep prior to the first night shift, and 6-hour day sleeps prior to the subsequent night shifts. Alertness due to process "S" decreases over successive night shifts for two reasons. First, the level of process "S" at sleep onset of the first day sleep is lower than normal because the period of prior wakefulness is extended. Secondly, this, and subsequent, day sleeps fail to fully restore process "S" due to their reduced duration. We have yet to test this prediction!

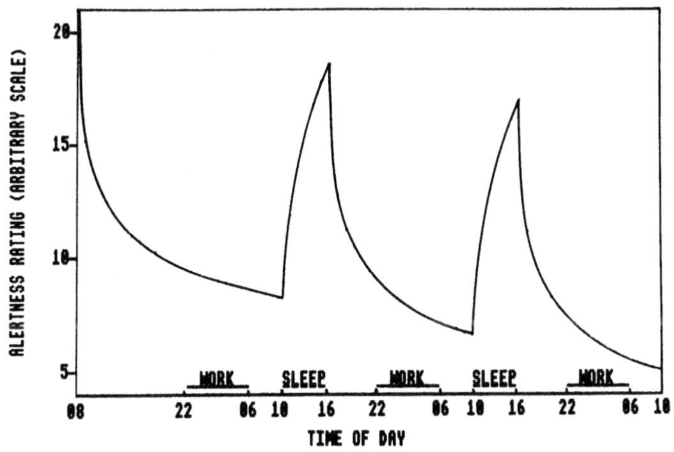

Figure 7. The predicted course of processes "S" and "S'" over three successive night shifts.

5. DISCUSSION

Despite the provisional nature of the model in its current form, it appears to be capable of "predicting" trends in alertness or fatigue under a variety of conditions and with considerable accuracy. At present the model is largely limited to the prediction of trends in group data. Although reliable fits have been found to the data from individual subjects, we are, as yet, unable to take account of the individual differences which may potentially occur in all three processes. This, and other factors such as the effects of prolonged work on fatigue, clearly need to be taken into consideration in future developments of the model.

The most obvious shortcoming of the model to date is its failure to predict the "post-lunch dip" in alertness during the second and third days of sleep deprivation apparent in the data of Froberg (1977) and shown in Figure 3. This dip in alertness (or more strictly increase in fatigue) occurred despite the fact that these subjects received no lunch as such, but identical snacks at three-hourly intervals, supporting the view that it is endogenous in origin. Further developments of the model might simulate this dip by building a 12h sinusoidal component into process "C", and allowing it to increase in its relative magnitude with increasing sleep deprivation. However, in order for it to also account for the dip observed in vigilance performance (Figure 4), we would also need to allow this 12h component to increase with "boredom" or lack of stimulation. Such an assumption might also account for the clear increase in sleep propensity observed at this time using the "multiple sleep latency test".

In sum, and despite the obvious need for further refinement, we feel justified in claiming to have made some progress towards a model for the prediction of alertness on abnormal sleep/wake schedules.

REFERENCES

Borbely, A.A., 1982, 'A two process model of sleep regulation.' Human Neurobiology, 1, 195-204.

Folkard, S., 1983, 'Diurnal Variation.' In G.R.J. Hockey (Ed.) Stress and fatigue in human performance. Chichester: John Wiley & Sons Ltd, pp 245-272.

Folkard, S. and Akerstedt, T., 1987, 'Towards a model for the prediction of alertness and/or fatigue on different sleep/wake schedules.' In A. Oginski, J. Pokorski and J. Rutenfranz (Eds) Contemporary advances in shiftwork research: Theoretical and practical aspects in the late eighties. Krakow: Medical Academy, pp 231-240.

Folkard, S., Hume, K.I., Minors, D.S., Waterhouse, J.M. and Watson, F.L., 1985, 'Independence of the circadian rhythm in alertness from the sleep/wake cycle.' Nature, 313, 678-679.

Froberg, J.E., 1977, 'Twenty-four-hour patterns in human performance, subjective and physiological variables and differences between morning and evening active subjects.' Biological Psychology, 5, 119-134.

Gates, A.I., 1916, 'Variations in efficiency during the day, together with practice effects, sex differences, and correlations.' Univ. Calif. Publs. Psychol., 2, 1-156.

Howell, W.H., 1897, 'A contribution to the physiology of sleep upon plethysmographic experiments.' J. Exp. Med., 2, 313.

Michelson, M., 1897, 'Ueber die Tiefe des Schlafes.' Psychol Arbeiten, 2, 84-117.

Thorndike, E., 1900, 'Mental fatigue.' Psychol. Review, 7, 466-482.

ACKNOWLEDGMENT. We gratefully acknowledge support from the CEC concerted action on "Breakdown of human adaptation 2: Performance decrement" under the leadership of Dr Hans-M Wegmann, DFVLR Institut fur Flugmedizin, Linder Hohe, D-5000 Koln 90. We also thank him, and other friends and colleagues, for allowing us to model their data.

Most of this work was conducted while the first author was a member of the MRC Perceptual and Cognitive Performance Unit, University of Sussex, Brighton, BN1 9QG, Sussex.

IV

PHARMACOLOGIE, VIGILANCE, PERFORMANCE

PHARMACOLOGY, VIGILANCE, PERFORMANCE

PRISE DE SOMNIFERE ET REACTIVITE AU BRUIT AU COURS DU SOMMEIL : AVANTAGES ET DANGERS

A. MUZET et J.P. LIBERT
Laboratoire de Physiologie
et de Psychologie Environnementales
21 rue Becquerel
67087 Strasbourg Cedex
France

SUMMARY

Sleep is a resting period that has clear restorative function. The sleep-wake cycle can be desorganized by several factors, such as ambient noise. The responsiveness of sleeping man depends on sleep stage, and also on the type and the signification of the noise. Thus sleeper will respond more easily to the call of his own name than to somebody else'name. The ability to respond in an adequate manner to an external stimulation also depends on sleep stage and on the acquisition of a behavioral response learned while awake. Recent works suggest that the consumption of sleeping pills is increased in noisy environments, as the drug is possibly used as a sleep protector. Perception of the noise and ability to act correctly when hearing it can be markedly impaired by using hypnotics. The implications of these findings are of importance for fire safety or response to any auditory emergency signal.

RESUME

Le rythme veille-sommeil peut être désorganisé par différents facteurs, dont notamment le bruit ambient. La réactivité de l'Homme endormi dépend du stade de sommeil dans lequel il est plongé, mais également de la nature et de la signification de la stimulation sonore. Sa capacité à répondre de façon adaptée à la stimulation perçue est fonction là encore du type de sommeil, mais également d'une éventuelle consigne acquise à l'état de veille. La perception, tout comme la capacité de répondre, peuvent être sensiblement réduites par l'absorption de somnifères. Les conséquences de tels effets peuvent être particulièrement néfastes chaque fois que le dormeur est placé dans une situation où il doit réagir à des signaux l'informant de la survenue imminente d'un danger.

A. Coblentz (ed.), Vigilance and Performance in Automatized Systems/ Vigilance et Performance de l'Homme dans les Systèmes Automatisés, 299–305.

INTRODUCTION

Le sommeil est l'état de repos naturel qui permet la restauration des capacités physiques et intellectuelles. L'équilibre du rythme veille-sommeil est de ce fait extrêmement important, car sa désorganisation entraîne une réduction des performances, une sensation de fatigue et des troubles divers des principales fonctions physiologiques. Plusieurs facteurs peuvent altérer ce rythme et affecter celui-ci au point de provoquer de graves désordres qui incitent la personne atteinte à utiliser soit des drogues éveillantes soit, le plus souvent, des drogues facilitant le sommeil.

Dans un certain nombre de cas c'est la perturbation sonore subie au cours du sommeil qui est la cause de l'emploi du somnifère ; celui-ci est alors utilisé comme un moyen de protection. Mais dans ce cas nous pouvons nous demander si l'emploi de tels produits ne présente pas des inconvénients supérieurs aux avantages escomptés ? Dans ce court exposé nous tâcherons d'apporter des réponses aux trois catégories de questions suivantes :

L'homme endormi perçoit-il l'environnement et ses stimulations de la même façon qu'à l'état de veille ? Est-il capable de réagir de façon rapide et adaptée aux signaux extérieurs ? Dans quelle mesure certains médicaments ne constituent-ils pas un obstacle à la bonne perception de ces signaux et mettent ainsi, quelques fois, la vie de l'individu en danger ?

REACTIVITE AU BRUIT AU COURS DU SOMMEIL NORMAL

Les réponses à la première question sont issues de travaux relativement anciens, datant d'une époque où l'on tentait de comprendre la nature et les fonctions du sommeil en s'opposant le plus souvent à son déroulement normal par l'utilisation de stimulations diverses visant soit à l'interrompre, soit à l'empêcher. C'est ainsi que bon nombre de travaux des années 1960-1970 utilisent des stimuli extéroceptifs (le plus souvent des sons de fréquence déterminée) pour analyser la "profondeur" du sommeil ou encore les capacités discriminatives de l'Homme endormi.

Dans une analyse des travaux portant sur le traitement de l'information au cours du sommeil, WILLIAMS, en 1973, souligne qu'en général avec des stimulations neutres, telles que clics, flashs, chocs électriques et équivalents, l'intensité nécessaire pour provoquer une réponse comportementale au cours du sommeil est plus élevée que celle nécessaire pour obtenir une réponse systématique au cours de la veille. Une telle différence pourrait être due à une augmentation des seuils sensoriels au cours du sommeil ou encore à une réduction des capacités de transmission des voies sensorielles afférentes. Cela est toutefois contesté par l'observation d'effets visibles en stades 1, 2 et 3, sur les composants électroencéphalographiques et, dans une moindre mesure, cardiovasculaires, d'une stimulation sonore ayant un niveau supérieur de 5 décibels seulement au seuil auditif de veille (WILLIAMS et al., 1964). KEEFE et al. (1971) ont étudié la hiérarchie des réponses électroencéphalographiques et végétatives à des stimulations sonores de 1000 Hz au cours de la veille et au cours du stade 2, des stades 3 et 4 regroupés (slow wave sleep ou SWS) et du sommeil paradoxal (REM sleep). Au cours de la veille le son calibré entraîne de façon statistiquement significative des réponses électroencéphlographiques, cardiaques, vasomotrices et électrodermales lorsque le niveau du seuil auditif est atteint. Il n'existe par contre aucune réponse lorsque le son est présenté avec une intensité inférieure à ce seuil. Au cours du sommeil la procédure expérimentale utilisée était la suivante : le son de 1000 Hz (5 secondes de durée) était présenté avec une intensité correspondant au seuil auditif de veille, puis avec une intensité augmentée de 5 dB en 5 dB jusqu'à l'obtention d'une réponse motrice adaptée (appuyer sur le bouton commandant l'arrêt de la diffusion du son de 1000 Hz) ou d'un changement de l'activité électroencéphalographique du dormeur indiquant une activation. Dans ces conditions on observe alors une dissociation dans la survenue des réponses au son. Ainsi on observe des signes de réponse électroencéphalographique pour des stimulations dont l'intensité est de 30 à 25 dB inférieure à celle provoquant une réponse motrice ou une activation. Les réponses vasomotrices (vasoconstrictions) sont observées pour des niveaux de 20 à 15 dB inférieurs au seuil d'activation et les réponses cardiaques (augmentation de la fréquence) sont visibles pour des niveaux de 20 à 5 dB plus bas que celui correspondant à ce seuil. Les réponses électrodermales, quant à elles, ne sont observables qu'au niveau sonore correspondant au déclenchement de l'activation ou de la réponse motrice adaptée. Par conséquent, il existe donc bien une hiérarchie dans l'ordre de survenue de ces réponses électroencéphalographiques et végétatives et cette hiérarchie est, selon KEEFE et al., constante pour les différents stades de sommeil explorés.

TRAITEMENT DE L'INFORMATION AU COURS DU SOMMEIL

En 1960, OSWALD et al. constatent qu'un dormeur présente des réponses électroencéphalographiques (complexes K) plus fréquentes à l'énoncé de son propre nom qu'à celui d'une autre personne. Un tel résultat suggère que l'homme endormi est susceptible de traiter, jusqu'à un certain degré, l'information qui lui parvient. Mc DONALD et al. (1975) arrivent à la même conclusion en présentant au cours du sommeil soit le nom du dormeur, soit le nom d'une autre personne, soit encore un stimulus sonore neutre (500 Hz). En fait, les sujets endormis répondent de façon différenciée à ces trois catégories de stimulations au cours du stade 2 et du sommeil paradoxal. Par contre ces différences n'existent plus en sommeil à ondes lentes (SWS). Dans le même article, Mc DONALD et al. notent également qu'un conditionnement discriminatoire acquis au cours de la veille persiste en stade 2 et en SWS alors qu'il ne persiste pas en sommeil paradoxal. La démonstration de l'effet selon lequel les dormeurs peuvent suivre des instructions données avant le sommeil afin de répondre de façon différentielle à des bruits considérés comme importants ou non importants, avait déjà été faite par ZUNG et WILSON en 1966.

Ainsi il apparaît évident que l'homme endormi dispose de moyens lui permettant de réagir à des stimulations externes, mais cette capacité de réaction dépend dans une certaine mesure du stade de sommeil dans lequel il est plongé. Cette capacité dépend aussi du moment de la nuit où la stimulation est exercée. WATSON et RECHTSCHAFFEN (1969) rapportent que les seuils d'activation observés en stade 2 et en sommeil paradoxal diminuent progressivement avec l'augmentation du temps passé en état de sommeil. L'influence du stade de sommeil est tout aussi visible immédiatement après l'éveil du sujet. Ainsi le temps de réaction est plus lent après réveil des sujets en stade 2 et en SWS qu'après leur réveil en sommeil paradoxal (GOODENOUGH et al., 1965 ; SCOTT, 1969). De même, BROUGHTON (1968) souligne que les sujets réveillés en SWS sont plus confus et incoordonnés que ceux réveillés en REM sleep.

REACTIVITE AU BRUIT EN CAS DE PRISE DE SOMNIFERE

De nombreux hypnotiques réduisent le nombre d'éveils nocturnes et induisent un sommeil subjectivement ressenti comme étant plus profond. C'est pourquoi certaines personnes particulièrement exposées au bruit la nuit sont amenées à utiliser le somnifère comme un moyen de protéger leur sommeil. Ceci constitue une des conclusions des enquêtes menées sur la consommation des somnifères et sédatifs dans des zones particulièrement exposées aux nuisances sonores (KNIPSCHILD 1977 ; LAMBERT et al., 1980). Il est intéressant de souligner toutefois qu'en dépit de jugements subjectifs tendant à laisser croire que les "mauvais dormeurs" ont un sommeil plus "léger", ces derniers présentent des seuils auditifs d'activation au cours de leur sommeil parfaitement comparables à ceux trouvés chez les "bons dormeurs" (JOHNSON et al., 1979).

L'emploi de certains somnifères entraîne une augmentation du seuil auditif d'activation (BONNET et al., 1979 ; JOHNSON et al., 1979) et une réduction du nombre de certaines modifications électrophysiologiques provoquées par le bruit (EHRENSTEIN et al., 1980 ; SALETU et al., 1980). Il est à remarquer que très peu d'études similaires ont été réalisées en dépit du volumineux dossier des travaux consacrés à l'étude de la structure du sommeil et de ses modifications sous l'effet d'une prise de somnifère. Dans une étude réalisée à Strasbourg (MUZET et al., 1983), trois groupes de 8 sujets se déclarant être "mauvais dormeurs" avaient été exposés à des bruits de passage d'avion compris entre 67 et 73 dB d'intensité de crête lors de trois nuits non consécutives faisant partie d'une série de 7 nuits passées au laboratoire. Lors de la première de ces trois nuits perturbées les sujets ont reçu, juste avant leur coucher, une capsule contenant du placebo alors que lors des 2 autres nuits perturbées ils ont absorbé, selon leur appartenance à l'un des trois groupes, soit du placebo, soit 0,25 mg de triazolam, soit 0,50 mg de triazolam. Dans le groupe ayant reçu du placebo, 12 à 16% des bruits d'avion diffusés entraînent un éveil électroencéphalographique ; il n'existe pas de différence sensible de ce pourcentage entre les 3 nuits perturbées. Dans le groupe ayant reçu 0,50 mg de triazolam lors des 2 dernières nuits perturbées, le pourcentage des bruits ayant entraîné un éveil électroencéphalographique passe de 24% en condition placebo à 5% lors de la première nuit sous somnifère et à 11% lors de la deuxième nuit. Dans le groupe ayant reçu la dose de 0,25 mg de triazolam les résultats vont dans le même sens que ceux observés pour le groupe précédent, mais cette fois-ci les différences ne sont pas statistiquement significatives. Par opposition, la réactivité cardiaque, mesurée par l'amplitude de la variation de fréquence cardiaque provoquée par le bruit, n'est que très modérément réduite par la plus forte dose de somnifère et cette réduction n'est observée que lors de la première nuit traitée seulement (LIBERT et al., sous presse).

Cette étude confirme donc que l'emploi du somnifère réduit la fréquence des éveils nocturnes provoqués par des stimulations sonores, mais elle montre aussi que cet effet de protection est labile puisque plus modéré après la 1ère administration. Cette étude permet également de souligner la dissociation qui existe entre les effets corticaux du bruit et les effets végétatifs, ces derniers étant manifestement moins affectés par l'emploi du somnifère.

La dernière étude discutée ici présente, quant à elle, des résultats qui ne sont pas surprenants à la lumière de ce qui vient d'être dit, mais dont les conséquences apparaissent de la plus grande importance. Dans cette étude (JOHNSON et al., 1987), trois groupes ("mauvais dormeurs") de 12 sujets chacun ont reçu lors de 5 nuits consécutives une capsule contenant soit du placebo, soit 0,25 mg de triazolam, soit 0,50 mg de triazolam. Lors de la 1ère et de la 4ème nuit un signal sonore d'alarme incendie était diffusé un certain nombre de fois selon un protocole précis. Le niveau sonore de cette alarme, mesuré à hauteur de l'oreiller, était de 78 dB ; le niveau de bruit de fond de la chambre étant compris entre 32 et 34 dB. Le signal d'alarme était diffusé pendant trois périodes successives de 1 minute chacune, à moins qu'il ne soit interrompu auparavant par le sujet qui avait pour consigne d'appuyer sur un bouton arrêtant ce signal dès qu'il le percevait. La présentation de l'alarme était faite au cours de trois périodes bien définies de la nuit :

- lors du stade 2 du début de nuit, c'est-à-dire 5 minutes après la survenue du 1er complexe K ou du 1er fuseau de sommeil ;
- lors du premier SWS, soit 20 minutes après le réendormissement consécutif au premier éveil ;
- lors du réveil matinal à 5 h 00 lorsque le sujet était toujours endormi.

Les résultats de cette étude confirment le fait que l'emploi du somnifère augmente clairement le seuil d'éveil. Lors de la 1ère nuit sous somnifère, 50% des sujets ayant reçu le triazolam ne se sont pas réveillés au cours des 3 alarmes successives de 1 minute chacune présentées en SWS. Lors de la même nuit 4 sujets sur les 24 ne se sont pas réveillés après les cinq premières minutes passées en stade 2. Par contre, tous les sujets ayant reçu le placebo ont été réveillés par toutes les alarmes. Lors de la deuxième nuit perturbée par ce signal d'alarme (4ème nuit sous traitement), 38% des sujets ayant reçu le somnifère ne se réveillent pas en SWS et 3 ne se réveillent pas lorsque l'alarme est diffusée après les cinq premières minutes de stade 2. Une fois encore tous les sujets ayant reçu le placebo sont réveillés à toutes les occasions où l'alarme est diffusée. Lorsque les sujets sont réveillés par l'alarme incendie on observe que la latence d'apparition de l'activation électroencéphalographique est la même quel que soit le groupe auquel appartient le sujet. Par contre la latence de la réponse comportementale adaptée (presser le bouton pour interrompre l'alarme) est plus longue chez les sujets ayant recu le somnifère que chez ceux ayant reçu le placebo.

Ainsi, bien que les sujets ayant pris le somnifère se réveillent aussi vite que ceux ayant reçu le placebo, le rappel de la consigne et la vitesse d'exécution motrice sont généralement ralentis chez les premiers.

CONCLUSION

La perception des stimulations sonores est diminuée chez le dormeur et ses capacités d'éveil dépendent notamment du stade de sommeil et de la nature des stimulations. L'emploi de somnifère réduit encore les capacités perceptives du dormeur, ce qui est souvent l'effet recherché dans des situations bruyantes ou lorsque la nouveauté du lieu de couchage (chambre d'hôtel par exemple) peut constituer un facteur aggravant la mauvaise qualité du sommeil. Les quelques recherches citées ici montrent à l'évidence que cette "protection" du sommeil est limitée et qu'elle peut présenter des effets aux conséquences particulièrement néfastes chaque fois que le dormeur est placé dans l'incapacité de réagir de façon adaptée à des signaux indiquant la survenue immminente d'un danger.

BIBLIOGRAPHIE

BONNET M.H., WEBB W.B., BARNARD G. 'Effect of flurazepam, pentobarbital, and caffeine on arousal threshold', Sleep, 1979, **1**: 271-279.

BROUGHTON R.J. 'Sleep disorders : Disorders of arousal ?' Science, 1968, **159**: 1070-1078.

EHRENSTEIN W., MULLER-LIMMROTH W., OPFERMANN M. 'The beneficial effects of a new benzodiazepine on the sleep disturbing effects of intensive noise produced by subsonic flyovers', in Sleep 1978, Karger, Basel, pp. 493-495, 1980.

GOODENOUGH D.R., LEWIS H.F., SHAPIRO A. 'Dream reporting following abrupt and gradual awakenings from different types of sleep', J. Personal. Social Psychol., 1965, **2**: 170-179.

JOHNSON L.C., CHURCH M.W., SEALES D.M., ROSSITER V.S. 'Auditory arousal thresholds of good and poor sleepers with and without flurazepam', Sleep, 1979, **1**: 259-270.

JOHNSON L.C., SPINWEBER C.L., WEBB S.C., MUZET A. 'Dose level effects of triazolam on sleep and response to a smoke detector alarm', Psychopharmacol., 1987, **91**: 397-402.

KEEFE F.B., JOHNSON L.C., HUNTER E.J. 'EEG and autonomic response pattern during waking and sleep stages', Psychophysiol., 1971, **8**: 198-212.

KNIPSCHILD P. 'Medical effects of aircraft noise', Int. Arch. Occup. Environ. Health, 1977, **40**: 185-204.

LAMBERT J., SIMONNET F., VALLET M. 'Comportement dans l'habitat soumis au bruit', Rapport INRETS n° 47, Bron, 1980.

LIBERT J.P., AMOROS C., MUZET A., EHRHART J., DI NISI J. 'Effects of triazolam on heart rate level and on phasic cardiac response to noise during sleep', Psychopharmacol., in press.

MC DONALD D.G., SCHICHT W.W., FRAZIER R.E., SHALLENBERGER H.D., EDWARDS D.J. 'Studies of information processing in sleep', Psychophysiol., 1975, **12**: 624-629.

MUZET A., WEBER LD., AMOROS C., EHRHART J., LIBERT J.P., TSAKONA C. 'Cardiovascular responses to noise during sleep. Effects of a benzodiazepine hypnotic, in Noise as a Public Health Problem, G. ROSSI (ed.), Centro Ricerche E Studi Amplifon, Milano, pp. 985-993, 1983.

OSWALD I., TAYLOR A.M., TREISMAN M. 'Discriminative responses to stimulation during human sleep', Brain, 1960, **83**: 440-453.

SALETU B., LIPPLER I., GRUNBERGER J. 'Nocturnal traffic noise and sleep. Effects and countereffects by oxazolam', Drug Res., 1980, **30**: 1231.

SCOTT J. 'Performance after abrupt arousal from sleep : comparison of a simple motor, a visual-perceptual, and a cognitive task', Proceedings, 77th Annual Convention, American Psychological Association, pp. 225-226, 1969.

WATSON R., RECHTSCHAFFEN A. 'Auditory awakening threshold and dream recall in NREM sleep', Percept. Motor Skills, 1969, **29**: 635-644.

WILLIAMS H.L., HAMMACK J.R., DALY R.L., DEMENT W.C., LUBIN A. 'Responses to auditory stimulation, sleep loss and the EEG stages of sleep', Electroencephal. clin. Neurophysiol., 1964, **16**: 269-279.

WILLIAMS H.L. 'Information processing during sleep', in Sleep, W.P. KOELLA and P. LEVIN (eds), Karger, Basel, pp. 36-43, 1973.

ZUNG, W.W.K., WILSON W.P. 'Attention, discrimination and arousal during sleep', Electroencephal. clin. Neurophysiol., 1966, **20**: 623.

PHARMACOLOGICAL BASIS OF WAKEFULNESS AND VIGILANCE

A. N. Nicholson and Peta A. Pascoe
Royal Air Force Institute of Aviation Medicine
Farnborough
Hampshire
United Kingdom

ABSTRACT. Although adequate sleep is essential to maintain the effectiveness of aircrew during military operations, the potential of drugs to sustain vigilance must also be considered. Beneficial changes in alertness may be achieved through the selective modulation of the activity of neurotransmitters involved in the control of sleep and wakefulness, though the influence of time of day is an important factor in determining the effects of centrally-acting drugs. In this context we have studied the effects of drugs which selectively modify monoaminergic or purinergic transmission on measures of alertness and performance throughout the 24h continuum. Electroencephalographic assessments of sleep and daytime alertness indicate that drugs which alter 5-HT transmission, those with dopamimetic activity and adenosine receptor antagonists promote wakefulness in man. However, with 5-HT uptake inhibitors enhanced alertness was associated with impaired performance, and so drugs such as pemoline or caffeine may be more appropriate for those involved in skilled activity.

1. INTRODUCTION

Although adequate sleep is essential to maintain the effectiveness of aircrew in military operations, the potential usefulness of drugs which sustain vigilance must also be considered. The states of sleep and wakefulness are controlled by the activity of many neurotransmitters and neuromodulators in the central nervous system, and beneficial changes in alertness might be achieved through the use of pharmacological agents which selectively modify one or more of these influences.

Much research has been carried out to investigate the role of established and putative transmitters in animals. However, though such studies are essential and have led to important advances in our understanding, extrapolation to man may be limited. The effects of centrally-acting drugs depend very much on the relative response of the mechanisms which regulate transmission, and there may be species differences, both in the response and in the mechanisms themselves.

A. Coblentz (ed.), Vigilance and Performance in Automatized Systems/ Vigilance et Performance de l'Homme dans les Systèmes Automatisés, 307–316.

Fenfluramine, for example, is believed to increase wakefulness in rats through stimulation of postsynaptic 5-HT receptors (Fornal and Radulovacki, 1983), but leads to sedation in man. Species differences in the nature of 5-HT receptors have been shown (Heuring et al, 1986; Hoyer et al, 1986; Hoyer et al, 1987), and may not only account for the differential effect on behaviour of drugs which modulate transmitter activity, but may also relate to the different patterns of sleep and wakefulness observed in animals and man.

One of the difficulties in reaching an understanding of the influence of neurotransmitters in the control of sleep and wakefulness in man has been the lack of specificity of drugs which alter their activity. This has been especially true of drugs which modulate monoaminergic transmission, as anticholinergic and antihistaminic activity frequently complicate the pharmacological profile. However, more recently, drugs with selective activity have become available. These include inhibitors of 5-hydroxytryptamine (5-HT) uptake, such as zimelidine, indalpine and fluoxetine, and pemoline, a heretocyclic amphetamine derivative which has dopamimetic activity.

A further complication in assessing the effects of drugs which modify transmission in the central nervous system is the change in the relative influence of neurotransmitters over the 24h continuum. Histamine H1 receptor antagonists, for example, lead to drowsiness and impaired performance during the day, but have little effect on wakefulness during nocturnal sleep (Nicholson et al, 1985). This raises the question whether some transmitters have a dominant role in the fundamental control of sleep and wakefulness, while others are more specifically involved in the regulation of vigilance. In each case the circadian variation in the activity of the transmitter may be present, but the influence of those selectively concerned with vigilance may only be evident against a particular background level of neurotransmitter activity or alertness.

It is, therefore, important to study the effects of centrally-acting drugs over the whole of the sleep-wakefulness continuum in man. Electroencephalographic studies of nocturnal sleep provide a preliminary indication of alerting or sedative properties against the background of neurotransmitter activity consistent with sleep. Drugs may also modify the pattern of performance usually observed overnight and characterised by a deterioration towards the early hours of the morning, and daytime studies of performance and alertness are equally important in determining the complete profile of a particular drug across the 24h continuum.

Using these models we have studied three groups of drugs which increase alertness, namely 5-HT uptake inhibitors, dopamimetic agents, and adenosine receptor antagonists. In the present paper, we discuss the effects of these drugs on sleep and wakefulness, with particular reference to the pharmacological mechanisms which may be involved in the modulation of alertness.

2. 5-HT UPTAKE INHIBITORS

Electroencephalographic studies on nocturnal sleep show that 5-HT uptake inhibitors, such as fluoxetine, indalpine and zimelidine, increase wakefulness and stage 1 (drowsy) sleep during the night, and reduce total sleep time (Figure 1) (Nicholson and Pascoe, 1986 and 1988).

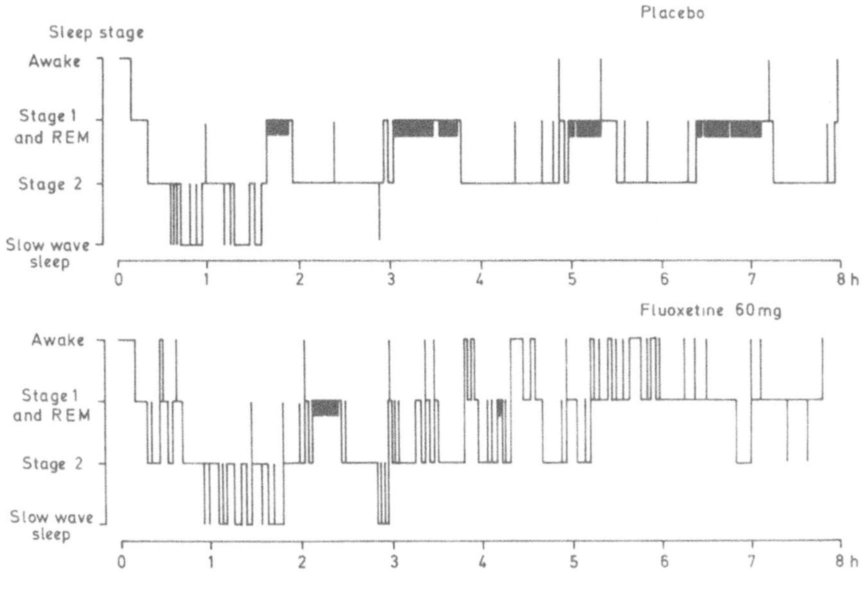

Figure 1. Effect of a 5-HT uptake inhibitor (fluoxetine 60mg) on nocturnal sleep.

To identify the relative influence which serotonergic mechanisms may have over the sleep-wakefulness continuum, studies with fluoxetine were extended to examine its effects on performance and alertness during the day (Nicholson and Pascoe, 1988). Performance was assessed at intervals after drug ingestion using digit symbol substitution and tests of sustained attention and recall, and alertness was assessed subjectively and electroencephalographically using the multiple sleep latency test (Carskadon and Dement, 1977).

Fluoxetine, ingested after the second test session, increased sleep latencies during the afternoon compared with placebo (Figure 2), indicating that subjects were more alert. However, this effect of fluoxetine was accompanied by subjective feelings of drowsiness and impaired coding ability. Saletu and Grunberger (1985) have also observed impaired performance with fluoxetine. However, subjective reports of sleepiness have been described with paroxetine, another 5-HT uptake inhibitor, at a dose which did not apparently impair performance (McClelland and Raptopoulos, 1986), and there are other studies which suggest that inhibition of 5-HT uptake is not associated with a decrement in performance. It is, therefore, difficult to be certain on this point, but it must be borne in mind that, although modulation of 5-HT transmission may increase alertness, this does not necessarily imply enhanced performance.

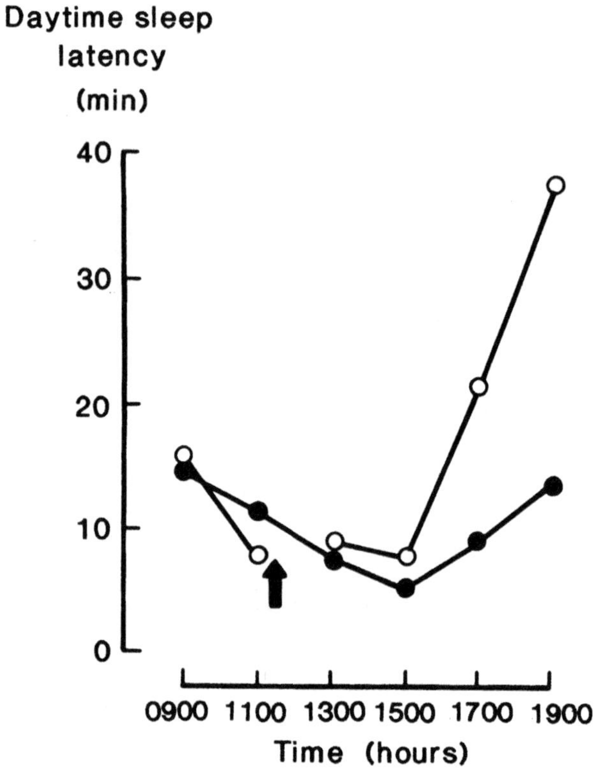

Figure 2. Daytime sleep latencies (back-transformed means for 6 subjects) after ingestion of placebo (filled circles) and after 60mg fluoxetine (open circles). Medication was given at 1130h (indicated by the arrow).

The role of monoamines in sleep and wakefulness is well recognised and 5-HT is considered necessary for the initiation and maintenance of sleep, possibly through the activity of other sleep factors (Jouvet, 1972; Sallanon et al, 1983). The alerting effects of 5-HT uptake inhibitors are, therefore, rather surprising as these drugs increase the level of 5-HT in the synapse by inhibiting the reuptake mechanism which normally terminates transmitter activity. However, inhibition of uptake may lead to modification of both postsynaptic and presynaptic influences on transmission, and raised levels of transmitter may increase presynaptic inhibition of transmitter release as well as increasing stimulation of postsynaptic receptors. Indeed, 5-HT uptake inhibitors lead to a rapid decrease in the synthesis of neurotransmitter with a decreased rate of neuronal firing (Carlsson and Lindqvist, 1978; Clemens et al, 1977; de Montigny et al, 1981), and to reduced 5-HT turnover and release (Fuller and Wong, 1977). Such a mechanism would account for the alerting profile of the 5-HT uptake inhibitors in man.

Our studies with fluoxetine confirm the pervasive influence of the serotonergic system as the alerting effect of fluoxetine was observed throughout the 24h continuum. It is of interest, however, that relatively small doses (20 - 60mg) of fluoxetine were needed to increase sleep latencies during the day compared with that which increased wakefulness during nocturnal sleep (60mg). This difference may have been due simply to the relative alertness of the subjects. However, many of the factors which determine the sensitivity of the 5-HT system are modified during wakefulness (Meyer and Quay, 1976; Wesemann et al, 1985; Mason, 1986) and in this way raised levels of 5-HT, coinciding with an enhanced responsiveness of the system, could lead to increased vigilance.

3. DOPAMIMETIC AGENTS

The ability to facilitate dopamine transmission contributes to the behavioural effects of many stimulants and, of the drugs we have studied, pemoline has the advantage of acting selectively on dopaminergic systems. Unlike amphetamine, pemoline does not modulate noradrenergic transmission, and so has no cardiovascular side-effects over a wide dose range. In the same way as the 5-HT uptake inhibitors, pemoline blocks transmitter reuptake but, in addition, stimulates the presynaptic release of dopamine. Again, the effects of increased availability of transmitter depend on the relative response of presynaptic and postsynaptic receptors, and it would seem that in the case of drugs which release as well as inhibit the uptake of transmitter there is continued stimulation of the postsynaptic receptor.

Studies with pemoline clearly demonstrate the role of the dopaminergic system in maintaining wakefulness. Pemoline leads to marked sleep disturbance, with increased awake activity and reduced total sleep time (Figure 3) (Nicholson and Stone, 1980; Nicholson et al, 1980), and to improved performance and increased alertness, as indicated by longer sleep latencies, during the day (Nicholson and Pascoe, unpublished data). On the other hand, pimozide, which at low doses is a potent and relatively specific antagonist of dopamine receptors, has little effect on nocturnal sleep but leads to drowsiness and impaired performance during the day (Sagales and Erill, 1975;

Nicholson and Pascoe, unpublished data).

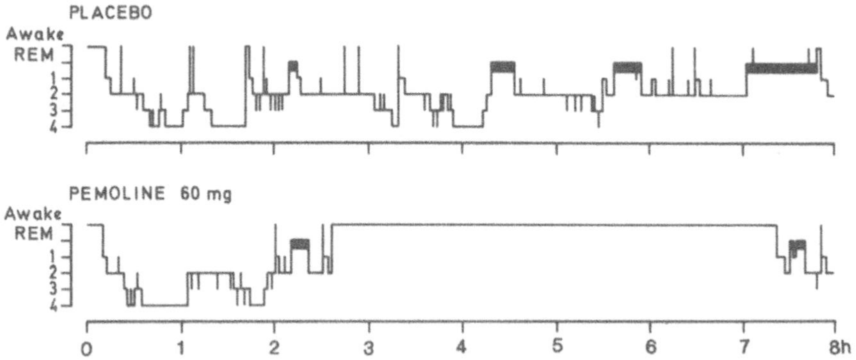

Figure 3. Effect of pemoline (60mg) on nocturnal sleep.

These studies provide some indication of the relative importance of dopamine activity over the 24h continuum. The limited effect of drugs which reduce dopaminergic transmission, together with the marked effect of dopamimetic agents, on nocturnal sleep are consistent with a decrease in the influence of the dopaminergic system during sleep. On the other hand, both dopamimetic agents and dopamine receptor antagonists modify alertness during the day, and this would suggest that dopamine activity is normally increased, but not necessarily maximal, during wakefulness. These observations in man are supported by studies in rats which indicate that the dopamine receptor antagonist α-flupenthixol is more effective in reducing wakefulness during periods of maximum spontaneous activity when dopamine turnover is reported to be increased (Lemmer and Berger, 1978; Wojcik and Radulovacki, 1982).

4. ADENOSINE RECEPTOR ANTAGONISTS

Unlike the 5-HT uptake inhibitors and pemoline, adenosine receptor antagonists, such as caffeine, do not directly modify monoaminergic transmission but modulate purinergic mechanisms by blocking adenosine receptors. There is increasing evidence that adenosine may be a regulator of central activity (Phillis and Wu, 1981; Dunwiddie, 1985). Adenosine and its metabolically stable analogues depress both the spontaneous and evoked firing of central neurones and, at the behavioural level, have sedative and hypnotic properties (Radulovacki, 1985). The inhibitory effect of adenosine on the release of other neurotransmitters, such as noradrenaline and dopamine, may well account for these effects, and in this way adenosine may function more as a neuromodulator rather than as a neurotransmitter. Indeed, though adenosine fulfils the requirements of a neurotransmitter with respect to its synthesis, release, reuptake, catabolism and receptors, purinergic neurones or specific synapses have not yet been identified in the brain (Snyder, 1985).

It is now widely accepted that the excitatory actions of caffeine and other methylxanthines are related to their antagonism of adenosine receptors rather than to inhibition of phosphodiesterase activity (Daly et al, 1981). The biochemical and neurophysiological effects of adenosine and its analogues are blocked by methylxanthine derivatives, and other studies have demonstrated antagonism of adenosine receptor-mediated behaviour, including sedation. In human brain tissue, caffeine antagonises the binding of adenosine receptor ligands at behaviourally active concentrations (Burg, 1975), and so this mechanism is most likely relevant to the stimulant effects of caffeine.

Caffeine leads to marked sleep disturbance (Figure 4) (Nicholson and Stone, 1980). It increases wakefulness and drowsy sleep, and reduces slow wave sleep and total sleep time.

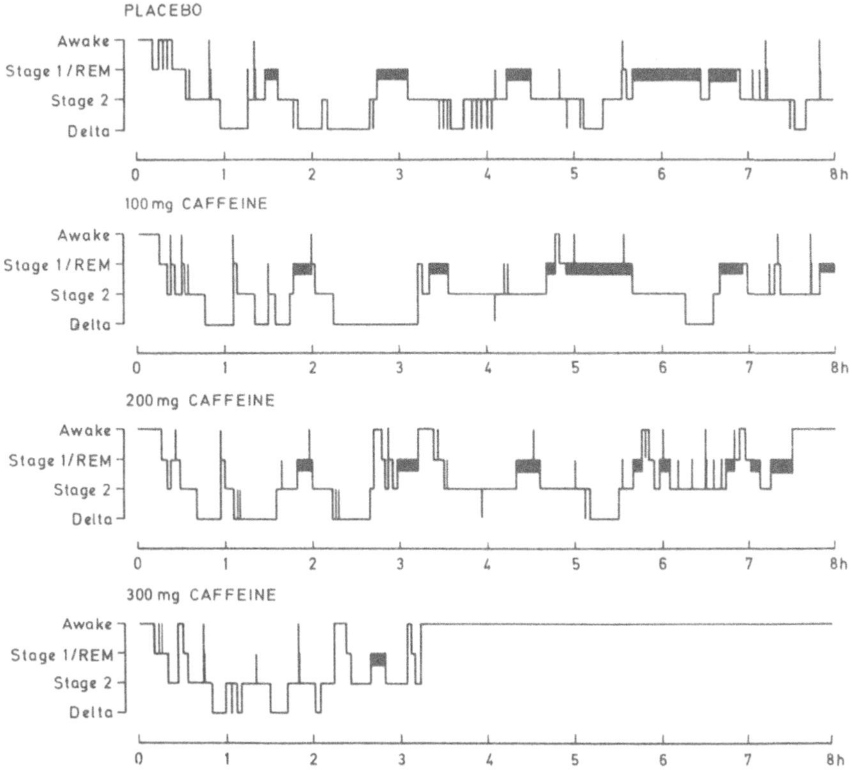

Figure 4. Effects of caffeine (100, 200 and 300mg) on nocturnal sleep.

Caffeine also attenuates the deterioration in overnight performance due to prolonged hours of work and the circadian fall in alertness (Borland et al, 1986), and improves performance and vigilance during the day (Nicholson and Pascoe, 1988). These studies suggest that adenosine may have a modulatory function throughout the 24h continuum. Indeed, circadian rhythms of adenosine receptors and adenosine levels in the rat brain indicate that adenosine may be involved in the control of sleep and wakefulness (Chagoya De Sanchez et al, 1983; Virus et al, 1984).

5. CONCLUSION

These studies suggest that alertness in man may be enhanced by modulation of at least three synaptic systems, though the possibility of adverse effects on performance due to the direct or indirect effect of the drugs cannot be ignored. Indeed, increased alertness does not necessarily imply improved vigilance, and even improved performance on one skill could be accompanied by impaired performance on another. Nevertheless, it is possible that drugs which have selective activity on the dopaminergic and purinergic systems are less likely to have adverse effects on behaviour than drugs which modulate noradrenergic and serotonergic transmission and such agents are, therefore, more likely to be useful as aids to wakefulness.

6. REFERENCES

Borland RG, Rogers AS, Nicholson AN, Pascoe PA, Spencer MB (1986) Performance overnight in shiftworkers operating a day-night schedule. Aviat Space Environ Med **57**:241-249.

Burg AW (1975) Physiological disposition of caffeine. Drug Metab Rev **4**:199-228.

Carlsson A, Lindqvist M (1978) Effects of antidepressant agents on the synthesis of brain monoamines. J Neural Trans **43**:73-91.

Carskadon MA, Dement WC (1977) Sleep tendency: An objective measure of sleep loss. Sleep Res **6**:200.

Chagoya De Sanchez V, Hernandez-Munoz R, Diaz-Munoz M, Suarez J, Vidrio S, Yanez L (1983) Circadian variations of adenosine and its physiological meaning in the energetic homeostasis of the cell and the sleep-waking of the rat. In: Sleep '82 (Koella WP, ed), p255, Karger, Basel.

Clemens JA, Sawyer BD, Cerimele B (1977) Further evidence that serotonin is a neurotransmitter involved in the control of prolactin secretion. Endocrinology **100**:692-698.

Daly JW, Bruns RF, Snyder SH (1981) Adenosine receptors in the central nervous system: Relationship to the central actions of methylxanthines. Life Sci **28**:2083-2097.

Dunwiddie TV (1985) The physiological role of adenosine in the central nervous system. Int Rev Neurobiol **27**:63-139.

Fornal C, Radulovacki M (1983) Sleep suppressant action of fenfluramine in rats. II. Evidence against the involvement of presynaptic serotonergic mechanism. J Pharmac Exp Ther **225**:675-681.

Fornal C, Wojcik WJ, Radulovacki M (1982) α-Flupenthixol increases slow-wave sleep in rats: Effects of dopamine receptor blockade. Neuropharmacology **21**:323-325.

Fuller RW, Wong DT (1977) Inhibition of serotonin reuptake. Fed Proc **36**:2154-2158.

Heuring RE, Schlegel JR, Peroutka SJ (1986) Species variations in RU 24969 interactions with non-5-HT1A binding sites. Eur J Pharmac **122**:279-282.

Hoyer D, Palacios JM, Pazos A, Probst A (1986) Serotonin (5-HT) recognition sites in human brain: Evidence for species differences. Br J Pharmac **87**:19P.

Hoyer D, Waeber C, Palacios JM (1987) Identification of a 5-HT1 recognition site in pig and human brain different from 5-HT1A, 5-HT1B and 5-HT1C. Br J Pharmac **92**:561P.

Jouvet M (1972) The role of the monoamines and acetylcholine-containing neurons in the regulation of the sleep-waking cycle. Ergebn Physiol **64**:163-307.

Lemmer B, Berger T (1978) Diurnal rhythm in the central dopamine turnover in the rat. Naunyn-Schmiedeberg's Arch Pharmac **303**:257-261.

Mason R (1986) Circadian variation in sensitivity of suprachiasmatic and lateral geniculate neurones to 5-hydroxytryptamine in the rat. J Physiol **377**:1-13.

McClelland GR, Raptopoulos P (1986) Paroxetine and amylobarbitone: Effects on psychomotor performance. Br J clin Pharmac **22**:227P-228P.

Meyer DC, Quay WB (1976) Hypothalamic and suprachiasmatic uptake of serotonin in vitro: Twenty-four-hour changes in male and proestrous female rats. Endocrinology **98**:1160-1165.

de Montigny C, Blier P, Caille G, Kouassi E (1981) Pre and post-synaptic effects of zimelidine and norzimelidine on the serotonergic system: Single cell studies in the rat. Acta Psychiat Scand **290**:79-90.

Nicholson AN, Pascoe PA (1986) 5-hydroxytryptamine and noradrenaline uptake inhibition: Studies on sleep in man. Neuropharmacology **25**:1079-1083.

Nicholson AN, Pascoe PA (1988) Studies on the modulation of the sleep-wakefulness continuum in man by fluoxetine, a 5-HT uptake inhibitor. Neuropharmacology **27**:597-602.

Nicholson AN, Pascoe PA, Stone BM (1985) Histaminergic systems and sleep: Studies in man with H1 and H2 antagonists. Neuropharmacology **24**:245-250.

Nicholson AN, Stone BM (1980) Heterocyclic amphetamine derivatives and caffeine on sleep in man. Br J clin Pharmac **9**:195-203.

Nicholson AN, Stone BM, Jones MMC (1980) Wakefulness and reduced rapid eye movement sleep: Studies with prolintane and pemoline. Br J clin Pharmac **10**:465-472.

Phillis JW, Wu PH (1981) The role of adenosine and its nucleotides in central synaptic transmission. Prog Neurobiol **16**:187-239.

Radulovacki M (1985) Role of adenosine in sleep in rats. Rev Clin Basic Pharmac **5**:327-339.

Sagales T, Erill S (1975) Effects of central dopaminergic blockade with pimozide upon the EEG stages of sleep in man. Psychopharmacologia (Berl) **41**:53-56.

Saletu B, Grunberger J (1985) Classification and determination of cerebral bioavailability of fluoxetine: Pharmacokinetic, pharmaco-EEG and psychometric analyses. J clin Psychiat **46**:45-52.

Sallanon M, Janin M, Buda C, Jouvet M (1983) Serotonergic mechanisms and sleep rebound. Brain Res **268**:95-104.

Snyder SH (1985) Adenosine as a neuromodulator. Ann Rev Neurosci **8**:103-124.

Virus RM, Baglajewski T, Radulovacki M (1984) Circadian variation of $(^3H)N^6$-(L-phenylisopropyl) adenosine binding in rat brain. Neurosci Lett **46**:219-222.

Weseman W, Rotsch M, Schulz E (1985) Effect of sleep deprivation and imipramine on the circadian rhythm of serotonin binding in rat brain. In: Sleep '84 (Koella WP, Ruther E, Schulz E, eds), pp 256-258, Gustav Fischer, New York.

LISTE DES AUTEURS

LIST OF CONTRIBUTORS

LISTE DES AUTEURS
LIST OF CONTRIBUTORS

INDEX DES MATIERES

SUBJECT INDEX

INDEX DES MATIERES

SUBJECT INDEX